The SkyX™ Workbook

D0781666

Thomas Jordan
Ball-State University

Scott Peters

BROOKS/COLE
CENGAGE Learning™

Australia • Brazil • Japan • Korea • Mexico • Singapore • Spain • United Kingdom • United States

© 2011 Brooks/Cole, Cengage Learning

ALL RIGHTS RESERVED. No part of this work covered by the copyright herein may be reproduced, transmitted, stored, or used in any form or by any means graphic, electronic, or mechanical, including but not limited to photocopying, recording, scanning, digitizing, taping, Web distribution, information networks, or information storage and retrieval systems, except as permitted under Section 107 or 108 of the 1976 United States Copyright Act, without the prior written permission of the publisher.

For product information and technology assistance, contact us at **Cengage Learning Customer & Sales Support, 1-800-354-9706**

For permission to use material from this text or product, submit all requests online at **www.cengage.com/permissions** Further permissions questions can be emailed to **permissionrequest@cengage.com**

ISBN-13: 978-0-538-73852-1
ISBN-10: 0-538-73852-9

Brooks/Cole
20 Channel Center Street
Boston, MA 02210
USA

Cover image: NGC 4755, The Jewel Box Cluster, courtesy of ESA/Hubble

Cengage Learning is a leading provider of customized learning solutions with office locations around the globe, including Singapore, the United Kingdom, Australia, Mexico, Brazil, and Japan. Locate your local office at: **www.cengage.com/global**

Cengage Learning products are represented in Canada by Nelson Education, Ltd.

To learn more about Brooks/Cole, visit **www.cengage.com/brookscole**

Purchase any of our products at your local college store or at our preferred online store **www.cengagebrain.com**

Printed in the United States of America
1 2 3 4 5 6 7 14 13 12 11 10

Table of Contents

© 2011 Cengage Learning. All Rights Reserved. May not be scanned, copied or duplicated, or posted to a publicly accessible website, in whole or in part.

Chapter 4 — Naming Objects in *TheSkyX* – *Continued*

Chapter 5 — Locating Celestial Objects in *TheSkyX*

© 2011 Cengage Learning. All Rights Reserved. May not be scanned, copied or duplicated, or posted to a publicly accessible website, in whole or in part.

Chapter 6 — Motions in *TheSkyX*

© 2011 Cengage Learning. All Rights Reserved. May not be scanned, copied or duplicated, or posted to a publicly accessible website, in whole or in part.

Chapter 7 — Keeping Time in *TheSkyX*

Chapter 8 — Seasons in *TheSkyX*

© 2011 Cengage Learning. All Rights Reserved. May not be scanned, copied or duplicated, or posted to a publicly accessible website, in whole or in part.

Chapter 8 — Seasons in *TheSkyX* (continued)

Chapter 9 — Phases and Eclipses In *TheSkyX*

© 2011 Cengage Learning. All Rights Reserved. May not be scanned, copied or duplicated, or posted to a publicly accessible website, in whole or in part.

Chapter 9 — Phases and Eclipses In *TheSkyX* – (continued)

© 2011 Cengage Learning. All Rights Reserved. May not be scanned, copied or duplicated, or posted to a publicly accessible website, in whole or in part.

Preface

The workbook you are about to use is a ***non-traditional*** workbook. By this, we mean that it is a stand-alone instruction manual. Many of the concepts presented in this workbook make using *TheSkyX* easier. It has been designed for use in an introductory observational astronomy course or for the causal observer.

Many traditional workbooks give brief and sometimes incomplete information about astronomical topics. They tend to focus more on how to use the software and less so on the concepts. This workbook employs an in-depth approach in discussing many astronomical topics. It gives detailed explanations of several difficult concepts. It is written with the assumption that most students who will use this workbook have little or no knowledge of the concepts used in observational astronomy.

The authors have given special attention to details in how to fully use the software, with many examples, exercises, and 'Sky' windows. All have been integrated into the chapters throughout the workbook. At the end of each chapter review, exercises and review questions help students test their understanding of the concepts in the chapter.

The first two chapters are written with the assumption that the reader has ***little*** or ***no*** computer experience. The authors wrote this workbook for a wide variety of backgrounds and disciplines so that everyone can enjoy *TheSkyX*. The first two chapters include detailed installation instructions for the necessary software. Computer-savvy individuals will probably want to skip these chapters and dive directly into the material.

When we decided to write this workbook to coincide with the Student Edition of *TheSkyX*, we thought there would be many sacrifices made in offering *TheSkyX* at a less sophisticated level. After using it, we feel that it is by far one of the best planetarium-type software products on the market today for both students and amateur astronomers.

Dr. Jordan has been using *Software Bisque* software in his observational astronomy class for many years. His students enjoy the hands-on experiences this software provides and have expressed how essential it has been in preparing them for the activities they later do in the observatory.

This workbook provides both instructors and students with a thorough understanding of how astronomers designate astronomical objects. Great care has been taken to explain confusing nomenclature that has been used historically and that is currently used in observational astronomy today. The exercises and questions are designed to assist students in gaining useful knowledge and insight of how astronomical objects are named, cataloged, and found. *TheSkyX* provides many images that are archived for your viewing pleasure.

TheSkyX software package is very user-friendly. It is an excellent and fun way to learn about stars, constellations, and coordinate systems used in observational astronomy. We have discussed coordinate systems in detail, beginning with a review of geographic coordinates. Students learn how the sky appears and how it changes when moving to a different location on Earth. Many major cities in the United States are included in *TheSkyX's* database, as well as many international cities and major observatories around the world. Students will enjoy taking imaginary trips to Earth's North Pole or to its Equator and all points in between.

We have thoroughly discussed both sky coordinates in this workbook. The horizon and equatorial coordinate systems are the coordinates that both amateur and professional astronomers use to locate objects in the sky. There are many examples and exercises included in this workbook to assist in their use. They are specifically designed to be used in conjunction with *TheSkyX* software. They help students and casual observers in learning where objects are in the sky and, most importantly, how to find them. Students have an opportunity to change their location on Earth and observe in real time how these changes affect the appearance and location of objects in the sky.

A troublesome concept that many students struggle with in astronomy involves the motions they observe in the sky. Many students have difficulty distinguishing the true motions of Earth from apparent motions seen in the sky. Readers know Earth rotates on its axis and revolves around the Sun, but many fail to understand what consequences arise from these two basic

© 2011 Cengage Learning. All Rights Reserved. May not be scanned, copied or duplicated, or posted to a publicly accessible website, in whole or in part.

motions of Earth. This workbook discusses in depth both the short-term and long-term motions of Earth and their effects on objects in the sky.

One advantage of *TheSkyX* is its demonstration of the effects of Earth's rotation on objects seen in the sky. Students and instructors can observe the continuous east-west motion of the sky over a 24-hour period. At the same instant, they can face any direction on the surface of Earth and observe this motion. Time increments (in seconds, minutes, hours, and days) can be manually set to observe short-term and long-term effects on the positions of objects. The remarkable thing is that you can make observations from *any* location on Earth.

Another interesting thing that can be done with *TheSkyX* is to look at the Sun's motion through the sky. This, of course, is caused by the revolution of Earth around the Sun. Users observe first-hand the location of the Sun relative to the stars and can correlate its position in the sky to the seasons on Earth. You can also observe short-term and long-term motions of the Moon and the planets. These motions puzzled many ancient astronomers for centuries and eventually led them to construct models of their "universe" that were incorrect.

TheSkyX provides several 'Sky' windows to illustrate different phenomena observed in the sky. The motions of the Sun, Moon, and the planets are just a few examples of interesting windows that you can open and view. The user simply opens a sky window, sits back, and observes! The authors have painstakingly prepared additional 'Sky' windows in addition to those provided in *TheSkyX's* database. These 'Sky' windows are intended to help demonstrate the concepts we discuss throughout the workbook. The windows are well integrated into the workbook for continuity. Using these 'Sky' windows helps to gain a better insight of the phenomena happening in the sky.

This workbook details planetary motion and configurations. It uses *TheSkyX* to show their orientation within the Solar System and their appearance as well as their observed motions in Earth's sky. This is very helpful when you are observing the planets for the first time.

The workbook also provides clear and concise explanations of planetary orientations with respect to Earth and the Sun. *TheSkyX* provides an excellent tool that displays three-dimensional views of the Solar System. It shows precise arrangements and alignments of all the planets for any specified time. This tool offers the user a clear understanding of where planets are located relative to Earth and Sun as well of as their appearance in Earth's sky.

Students are provided with many hands-on exercises while using *TheSkyX* software. The software allows users to move ahead or backward in time. They may observe the sky as it was when Julius Caesar was in power, or perhaps, as it will appear thousands of years in the future. By moving ahead or back in time, one will appreciate the effects of Earth's precession on the orientation of the sky and the effect this precession has on the coordinates of objects in it. This is something you would normally have to go to a planetarium to view.

Perhaps one of the most difficult concepts to convey to students and people in general, is that of time. A whole chapter is devoted to the concept of time and the different time systems in use throughout the world and in observational astronomy. It outlines how time is determined for the world, for different locations on Earth, and most importantly locally. It is certainly true that time *is relative* but it is more important to understand how your location on Earth affects the observed time of astronomical events in the sky.

We have specifically designed exercises to show how one's location on Earth affects the time of observed events. Using *TheSkyX* allows anyone to determine the time for the beginning of the seasons for any year (past, present, or future). It helps users determine the local time of astronomical events, such as the apparent sunrise and sunset, for any location on Earth.

The most important time used by astronomers and in observational astronomy is *sidereal time*, time by the stars. Indeed, it is the time system students find most exasperating to learn and comprehend. We have taken great pains here to explain simply how astronomers keep track of this time.

Sidereal time is perhaps the most important time system in observational astronomy to master. Detailed observations are usually planned accordingly on how the sidereal time is computed. The exercises dealing with sidereal time provide students a comprehensive explanation of this complicated time system. It also helps them understand how to calculate it. Time calculation is one of the fundamental tasks in observational astronomy that students must master in order to plan constructive and effective observing sessions.

© 2011 Cengage Learning. All Rights Reserved. May not be scanned, copied or duplicated, or posted to a publicly accessible website, in whole or in part.

Several other tools are available in *TheSkyX* to demonstrate different types of phenomena. One interesting tool is the lunar phase calendar. It provides an up-to-date Moon calendar, for planning observation sessions.

The phasing of the Moon is a phenomenon we discuss at length in the last chapter of this workbook. After completing the exercises, observers should have a clearer understanding of why the Moon appears the way it does in the sky. Diagrams in textbooks and on many Web pages are often confusing. The depiction of lunar phases is difficult for many to comprehend at a glance. We have tried hard in this workbook to make sure the diagrams are correct and easy to follow. They depict the Moon's motion as well as its position relative to Earth and the Sun. In fact, after using *TheSkyX* and doing the exercises, students should be able to determine the Moon's orbital position from simply observing its appearance in the sky. With experience, you will even be able to estimate the Moon's apparent rising and setting times as well. *TheSkyX* depicts lunar phases, apparent rising and setting times, and motions very accurately.

Another tool in *TheSkyX* is the eclipse tool. This tool can be used to determine the time and location of eclipses. Eclipses of the Sun and Moon are somewhat rare phenomena that people usually read or hear about after they have happened. They often occur in remote or inaccessible locations on Earth. *TheSkyX* allows you the luxury of looking for and observing these eclipses from those mysterious and distant locations from the comfort of your own desktop.

This workbook provides a detailed description of the types of eclipses observed. It discusses the geometry of eclipses and the conditions necessary for eclipses to occur. We have clearly explained the connection between lunar phases and eclipses. In fact, with the understanding of how eclipses occur that *TheSkyX* provides, it helps users prepare for solar and lunar eclipses that will occur over the next few millennia.

The main objective of this workbook is to use this newly found knowledge to be able to go outside and have a greater appreciation for the wonders that many have taken for granted over the years. It is our intention to promote an easy hands-on approach to viewing the sky with *TheSkyX*. After completing this workbook and its exercises, students and instructors alike will become very proficient in the use of *TheSkyX*. The possibilities are limitless!

I (Dr. J.) have been associated with the teaching profession for over 32 years. I have always considered myself a 'disciple of astronomy', spreading the 'good news' about what is going on in the sky above. I have been conveying these ideas and concepts to students in my astronomy classes for years. Moreover, I especially would like everyone to be able to appreciate (as I have for many years) the beauties and wonders of the sky and the Universe.

We decided to write this book to provide you, the reader, an opportunity to appreciate the wonders in the night skies as we do. With everything that this software package provides you too will be astounded by the large variety of *"astronomical creatures"* that exist and the phenomena that occur in nature.

With *TheSkyX*, enjoy the sky, as you never have before…may you always have clear skies!!

Tom Jordan
Scott Peters

© 2011 Cengage Learning. All Rights Reserved. May not be scanned, copied or duplicated, or posted to a publicly accessible website, in whole or in part.

Chapter 1

Introduction

We humans have observed the stars for thousands of years. Yet today some of their names and meanings are difficult to track down. We attribute the names given to the bright stars to our ancestors to whom they had a special significance. Our forebears have also given us the names of larger groups of stars called constellations. They named constellations after mythological beasts, gods, and humans with supernatural attributes, and even some common everyday objects.

The stories told long ago about these constellations rival any story told around a blazing campfire today. The ancients used these constellations to determine seasons and to remember different areas of the sky. It seems that humans can remember patterns much better than thousands of individual names. In these patterns, many names have become very familiar after many years of observing.

Astronomers today continue to use the names of the constellations first plotted by the Greek astronomer Hipparchus over 2000 years ago. In 1927, by international agreement, definite boundaries were assigned to the constellations in the sky. Today, astronomers still use the constellations to locate many stellar and non-stellar objects in the sky.

There are several ways to locate celestial objects. Using a star chart is perhaps the easiest. Nevertheless, to do so you must have some knowledge of coordinate systems used in the sky. That is what this workbook is all about. It introduces you to astronomical coordinate systems used in observational astronomy and helps you locate any naked-eye or telescopic object with a click of a mouse button.

This workbook guides you through many exercises so that you can fully use the software package. *TheSkyX* displays the bright "naked-eye" stars and those down to the limit of the Hubble Space Telescope. It locates planets and many other Solar System objects as well as thousands of deep sky objects.

Before starting your adventure, however, you will need to install *TheSkyX* on your PC or Mac. Listed below are the **Minimum System Requirements** necessary for the program to run effectively. Once you've determined that your PC or Mac meets or exceeds these minimum requirements, a detailed step-by-step procedure follows in order that the necessary files are installed properly.

Minimum System Requirements

TheSkyX program requires the following minimum system requirements:

Windows Users

➢ 1.5 GHz or faster processor

➢ Intel Pentium 4, Pentium M, Pentium D or better, or AMD K-8 (Athlon) or better

➢ OpenGL 1.5 and later

➢ Windows Vista, Windows XP Professional or Windows XP Home

➢ 1024 x 768 display resolution with true color

➢ 128 MB (minimum) video RAM

➢ 520 MB free disc space

➢ CD-ROM drive

© 2011 Cengage Learning. All Rights Reserved. May not be scanned, copied or duplicated, or posted to a publicly accessible website, in whole or in part.

➢ Mouse or other pointing device

➢ Keyboard

Macintosh Users

➢ 1.28 GHz G4 PowerPC or faster processor or 2 GHz Intel Core Duo or faster processor

➢ Macintosh OS X version 10.4.8 or later

➢ 512 MB RAM

➢ 64 MB video RAM

➢ 1 GB disc space

➢ CD-ROM or DVD drive

➢ Mouse or other pointing device

➢ Keyboard

Installation of *TheSkyX*

If your PC or Mac meets or exceeds these system requirements, then you are ready for an incredible journey in *TheSkyX*. Before you start, however, it is wise to read through these installation instructions before you attempt to install *TheSkyX* software. The following only provides some suggestions that will hopefully minimize any frustrations that might occur later on during the installation process.

Windows 7:

1. Logon as an administrator. *TheSkyX* requires administrative privileges to be installed under Windows 7.

2. Insert the CD-ROM in the CD-ROM drive or DVD drive.

3. If the AutoPlay window in **Figure 1-3** does not appear, Click **Start** in the lower left corner of your desktop window, then **Computer** as shown in **Figure 1-1**. If the AutoPlay window appears, proceed to step 6 below.

© 2011 Cengage Learning. All Rights Reserved. May not be scanned, copied or duplicated, or posted to a publicly accessible website, in whole or in part.

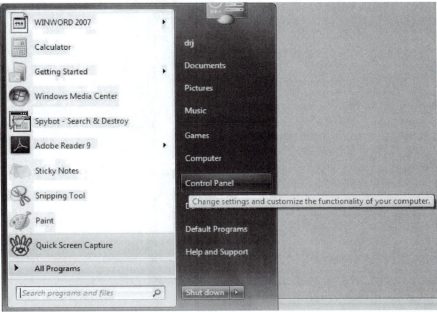

Figure 1-1 The Start Menu in Windows 7

4.   On the Computer window, select the CD-ROM drive or DVD drive that holds *TheSkyX* disc as shown in **Figure 1-2**.

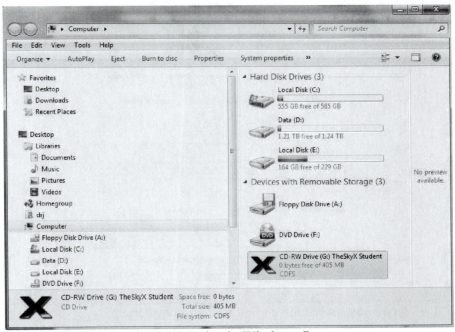

Figure 1-2 Selecting the DVD-Drive in Windows 7

5.   Click the **AutoPlay** button located near the top-left of the window in **Figure 1-2**.

6.   This opens the AutoPlay window. Click **Run Readme.htm** as shown in **Figure 1-3**. This connects the user with the installation link on the Software Bisque disc as shown in **Figure 1-4**.

© 2011 Cengage Learning. All Rights Reserved. May not be scanned, copied or duplicated, or posted to a publicly accessible website, in whole or in part.

Figure 1-3 The AutoPlay Menu in Windows 7

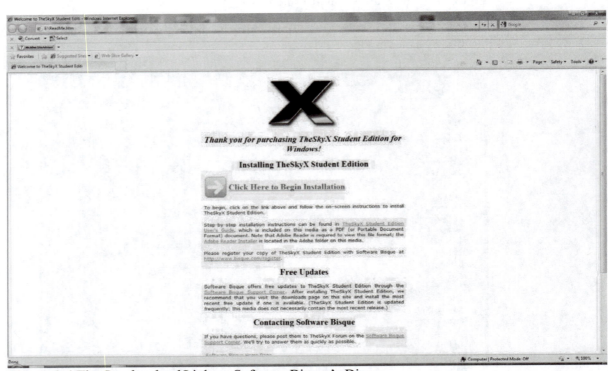

Figure 1-4 The *Readme.html* Link on Software Bisque's Disc

7. After carefully reading the instructions in the ReadMe file, click the "**Click Here to Begin Installation**" link.

8. Follow the on-screen instructions to complete the installation.

Windows Vista:

1. Logon as an administrator. *TheSkyX* requires administrative privileges to be installed under Vista.

© 2011 Cengage Learning. All Rights Reserved. May not be scanned, copied or duplicated, or posted to a publicly accessible website, in whole or in part.

2. Insert the CD-ROM in the CD-ROM drive or DVD drive.

4. If the AutoPlay window in **Figure 1-7** does not appear, Click **Start** in the lower left corner of your desktop window, then **Computer** as shown in **Figure 1-5**. If the AutoPlay window appears, proceed to step 6 below.

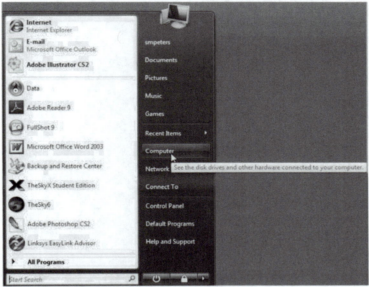

Figure 1-5 The Start Menu in Windows Vista

4. On the Computer window, select the CD-ROM drive or DVD drive that holds *TheSkyX* disc as shown in **Figure 1-2**.

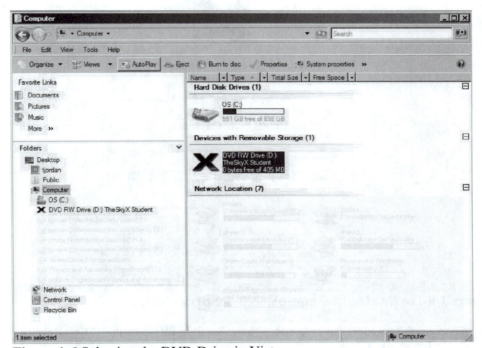

Figure 1-6 Selecting the DVD-Drive in Vista

5. Click the **AutoPlay** button located near the top-center of the window in **Figure 1-6**.

© 2011 Cengage Learning. All Rights Reserved. May not be scanned, copied or duplicated, or posted to a publicly accessible website, in whole or in part.

6. This opens the AutoPlay window. Click **Run Readme.htm** as shown in **Figure 1-7**. This connects the user with the installation link on the Software Bisque disc as shown in **Figure 1-8**.

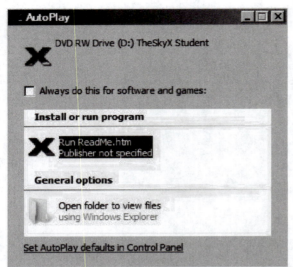

Figure 1-7 The AutoPlay Menu in Windows Vista

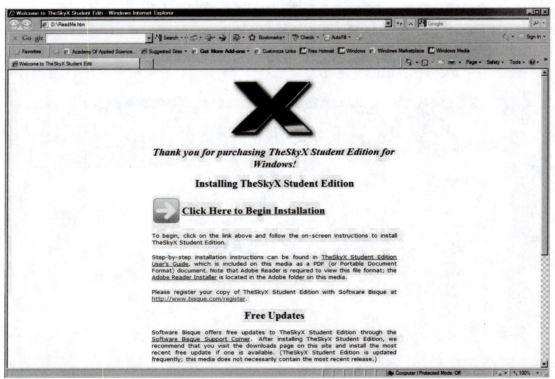

Figure 1-8 The *Readme.html* Link on Software Bisque's Disc

7. After carefully reading the instructions in the ReadMe file, click the "**Click Here to Begin Installation**" link.

8. Follow the on-screen instructions to complete the installation.

© 2011 Cengage Learning. All Rights Reserved. May not be scanned, copied or duplicated, or posted to a publicly accessible website, in whole or in part.

Windows XP:

1. Log on as an administrator. *TheSkyX* requires administrative privileges to be installed under XP.

2. Insert the CD-ROM in the CD-ROM or DVD drive

3. Wait for the ReadMe file, like the one in **Figure 1-8,** to appear in a browser window.

4. If Windows XP Auto Run is not active, then click **Start** in the lower left corner of your desktop window. Click on **My Computer**, as shown in **Figure 1-9**.

Figure 1-9 The Start Menu in Windows XP

5. Now, right-click on the CD-ROM drive or DVD-drive that holds *TheSkyX* media, as shown in **Figure 1-10**.

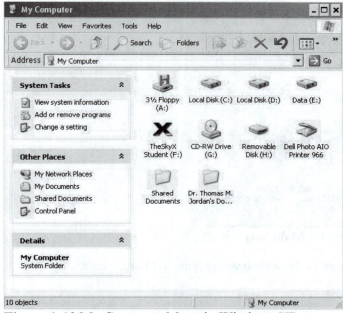

Figure 1-10 My Computer Menu in Windows XP

6. Next, click **AutoPlay** that is shown in the menu displayed in **Figure 1-11**.

© 2011 Cengage Learning. All Rights Reserved. May not be scanned, copied or duplicated, or posted to a publicly accessible website, in whole or in part.

Figure 1-11 The AutoPlay Option

7. This connects the user with the installation link on the Software Bisque disc shown in **Figure 1-8.**

8. After carefully reading the instructions in the **ReadMe file**, click the "<u>**Click Here to Begin Installation**</u>"link on Software Bisque's disc as shown in **Figure 1-8**.

9. Follow the on-screen instructions to complete the installation.

Detailed Installation Instructions and Menus for PCs

Once the user clicks "**Begin Installation**" they are prompted by several menu windows. The following details the installation process. There are several files that are downloaded from Software Bisque's website. At first, your PC may not recognize the installation protocol as being a "friendly" file. It will probably warn you of a potentially malicious file download as displayed in **Figure 1-12**. The user has the option to either Save the file, Run the file, or to Cancel it. Just click "Run."

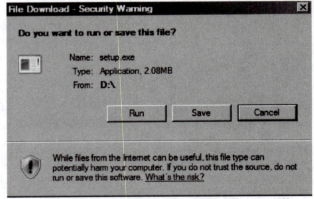

Figure 1-12 Security Warning for Malicious File

After clicking Run, a window such as the one shown in **Figure 1-13** opens indicating what is being downloaded from the website.

© 2011 Cengage Learning. All Rights Reserved. May not be scanned, copied or duplicated, or posted to a publicly accessible website, in whole or in part.

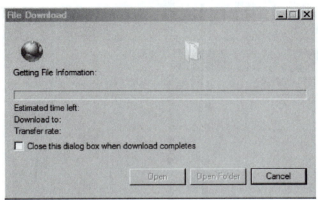

Figure 1-13 The File Download Window

Once this file is downloaded *TheSkyX* Student Edition InstallShield Wizard that installs *TheSkyX* program opens. It is shown in **Figure 1-14**. The program is installed while this window is displayed.

Figure 1-14 *TheSkyX* InstallShield Window

As one can see, this may take several minutes depending on your operating system. New files are then copied to the user's PC as shown in **Figure 1-15**.

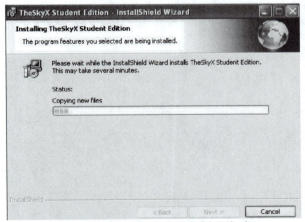

Figure 1-15 *TheSkyX* InstallShield Window

© 2011 Cengage Learning. All Rights Reserved. May not be scanned, copied or duplicated, or posted to a publicly accessible website, in whole or in part.

Once the program is successfully installed by the Wizard the user is then notified by the InstallShield Wizard as shown in **Figure 1-16**.

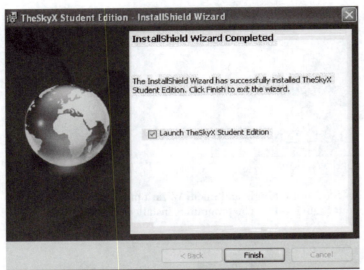

Figure 1-16 InstallShield Wizard Successfully Installed *TheSkyX*

At this point, one has the option to "Launch *TheSkyX* Student Edition or not. If one chooses not to run the program, then uncheck the box next to "Launch *TheSkyX* Student Edition and click "Finish."

When you decide to run *TheSkyX* program at a later time simply click the icon on the desktop window. If the program fails to start, then simply click **Start** at the bottom left of your desktop, then **All Programs, Software Bisque, TheSkyX Student Edition,** *TheSkyX Student Edition*.

Name and Serial Number Registration

The first time *TheSkyX* is launched, you'll be prompted to enter your name and serial number as displayed in **Figure 1-17**. The serial number is located on the outside of the CD case.

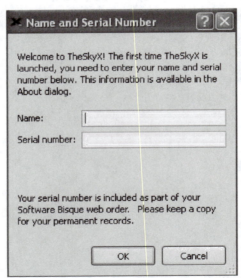

Figure 1-17 Name and Serial Number Window

© 2011 Cengage Learning. All Rights Reserved. May not be scanned, copied or duplicated, or posted to a publicly accessible website, in whole or in part.

Figure 1-18 displays the Name and Serial Number window that has been completed by Dr. Jordan.

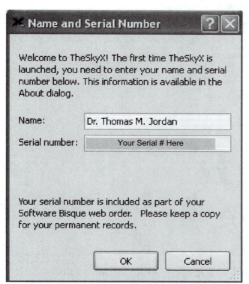

Figure 1-18 The Name and Serial Number Window

After the installation and registration are completed *TheSkyX* program runs for the first time. While loading, a window like the one displayed in **Figure 1-19** appears on your desktop.

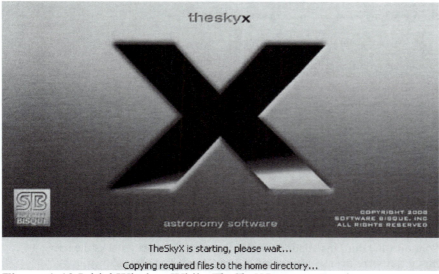

Figure 1-19 Initial Window While *TheSkyX* Program Loads

MacIntosh:

1. To install most applications made for Mac OS X, simply insert the disc. For applications downloaded from the Internet, double-click the disk image file. Then open the installer provided and follow onscreen instructions. *TheSkyX* requires administrative privileges to be installed under Macintosh OS X version 10.4.8 or later.

2. Insert the CD-ROM in the CD-ROM drive or DVD drive. A window appears like the one shown in **Figure 1-20**. This window is the Introduction window. It describes where the software is going to be installed on your Mac.

© 2011 Cengage Learning. All Rights Reserved. May not be scanned, copied or duplicated, or posted to a publicly accessible website, in whole or in part.

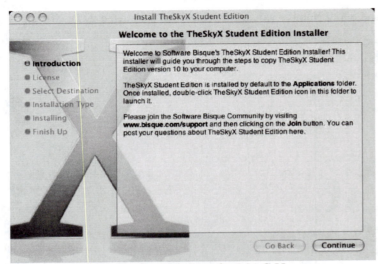

Figure 1-20 Introduction Window for MAC Users

3. Click "Continue." A window opens like the one shown in **Figure 1-21** that displays the Software License Agreement. Click "Continue" again.

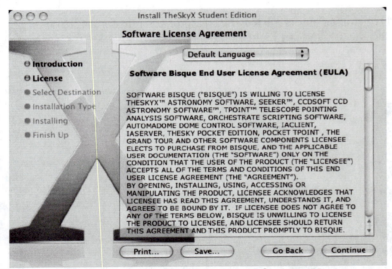

Figure 1-21 Software License Agreement for Mac Users

4. After clicking Continue, the third installation window opens. The installer asks the user if they agree with the software license agreement. Click "Agree." This is shown in **Figure 1-22**.

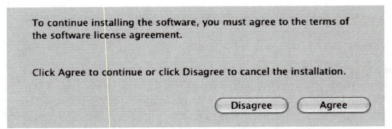

Figure 1-22 Third Installation Window for Mac Users

© 2011 Cengage Learning. All Rights Reserved. May not be scanned, copied or duplicated, or posted to a publicly accessible website, in whole or in part.

5. After clicking Continue, the installer prompts the user to "Select the Destination" where the program is to be installed. Click "Continue." The default destination is the Mac's hard drive in the Applications folder as shown in **Figure 1-23**.

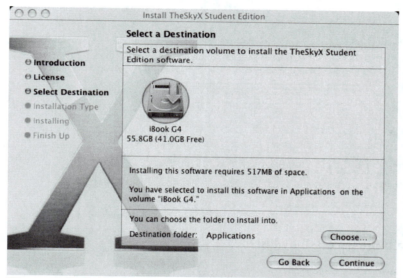

Figure 1-23 Select Destination Window for Mac Users

6. After selecting "Continue," the installation process begins. **Figure 1-24** displays that 50% of the program files have been written to the hard drive.

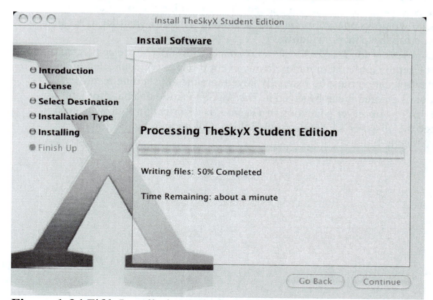

Figure 1-24 Fifth Installation Window for Mac Users

7. Once the program is installed the installer informs the user that the installation was a success as is displayed in **Figure 1-25**.

© 2011 Cengage Learning. All Rights Reserved. May not be scanned, copied or duplicated, or posted to a publicly accessible website, in whole or in part.

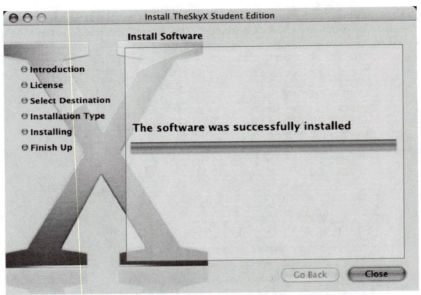

Figure 1-25 Software Successfully Installed

8. Click "Close."

Once *TheSkyX* program is loaded a window, like the one shown in **Figure 1-26**, appears on your desktop. This is the default window in *TheSkyX* program. The user may now select how to display the sky from the command menus that are opened. The first thing that one notices is that there are a sundry of menu windows opened in *TheSkyX's* desktop window. In order to see the desktop window clearly, close these menu windows.

The default location setting for *TheSkyX* desktop is *your* location if you are connected to the Internet. If not, it may be set to Golden, Colorado. In any event if one is not connected to the Internet, then the observer's location must be manually keyed into the appropriate menu.

Once this information is entered it can be saved in *TheSkyX's* location database for later use. The procedure to set or reset an observer's location in *TheSkyX* is described in following section. This exercise will change the observer's location from Golden, Colorado to the Ball State Observatory located in Muncie, Indiana.

© 2011 Cengage Learning. All Rights Reserved. May not be scanned, copied or duplicated, or posted to a publicly accessible website, in whole or in part.

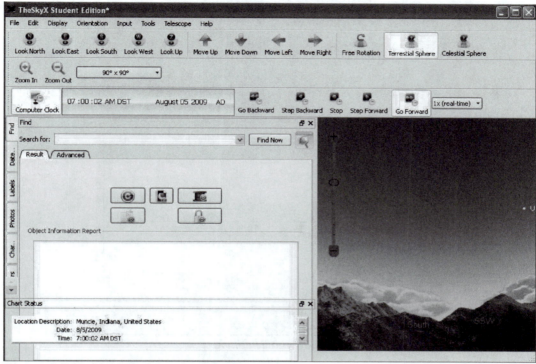

Figure 1-26 Initial Window after *TheSkyX* Loads

Setting Your City's Latitude and Longitude in *TheSkyX*

1. Close all the menu windows that are open in *TheSkyX* window.

2. Go to ***Input*** on the Standard toolbar at the top of *TheSkyX* window as shown in **Figure 1-27**.

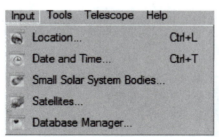

Figure 1-27 The Input Menu Window in *TheSkyX*

3. Click ***Location*** in the Input Menu. The Location menu window appears such as the one shown in **Figure 1-28**.

4. One may now select any location in Germany, an International location, the location of any Observatory world-wide, Star Parties locations, or any city in the United States. In order to do this you must expand the list. Click on ⊞ next to one of the location sites such as the United States as shown in **Figure 1-29**. Scroll down the list until you find your location.

© 2011 Cengage Learning. All Rights Reserved. May not be scanned, copied or duplicated, or posted to a publicly accessible website, in whole or in part.

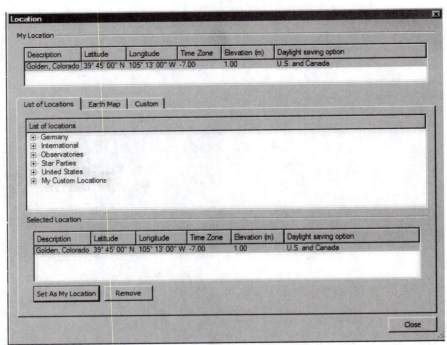

Figure 1-28 The Location Menu Window in *TheSkyX*

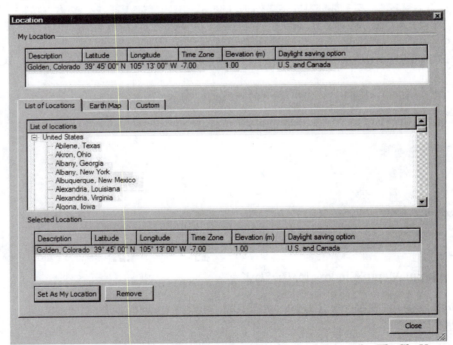

Figure 1-29 Expanded Location List for The United States in *TheSkyX*

5. If your location is not in any of these lists, then you may customize your location by clicking on the *Custom* tab as shown in **Figure 1-30**.

© 2011 Cengage Learning. All Rights Reserved. May not be scanned, copied or duplicated, or posted to a publicly accessible website, in whole or in part.

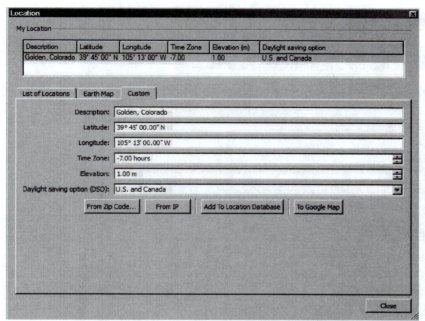

Figure 1-30 Customizing One's Location in *TheSkyX*

6. Type in the name of your location in the Description line, then the Latitude and Longitude. Next, type in your Time Zone (-5 hours – Eastern Time Zone; -6 hours – Central Time Zone; -7 hours – Mountain Time Zone; -8 hours – Pacific Time Zone). Next, your Elevation if known. Finally, if your location observes Daylight Savings Time then you must choose the Daylight saving option (DSO) to be the U.S. and Canada as displayed in **Figure 1-31**. If you do not observe Daylight Savings Time, then choose "Not Observed" at the top of the pull down menu shown in **Figure 1-31**.

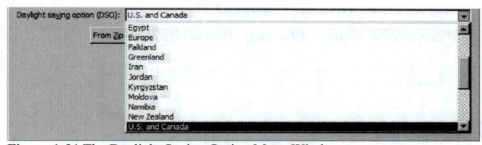

Figure 1-31 The Daylight Saving Option Menu Window

7. **Figure 1-32** displays the change in the observer's location from Golden, Colorado to the Ball State Observatory located in Muncie, Indiana. Once the change has been made, click

    Add to the Location Database (Add To Location Database) bar.

© 2011 Cengage Learning. All Rights Reserved. May not be scanned, copied or duplicated, or posted to a publicly accessible website, in whole or in part.

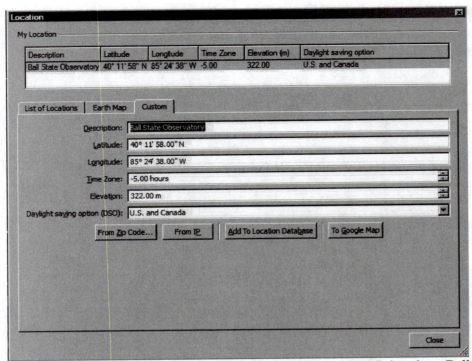

Figure 1-32 Changing the Observer's Location from Golden, Colorado to Ball State Observatory in Muncie, Indiana

8. Now, click on the **List of Locations** tab and expand the **My Custom Locations** by clicking on the ⊞ again.

9. Ball State Observatory should now be on the "My Custom Locations" list and may be accessed at any future time. It is shown in **Figure 1-33**.

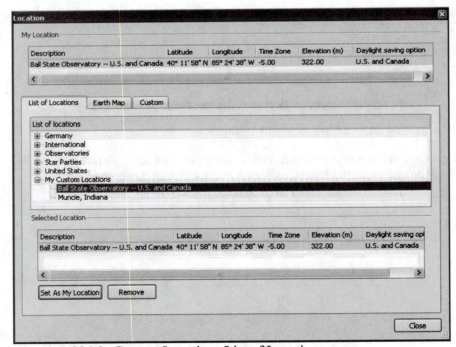

Figure 1-33 My Custom Locations List of Locations

10. Click **Set As My Location** to finish the change.

© 2011 Cengage Learning. All Rights Reserved. May not be scanned, copied or duplicated, or posted to a publicly accessible website, in whole or in part.

There is one more task to be performed before using *TheSkyX* to its fullest. In order to use the *Sky Files* discussed in later chapters, a file folder entitled **Sky Files**, must be created in *TheSkyX* folder, using Windows Explorer. The following procedure describes how to create this folder.

Creating the 'Sky Files' File Folder for the PC

Windows 7:

1. Click on the Windows **Start** button ( ) as displayed in **Figure 1-34.**

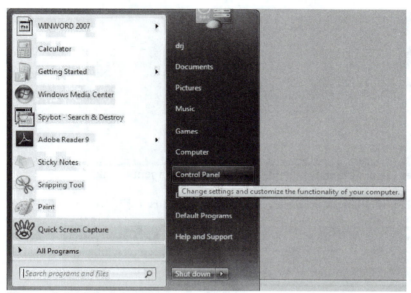

Figure 1-34 Windows Start Menu

2. Next, click on **Computer**. A window similar to **Figure 1-35** appears.

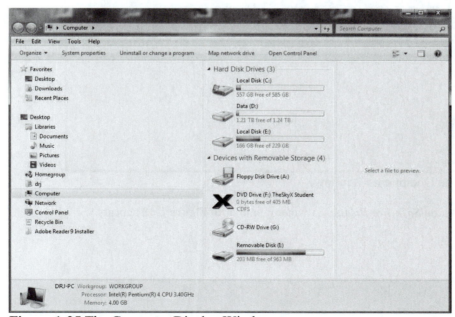

Figure 1-35 The Computer Display Window

© 2011 Cengage Learning. All Rights Reserved. May not be scanned, copied or duplicated, or posted to a publicly accessible website, in whole or in part.

3. Click on **Documents** in the upper left as shown in **Figure 1-36**.

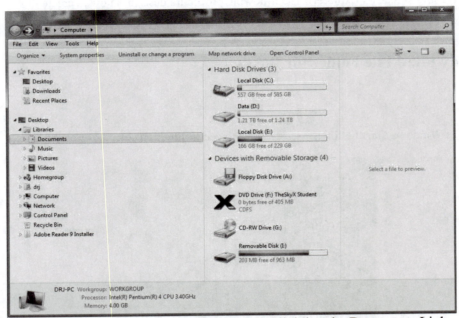

Figure 1-36 Computer Display Window Highlighting the Documents Link.

4. The **Documents** folder should now be displayed as shown in **Figure 1-37**.

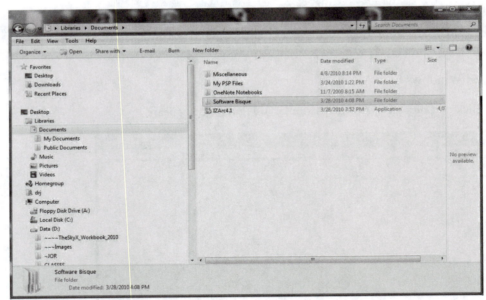

Figure 1-37 The Documents Window

5. Double-Click on **Software Bisque**. A window similar to **Figure 1-38** appears.

© 2011 Cengage Learning. All Rights Reserved. May not be scanned, copied or duplicated, or posted to a publicly accessible website, in whole or in part.

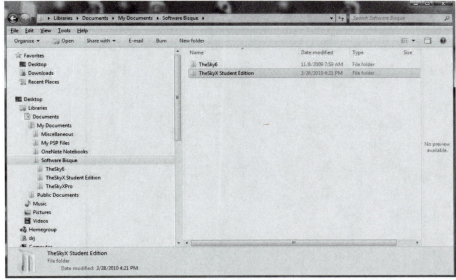

Figure 1-38 The Software Bisque Folder

6. Double-click on ***TheSkyX Student Edition*** folder. A window similar to **Figure 1-39** appears.

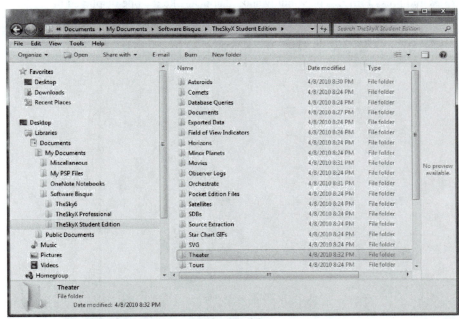

Figure 1-39 *TheSkyX* Student Edition Folder

7. Click on ***New Folder*** in the upper left corner of the window. A window similar to **Figure 1-40** appears.

© 2011 Cengage Learning. All Rights Reserved. May not be scanned, copied or duplicated, or posted to a publicly accessible website, in whole or in part.

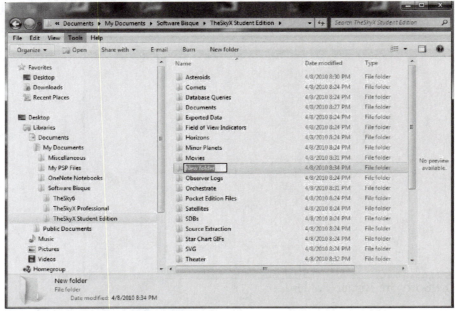

Figure 1-40 The New Folder Window

8.   Type ***Sky Files*** and press ***Enter***.   The window should appear like the one displayed in Figure **1-41**.

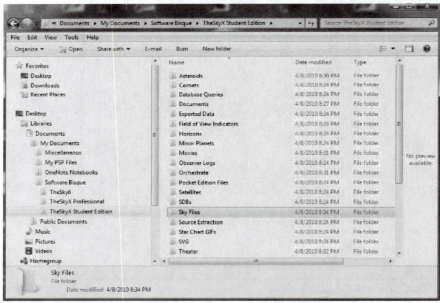

Figure 1-41 Addition of the 'Sky Files' File Folder Complete

9.   The 'Sky Files' file folder has now been created.   The complete path for this folder should be:

C:\Users\Your Username\Documents\Software Bisque\TheSkyX Student Edition\
Sky Files

Later on, you will need to access and download these Sky Files via the Internet.   It is wise to ***organize*** your file folders by ***chapters***.   This will make the download process easier and

© 2011 Cengage Learning. All Rights Reserved. May not be scanned, copied or duplicated, or posted to a publicly accessible website, in whole or in part.

more efficient. These 'Sky Files' may be downloaded from the webpage of Brooks-Cole Publishing Company. Their *URL* is http://www.cengage.com/astronomy/jordan/skyx.

Windows Vista:

1. Click on the Windows *Start* button (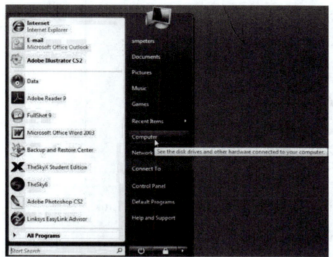) as displayed in **Figure 1-42.**

Figure 1-42 Windows Start Menu

2. Next, click on *Computer*. A window similar to **Figure 1-43** appears.

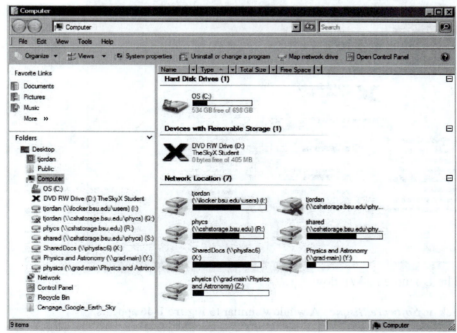

Figure 1-43 The Computer Display Window

3. Click on *Documents* in the upper left as shown in **Figure 1-44**.

© 2011 Cengage Learning. All Rights Reserved. May not be scanned, copied or duplicated, or posted to a publicly accessible website, in whole or in part.

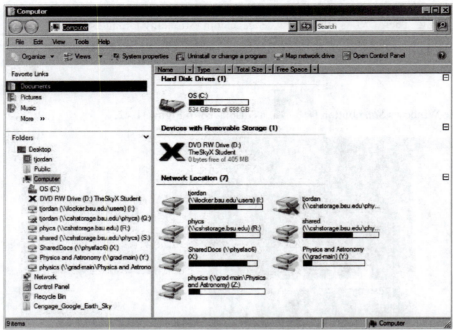

Figure 1-44 Computer Display Window Highlighting the Documents Link.

4.   The ***Documents*** folder should now be displayed as shown in **Figure 1-45**.

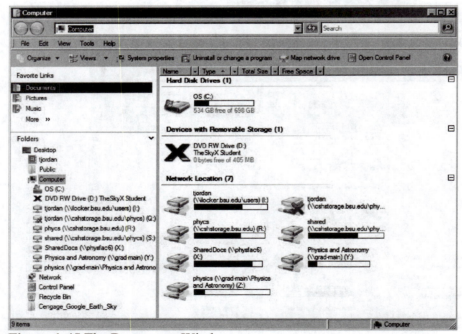

Figure 1-45 The Documents Window

5.   Double-Click on ***Software Bisque***. A window similar to **Figure 1-46** appears.

© 2011 Cengage Learning. All Rights Reserved. May not be scanned, copied or duplicated, or posted to a publicly accessible website, in whole or in part.

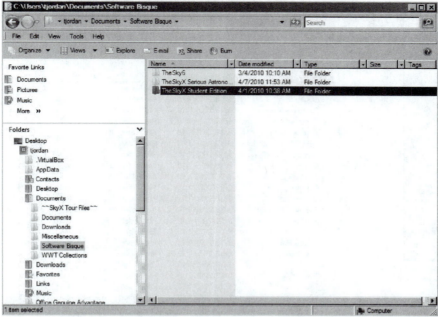

Figure 1-46 The Software Bisque Folder

6. Double-click on *TheSkyX Student Edition* folder. A window similar to **Figure 1-47** appears.

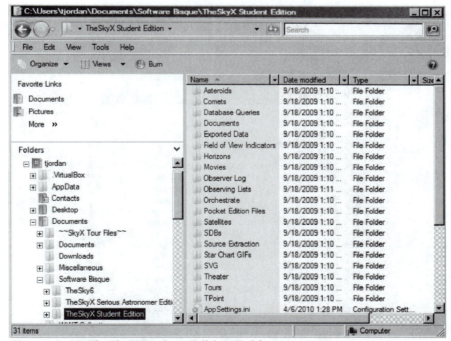

Figure 1-47 *TheSkyX* Student Edition Folder

7. Click on *Organize* in the upper left corner of the window. A drop-down menu appears like in **Figure 1-48**.

© 2011 Cengage Learning. All Rights Reserved. May not be scanned, copied or duplicated, or posted to a publicly accessible website, in whole or in part.

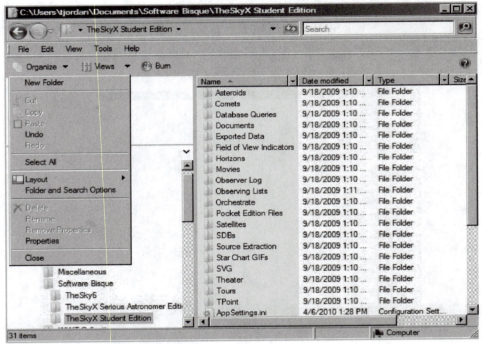

Figure 1-48 The Organize Menu.

8. Click *New Folder*. The window now appears like **Figure 1-49**.

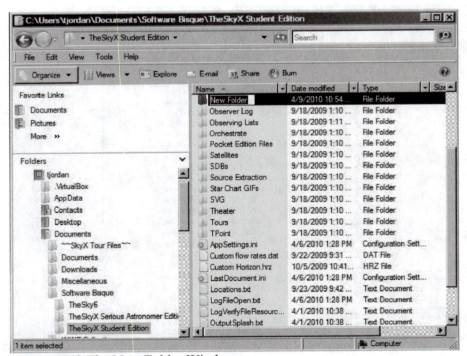

Figure 1-49 The New Folder Window

9. Type *Sky Files* and press *Enter*. The window should appear like the one displayed in Figure **1-50**.

© 2011 Cengage Learning. All Rights Reserved. May not be scanned, copied or duplicated, or posted to a publicly accessible website, in whole or in part.

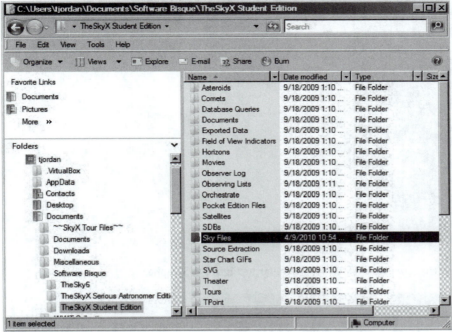

Figure 1-50 Addition of the 'Sky Files' File Folder Complete

10. The 'Sky Files' file folder has now been created. The complete path for this folder should be:

**C:\Users*Your Username*\Documents\Software Bisque\TheSkyX Student Edition\\
Sky Files**

Later on, you will need to access and download these Sky Files via the Internet. It is wise to *organize* your file folders by *chapters*. This will make the download process easier and more efficient. These 'Sky Files' may be downloaded from the webpage of Brooks-Cole Publishing Company. Their *URL* is http://www.cengage.com/astronomy/jordan/skyx.

Windows XP:

1. Click on the Windows **Start** button as displayed in **Figure 1-51.**

© 2011 Cengage Learning. All Rights Reserved. May not be scanned, copied or duplicated, or posted to a publicly accessible website, in whole or in part.

Figure 1-51 Windows Start Menu

2. Next, click on *My Computer*. A window similar to **Figure 1-52** appears.

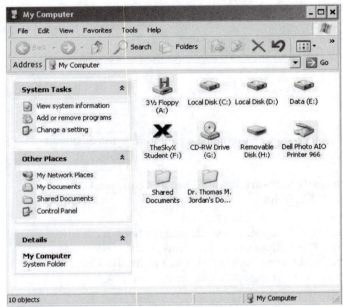

Figure 1-52 My Computer Display Window

3. Double-click on [username] *Documents* in the upper right.

4. The *My Documents* folder should now be displayed as shown in **Figure 1-53**.

© 2011 Cengage Learning. All Rights Reserved. May not be scanned, copied or duplicated, or posted to a publicly accessible website, in whole or in part.

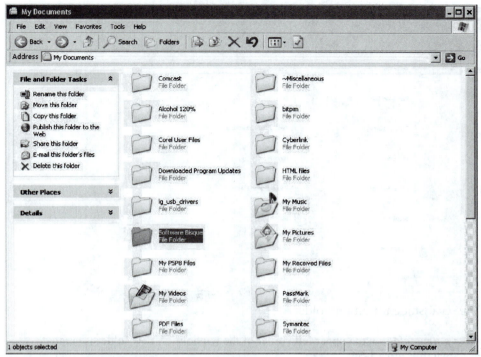

Figure 1-53 The My Documents Window

5. Double-Click on **Software Bisque**. A window similar to **Figure 1-54** appears.

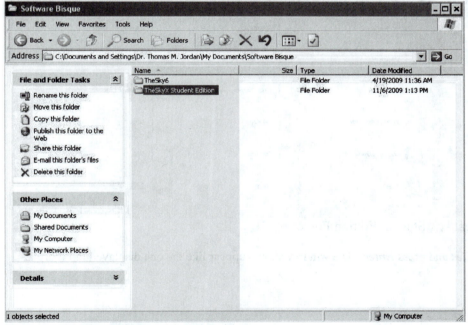

Figure 1-54 The Software Bisque Folder

6. Double-click on **TheSkyX Student Edition** folder. A window similar to **Figure 1-55** appears.

© 2011 Cengage Learning. All Rights Reserved. May not be scanned, copied or duplicated, or posted to a publicly accessible website, in whole or in part.

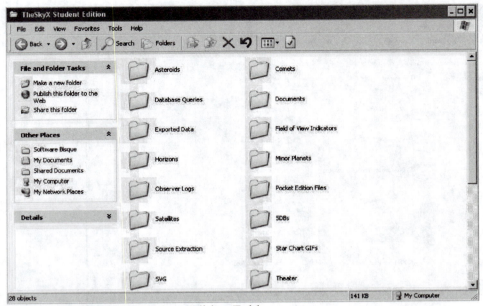

Figure 1-55 *TheSkyX* Student Edition Folder

7. Click on ***Make a new folder*** in the upper left corner of the window. A new folder is created like the one shown in **Figure 1-56**.

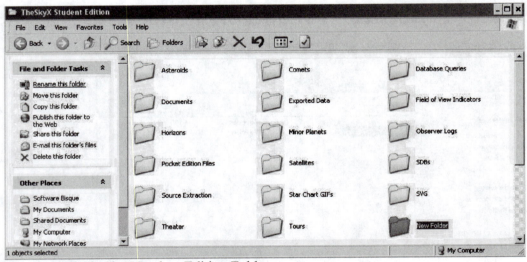

Figure 1-56 *TheSkyX* Student Edition Folder.

8. Type ***Sky Files*** and press ***Enter***. The window should appear like the one displayed in Figure **1-57**.

© 2011 Cengage Learning. All Rights Reserved. May not be scanned, copied or duplicated, or posted to a publicly accessible website, in whole or in part.

Figure 1-57 Addition of the 'Sky Files' File Folder Complete

9. The 'Sky Files' file folder has now been created. The complete path for this folder should be:

**C:\Documents and Settings*Your Username*\\My Documents\Software Bisque\\
TheSkyX Student Edition\Sky Files\\.**

Later on, you will need to access and download these Sky Files via the Internet. It is wise to *organize* your file folders by *chapters*. This will make the download process easier and more efficient. These 'Sky Files' may be downloaded from the webpage of Brooks-Cole Publishing Company. Their *URL* is http://www.cengage.com/astronomy/jordan/skyx.

Creating the 'Sky Files' File Folder for the Mac

MacIntosh:

1. First, open Finder and click on User (dr.j.) as shown in **Figure 1-58**.

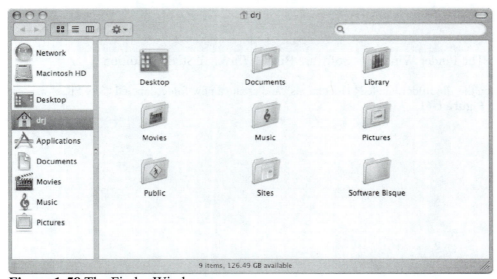

Figure 1-58 The Finder Window

© 2011 Cengage Learning. All Rights Reserved. May not be scanned, copied or duplicated, or posted to a publicly accessible website, in whole or in part.

2. Next, open the file folder labeled "Software Bisque" as shown in **Figure 1-59**.

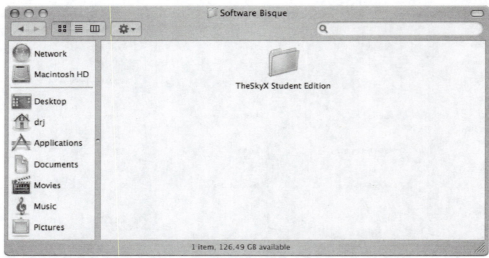

Figure 1-59 The Finder Window – Software Bisque

3. Next, open the file folder labeled *TheSkyX Student Edition* as shown in **Figure 1-60**.

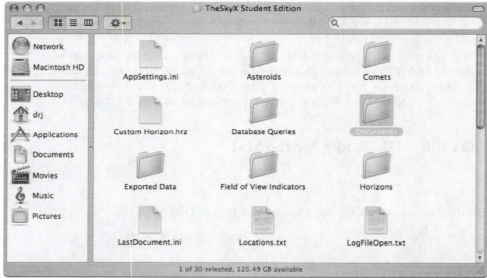

Figure 1-60 The Finder Window – Software Bisque*TheSkyX* Student Edition

4. Next, open the file folder labeled "Documents" and create a new folder labeled "Sky Files" as shown in **Figure 1-61**.

© 2011 Cengage Learning. All Rights Reserved. May not be scanned, copied or duplicated, or posted to a publicly accessible website, in whole or in part.

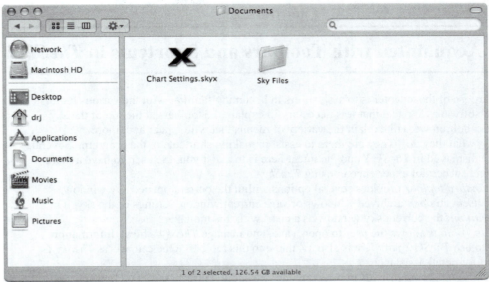

Figure 1-61 The Finder Window – Software Bisque*TheSkyX* Student Edition\\Documents

5. It will be necessary to create several other file folders in the "Sky Files" folder. It is suggested that you create these folders according to chapters in the *TheSkyX* Workbook as shown in **Figure 1-62**.

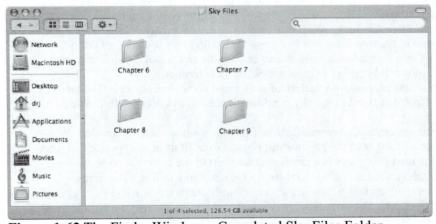

Figure 1-62 The Finder Window – Completed Sky Files Folder

The location of your Sky files file folder is:

User\\Software Bisque\\TheSkyX Student Edition\\Documents\\Sky Files\\.

Later on, you will need to access and download these Sky Files via the Internet. Since you have **organized** your file folders by **chapters** this will make the download process easier and more efficient. These 'Sky Files' may be downloaded from the webpage of Brooks-Cole Publishing Company. Their *URL* is http://www.cengage.com/astronomy/jordan/skyx.

© 2011 Cengage Learning. All Rights Reserved. May not be scanned, copied or duplicated, or posted to a publicly accessible website, in whole or in part.

Chapter 2

Getting Acquainted with Toolbars and Shortcuts in *TheSkyX*

The purpose of this chapter is to assist users in becoming familiar with the various toolbars in *TheSkyX* software. We feel that it is necessary to explain the toolbars at the top of the sky window. In addition we will explain the variety of menus that will appear and, more importantly, what they do! They are there to assist in making shortcuts in the program. Several toolbars are displayed in *TheSkyX* and the intent here is to assist you, the user, to have a more enjoyable and successful experience in using *TheSkyX*.

The *Standard toolbar* provides several options within the current opened sky window. With it, the user can *open* archived windows or *save* current window settings of the sky. One can even *print out* the current sky window as a chart with this toolbar.

The *File Toolbar* allows the user to open, save, and manage *TheSkyX* files. Information about how to use *TheSkyX* software is also available in this toolbar. One can access *TheSkyX* user manual for detail assistance.

The *Orientation toolbar* helps in orienting the sky and the direction in which the user is looking. The user may choose either the celestial pole or the zenith to be up in either of these two modes.

The *View toolbar* provides shortcuts to finding the coordinates of celestial objects. It allows users to display times that objects appear to rise, set, or transit the local celestial meridian. It lets them change their view of the sky from the night sky mode to a day sky mode and vice versa. It allows the toggling on and off of names of objects and constellation lines and boundaries. It will even let users change the sky window to a **night vision** mode.

The *Field of View toolbar* lets users zoom in or out on a particular region in the sky. One can zoom from a horizon to horizon view to a telescopic one with just a touch of a button.

The *Field of View Chooser toolbar* allows one to change the sky desktop window to any particular field of view not exhibited in the regular field of view toolbar.

The *Objects toolbar* lets users view a variety of stellar and nonstellar objects. The types of objects can be selected and displayed in the sky window. It also allows users to find objects in the sky.

The *Time Skip toolbar* permits incrementing time flow steps while viewing the sky. With this toolbar, you can view changes in the positions of objects over short or long periods of time.

What follows in the next few pages is a detailed discussion of the information about these toolbars and what one might want to include in them on the desktop window. The items displayed on the various toolbars that follow are not necessarily items that are shown after in the default installation of *TheSkyX*. These toolbars may be customized by the user to include more or less shortcuts at any time.

Using the Standard Toolbar

The first toolbar is the *Standard toolbar*. It contains buttons for creating new documents, opening existing documents, saving active documents, copying, pasting, printing information about *TheSkyX*, and getting help. It is displayed in **Figure 2-1**.

File	Edit	Display	Orientation	Input	Tools	Telescope	Help

Figure 2-1 The Standard Toolbar in *TheSkyX*

Below the *Standard toolbar* provide a variety of *toolbars* that one may use in *TheSkyX* that are displayed like icons. One such toolbar is the *File toolbar*.

© 2011 Cengage Learning. All Rights Reserved. May not be scanned, copied or duplicated, or posted to a publicly accessible website, in whole or in part.

Using the File Toolbar

The *File toolbar* can host a variety of commands at the user's disposal. The authors suggest a configuration similar to the one shown in **Figure 2-2**. The default File toolbar, however, does not display all that is shown in **Figure 2-2**. The customized *File toolbar* is displayed in **Figure 2-2** and discussed in detail in **Appendix E**. The user essentially manages *TheSkyX* files with this menu.

Figure 2-2 The File Toolbar

Allows the user to open sky files in *TheSkyX*.

Allows the user to save sky files in *TheSkyX*.

Allows the user to rename sky files in *TheSkyX*.

Allows the user to edit toolbars in *TheSkyX*

Allows the user to print the desktop window as a chart in *TheSkyX*.

Allows the user to modify the printer setup for charts in *TheSkyX*.

Allows the user to play animated sky windows that have been created so that the user may appreciate the fascinating things that one can see in the sky.

Allows the user to open the User's Guide PDF file.

Allows the user access to information about *TheSkyX Student Edition*.

© 2011 Cengage Learning. All Rights Reserved. May not be scanned, copied or duplicated, or posted to a publicly accessible website, in whole or in part.

Allows the user to exit sky files in *TheSkyX*.

Another useful toolbar is the Orientation toolbar. It is in this toolbar that allows the user to view the sky in many different ways.

Using the Orientation Toolbar

The *Orientation toolbar* contains many buttons that control where the observer is looking in the sky and is displayed in **Figure 2-3**.

Figure 2-3 The Orientation Toolbar

With them, one can move view the sky in any direction even overhead. The directional buttons are clearly marked and control the direction in which *TheSkyX* is plotted on the computer monitor and the direction that the user is facing. In addition, one can move the sky left, right, up, and down. In addition to these directions, it permits the user to select different orientations of the sky.

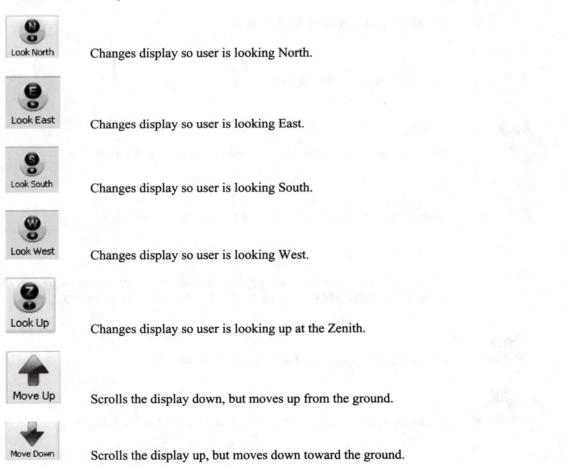

Changes display so user is looking North.

Changes display so user is looking East.

Changes display so user is looking South.

Changes display so user is looking West.

Changes display so user is looking up at the Zenith.

Scrolls the display down, but moves up from the ground.

Scrolls the display up, but moves down toward the ground.

© 2011 Cengage Learning. All Rights Reserved. May not be scanned, copied or duplicated, or posted to a publicly accessible website, in whole or in part.

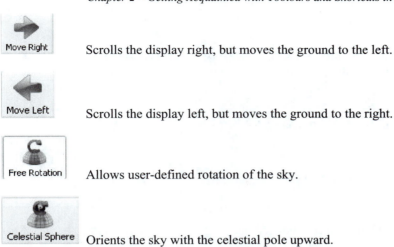

Scrolls the display right, but moves the ground to the left.

Scrolls the display left, but moves the ground to the right.

Allows user-defined rotation of the sky.

Orients the sky with the celestial pole upward.

Orients the sky with the zenith upward.

Using the Display Toolbar

The *View toolbar* contains buttons to configure *TheSkyX* in a variety of ways and is displayed in **Figure 2-4**. It allows users to view the Solar System in a three-dimensional mode.

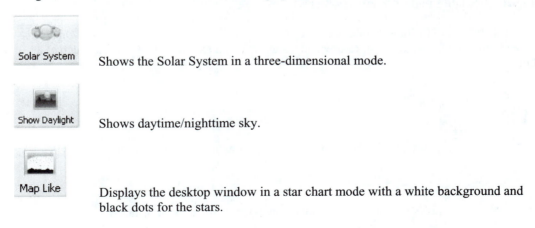

Figure 2-4 The Display Toolbar

It lets users view the sky in a daytime or nighttime mode. It allows users to display the desktop window in either a "Map-Like" or "Photo-Like" appearance. It also allows users to display the horizon or equatorial grid onto the sky. These grids are useful in determining where astronomical objects are located in the sky. You can display constellation lines and boundaries using this toolbar. Several other screen-viewing options are also available to users.

Shows the Solar System in a three-dimensional mode.

Shows daytime/nighttime sky.

Displays the desktop window in a star chart mode with a white background and black dots for the stars.

© 2011 Cengage Learning. All Rights Reserved. May not be scanned, copied or duplicated, or posted to a publicly accessible website, in whole or in part.

Displays the desktop window in a photo-like mode with a color rendered Milky Way. The stars appear to have color to simulate their temperatures.

Displays or hides the horizon grid.

Displays or hides the equatorial grid.

Displays or hides the constellation lines.

Displays or hides the constellation boundaries.

Displays or hides the common names of objects.

Changes the color scheme of the screen in order to preserve the user's night vision.

The *Field of View* (FOV) *toolbar* has just two buttons when installed...Zoom In and Zoom Out. The observer, however, has various options to choose from in *TheSkyX*. as shown in **Figure 2-5**. In order to change the field of view one simply clicks on any of the items displayed or use the appropriate set of keys on the keyboard. See **Appendix E** for further discussion.

Using the Field of View Toolbar

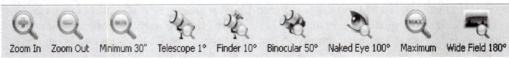

Figure 2-5 The Field of View Toolbar

This like all of the other toolbars can be customized to the needs of the observer. One can zoom in to a field of view of 30″ (arcseconds) and out to a maximum of 235° at the touch of a button.

Allows the user to zoom in on an object in the desktop window.

Allows the user to zoom out on an object in the desktop window.

© 2011 Cengage Learning. All Rights Reserved. May not be scanned, copied or duplicated, or posted to a publicly accessible website, in whole or in part.

Allows the user to zoom in to a field of view of 30″ (arcseconds) in the desktop window.

Allows the user to zoom to a typical telescope finder field of view of 1° in the desktop window.

Allows the user to zoom to a typical telescope finder field of view of 10° in the desktop window.

Allows the user to zoom to a typical binocular finder field of view of 50° in the desktop window.

Allows the user to zoom to the typical naked-eye field of view of 100° in the desktop window.

Allows the user to zoom to a horizon to horizon view. A field of view of 180° in the desktop window.

Allows the user to zoom to maximum field of view of 235° in the desktop window.

If none of the options in this toolbar are suitable, the observer *may* customize the field of view in order to display the desktop window to match their needs. By clicking on the *Field of View Chooser* toolbar, the user can easily change the field of view in the desktop window. The field of view chooser bar is displayed in **Figure 2-6**.

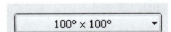

Figure 2-6 The Field of View Toolbar

Using the Field of View Chooser Toolbar

After clicking the field of view chooser toolbar icon, a window opens such as the one shown in **Figure 2-7**. there you will find all of the options previously mentioned. If you wish to change the FOV to something other than what is listed in **Figure 2-7**, then click the "Custom Fields of View" option.

© 2011 Cengage Learning. All Rights Reserved. May not be scanned, copied or duplicated, or posted to a publicly accessible website, in whole or in part.

	Custom...	
	Minimum 30"	Ctrl+Alt+End
	Telescope 1°	Ctrl+Alt+T
	Finder 10°	Ctrl+Alt+F
	Binocular 50°	Ctrl+Alt+B
	Naked Eye 100°	Ctrl+Alt+Y
	Wide Field 180°	Ctrl+Alt+I
	Maximum	Ctrl+Home

Figure 2-7 The Field of View (FOV) Menu

After clicking the "Custom Fields of View" option, a window opens such as the one displayed in **Figure 2-8**. It is in this window that the user may set the field of view to any desired setting! Once these desired fields of view are entered they can be added to *TheSkyX* database to use at a later time. They will be listed in the "My Fields of Views" window.

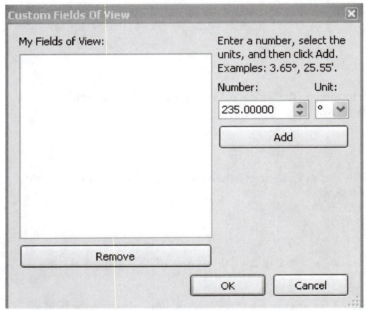

Figure 2-8 The Custom Fields of View Menu

The *Objects toolbar* has just a few buttons on it and is displayed in **Figure 2-9**. It lets users find many astronomical objects in *TheSkyX* database. Nonstellar objects such as galaxies, star clusters, and nebulae can be toggled on and off with a single mouse click.

Using the Objects Toolbar

Figure 2-9 The Object Toolbar

© 2011 Cengage Learning. All Rights Reserved. May not be scanned, copied or duplicated, or posted to a publicly accessible website, in whole or in part.

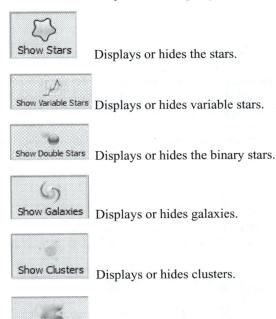

Displays or hides the stars.

Displays or hides variable stars.

Displays or hides the binary stars.

Displays or hides galaxies.

Displays or hides clusters.

Displays or hides nebulae.

The *Time Skip toolbar* allows users to control or input increments of time that can be skipped either ahead or behind in the sky and is displayed in **Figure 2-10**. It also controls tracking features in the software. The default setting is the current time which uses the observer's Computer Clock. By changing the settings in the *Time Skip toolbar*, users can observe short-term motions of objects such as the east–west movement of the Sun, Moon, planets, and stars. One can also observe long-term motions of the Sun, Moon, and planets.

Using the Time Skip Toolbar

Figure 2-10 The Time Skip Toolbar

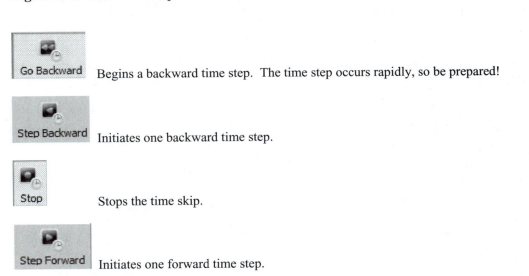

Begins a backward time step. The time step occurs rapidly, so be prepared!

Initiates one backward time step.

Stops the time skip.

Initiates one forward time step.

© 2011 Cengage Learning. All Rights Reserved. May not be scanned, copied or duplicated, or posted to a publicly accessible website, in whole or in part.

Begins a forward time step (play). The time step occurs rapidly, so be prepared!

The *Time Flow* controls the input time increment as shown in **Figure 2-11**. This increment can be set to anything the user desires such as 1 second, 1 minute, 1 hour, 1 day, sunrise, sunset, a synodic day, a sidereal day, or even 10,000 years. The time increment window is used to set the passage of time. It is also used when setting the date and time ahead and backwards in time. The user has the versatility to set the time flow to any time they wish. The default setting is 1x.

Figure 2-11 The Time Flow Toolbar

If you click on the *Time Flow Toolbar* in **Figure 2-11**, a menu opens such as the one displayed in **Figure 2-12**. Changing the Custom Time Flow Increment to that of 2 days is shown in **Figure 2-12**.

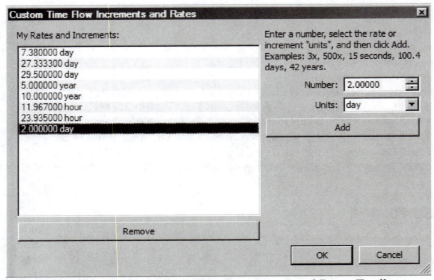

Figure 2-12 The Custom Time Flow Increments and Rates Toolbar

Now, try the exercises on the following pages.

© 2011 Cengage Learning. All Rights Reserved. May not be scanned, copied or duplicated, or posted to a publicly accessible website, in whole or in part.

Chapter 2

TheSkyX Review Exercises

TheSkyX Exercise 1: Setting Your Location, Date, and Time in *TheSkyX*

1. Click "Input" on the Standard toolbar at the top of the sky window, then "Location." Then click the "List of Locations" tab. Make certain that the information displayed on your monitor is identical to that shown in **Figure 2-13**.

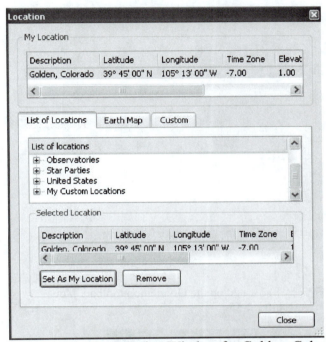

Figure 2-13 The Location Window for Golden, Colorado

2. If the information is not identical, then click the "+ sign" next to the United States to expand the list of cities. Scroll down the menu and click on "Golden, Colorado" in the list of U.S. cities. Click the "Set As My Location" button.

    Click on "Input" again in the Standard toolbar and then on "Date and Time." The default setting is the user's "Computer Clock." Click the "Stop Clock" (Stop) button and set the date to January 1, 2023 and the time to 8:00 PM STD. The window should look like the one that is illustrated in **Figure 2-14**. Click the "Close ( ) button.

© 2011 Cengage Learning. All Rights Reserved. May not be scanned, copied or duplicated, or posted to a publicly accessible website, in whole or in part.

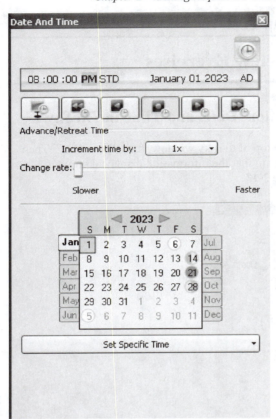

Figure 2-14 Setting the Date and Time in the Command Center Window

4. Click the South shortcut button (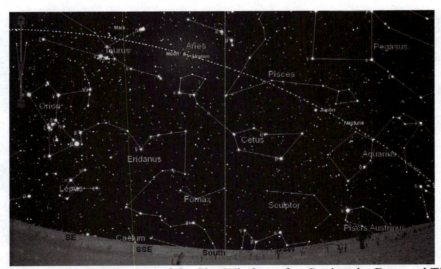 Look South) located on the Orientation toolbar. Your screen now looks like the one shown in **Figure 2-15**.

Figure 2-15 Appearance of the Sky Window after Setting the Date and Time

5. Click the stars shortcut button (Show Stars) on the Objects toolbar. Your screen will appear like the one in **Figure 2-16**.

© 2011 Cengage Learning. All Rights Reserved. May not be scanned, copied or duplicated, or posted to a publicly accessible website, in whole or in part.

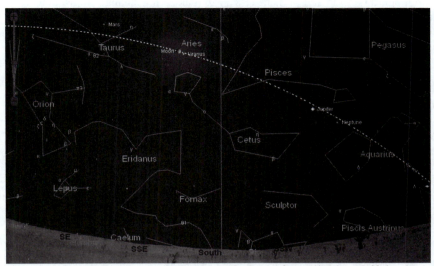

Figure 2-16 Appearance of the Sky Window after Clicking Off the Stars

6. If you click the [Show Stars] button again, the stars return to the sky window.

7. Now, try using some of the other buttons on your own.

TheSkyX Exercise 2: Changing Your Location, Date, and Time in *TheSkyX*

1. Click "Input" on the Standard toolbar at the top of the sky window, then "Location." Expand the list of cities in the "United States" once again by clicking on the "+ sign."

2. Scroll down the menu and click on "Muncie, Indiana." The location window should appear like the one displayed in **Figure 2-17**.

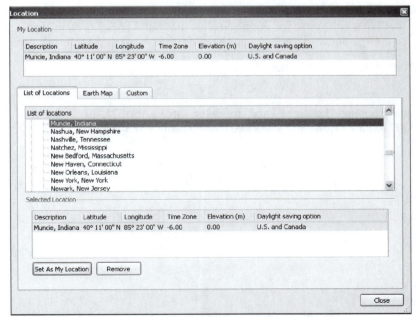

Figure 2-17 Changing One's Location to Muncie, Indiana

© 2011 Cengage Learning. All Rights Reserved. May not be scanned, copied or duplicated, or posted to a publicly accessible website, in whole or in part.

However, due to modern GPS devices, it is necessary to make a few changes to Muncie's latitude, longitude, and elevation.

3. To make changes and update any city's location in *TheSkyX* database one must do the following:

After choosing the location, click on the "Custom" tab. A window like the one displayed in **Figure 2-18** opens. One then can make the necessary changes in coordinates or any other of the location's parameters. The Latitude, Longitude, Time Zone, and Elevation have been changed for Muncie in **Figure 2-18**.

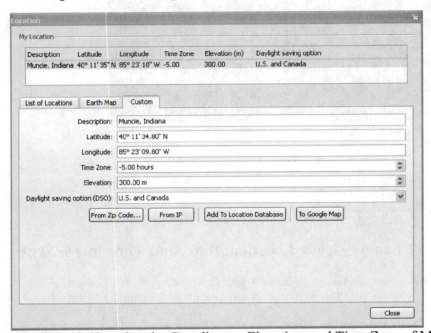

Figure 2-18 Changing the Coordinates, Elevation, and Time Zone of Muncie, Indiana

4. Once these parameters have been changed click the "Add To Location Database" bar. Now, click the "List of Locations" tab once again and click the "Set As My Location" bar. Make certain that the information displayed on your monitor is the same as the information shown in **Figure 2-18**. Click "Close" or the ✖ button.

5. Next, click on "Input" and then "Date and Time" to set the date to April 25, 2015, and the time to 10:00 PM DST. The window now looks like the one displayed in **Figure 2-19**.

© 2011 Cengage Learning. All Rights Reserved. May not be scanned, copied or duplicated, or posted to a publicly accessible website, in whole or in part.

Figure 2-19 Setting the Date and Time to April 25, 2015, at 10:00 PM DST

6.   Click the South shortcut button (Look South) located on the Orientation toolbar.

7.   Click the up blue button (Move Up) a few times. The screen now looks like the one illustrated in **Figure 2-20**. Notice that the constellation Leo is due south on the local celestial meridian in this sky window.

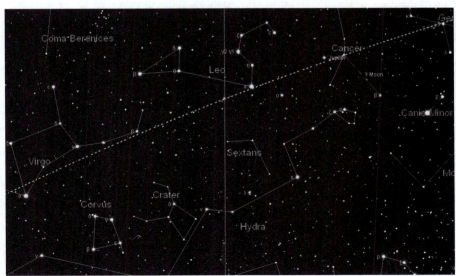

Figure 2-20 Sky's Appearance Looking South on April 25, 2015, from Muncie, IN

© 2011 Cengage Learning. All Rights Reserved. May not be scanned, copied or duplicated, or posted to a publicly accessible website, in whole or in part.

8. Now click the "Toggle Labels" () button on the *View toolbar*. Notice that the names of the objects disappear. The sky window looks just like the one in **Figure 2-21**.

Figure 2-21 The Sky Looking South after Toggling Off the Labels

9. Notice that only the names disappeared, not the objects.

10. Click the (Toggle Labels) button again and the labels return to the sky window.

It should be noted here that some of the features in the *Normal.skyx* window file have been changed since you started these exercises. A ***word of caution is warranted here!*** On exiting the program, **if any changes** have been made in the *Normal.skyx* window, the program ***automatically saves*** the *Normal* sky file and the configurations at the time you exit the program. In other words, the next time you start up *TheSkyX* software it will be just like the window that you just exited. *TheSkyX* does **not** prompt the user *"Do you want to save your changes"* as it did in previous versions!!

It is probably wise to make changes in the *Normal.skyx* file such as setting the time to that of the Computer Clock and the location to that of the user and then save the file before exiting.

The reason being that the next time *TheSkyX* program is run the user will not have to reset the location or the time. If you save the *Normal.skyx* file at any time, then it will always open to the sky window settings that you have previously used! Once your location and personal preferences have been set for *TheSkyX* window then it is time to **save** the file.

We recommend that you make a backup copy of the *Normal.skyx* file. You may copy this file to a thumb disk or perhaps to another folder on your PC. If you wish to keep this file *TheSkyX* program folder, it must be renamed. It is appropriate to rename this backup file as ***Normal.bak.skyx***. It then may be copied to the following folder:

C:\Program Files\Software Bisque\TheSkyX Student Edition.

Just be sure to keep this file in a safe place for future use!

© 2011 Cengage Learning. All Rights Reserved. May not be scanned, copied or duplicated, or posted to a publicly accessible website, in whole or in part.

Chapter 3

Setting Your Location and Time in *TheSkyX*

The purpose of this chapter is to help users set up their site information and other defaults in *TheSkyX* program. *TheSkyX* uses your latitude and longitude to accurately display objects that can be seen from any location on Earth. This information, of course, is different for each observer. Some major cities have been included in *TheSkyX's* location database, but it is best to use *your* latitude and longitude to make the fullest of *TheSkyX* experience. This can be done in two ways; either by accessing the Internet then searching for your location's latitude and longitude or by using a GPS device.

Exercise A: Setting the Site Information

1. Run *TheSkyX*.

2. Click "Input" on the Standard toolbar at the top of the sky window.

3. Click on "Location." This opens a window like the one shown in **Figure 3-1**.

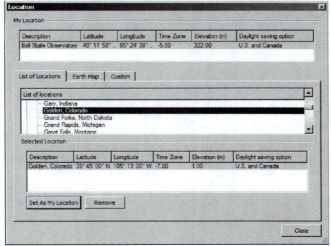

Figure 3-1 Location Window

4. The current city listed in the Description Line is Golden, Colorado. This is usual default setting in *TheSkyX* if you do not have Internet access.

5. To change the location setting to your location, click on the "Custom" tab. This opens a window such as the one displayed in **Figure 3-2**. Now, delete Golden, Colorado.

6. You have three options to add your location to *TheSkyX's* location database. You may use your latitude and longitude, your zip code, or the IP Address of your PC or MAC.

7. We will use the latitude and longitude method in this exercise. In the "Description" line type the name of your location (no more than 30 characters). In the description line type "Ball State Observatory."

© 2011 Cengage Learning. All Rights Reserved. May not be scanned, copied or duplicated, or posted to a publicly accessible website, in whole or in part.

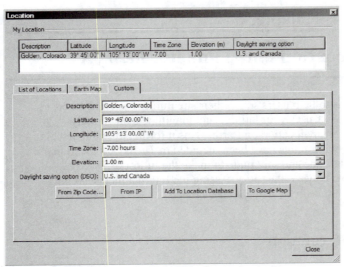

Figure 3-2 The Custom Tab Window

8. The "Latitude" and "Longitude" coordinates will be changed to those of Ball State Observatory.

9. *Highlight **only*** the numbers in "Latitude" and type in those for Ball State Observatory. Type 40° 11′ 58″ N for the latitude for Ball State Observatory.

10. *Highlight **only*** the numbers in "Longitude" and type in those for Ball State Observatory. Type 85° 24′ 38″ W for the longitude for Ball State Observatory.

11. Enter the "Time Zone" and "Elevation" (in meters) of Ball State Observatory's location. The time zone at the Ball State Observatory is -5.00, and the elevation is approximately 322 meters. The "Daylight saving option" (DSO) most likely changes throughout the year for most locations in the United States. Set it to "U.S. and Canada" option for now. The exception to this, of course, is Arizona and Hawaii.

12. Since this location is *not* listed in the indexed menu click the "Add To Location Database" button to save it to *TheSkyX's* location database. This location is saved in the Custom database. The Location window now looks like the one shown in **Figure 3-3**. This location will be used later on in many of the exercises in this workbook.

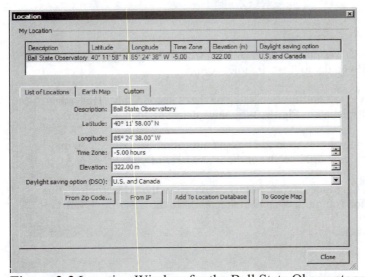

Figure 3-3 Location Window for the Ball State Observatory

© 2011 Cengage Learning. All Rights Reserved. May not be scanned, copied or duplicated, or posted to a publicly accessible website, in whole or in part.

13. Click "Close." The sky automatically changes the desktop display to this location.

14. Now, click "Input" again on the Standard toolbar.

15. Click "Date and Time." This menu is used to set the date and time of your sky desktop window. **Figure 3-4** displays the current date and time in the "Date and Time" window.

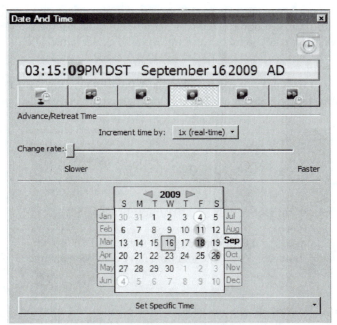

Figure 3-4 The Date and Time Window

16. Let's set the date to December 31, 2017 at 10:45 AM STD.

17. First the date…click the "Dec" month tab for December; then click "31" for the day, and advance the year by clicking the year advance button ▷ to 2017 or one may use the up arrow on the keyboard. **Figure 3-5** displays the date as December 31, 2017.

18. Next the time…*highlight* the numbers *only* on this line and then use the scroll wheel on the mouse or the up/down arrows on the keyboard to change the hours to **10**, the minutes to **45**, and the seconds to **00**. **Figure 3-5** shows the results of the date and time change.

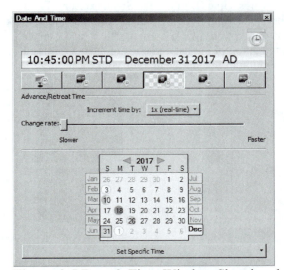

Figure 3-5 Date & Time Window Showing the Change to December 31, 2017

© 2011 Cengage Learning. All Rights Reserved. May not be scanned, copied or duplicated, or posted to a publicly accessible website, in whole or in part.

19. There are a couple options that one can use to set the date or time different from the computer clock but the previous method is perhaps the easiest. Before changing the time in *TheSkyX* window, you must first *always* turn off the computer clock by clicking the "Stop Clock" ( Stop) button.

20. Once the computer clock is stopped, you can change the current date and time to any date and time. The first method was just described, that is, highlighting the numbers and using the up or down keys on the keyboard or scrolling with the center wheel (PC) on your mouse.

21. You try it! Change the date and time to December 31, 2017 at 9:00 PM STD. **Figure 3-6** displays the time change to 9:00 P.M. Note the "00" is still highlighted in the seconds' window.

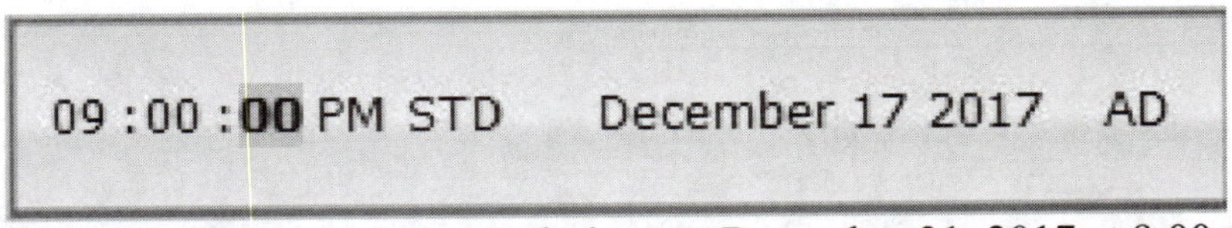

09 : 00 : **00** PM STD   December 17 2017   AD

Figure 3-6 Date and Time Window on December 31, 2017 at 9:00 PM EST

22. Your location has now been changed from Golden, Colorado, to the Ball State Observatory, located in Muncie, Indiana. Your date and time have also been changed to December 31, 2017, at 9:00 PM STD.

23. Click the Close button (⊠). The desktop window of your *TheSkyX* is now displaying the night sky at the Ball State Observatory for December 31, 2017, 9:00 PM EST.

24. A change in the time setting may be necessary if your location ***does not*** observe Daylight Savings Time. *TheSkyX* software assumes that you do observe it and sets the default Time Zone setting in the program to the "U.S. and Canada" zone. This means that your time will be changed accordingly to Daylight Savings Time when it is in effect. For the Ball State Observatory Daylight Savings Time is observed from late spring into the early autumn months.

25. In some of the exercises that follow in later chapters, it was convenient to change the Time Zone to not observing it. When this was done the Time Zone option was changed to "Not observed." To change from Daylight Savings Time (DST) to the Standard Zone Time (STD) click on the "Input" menu on the Standard toolbar then on "Location." Next, click the "Custom" tab and then "Daylight saving option (DSO)." On the "U.S. and Canada" line as shown in **Figure 3-7**, scroll up to the top of the list and click on "Not Observed." This will change the Daylight Savings Option to Standard Time for your location.

© 2011 Cengage Learning. All Rights Reserved. May not be scanned, copied or duplicated, or posted to a publicly accessible website, in whole or in part.

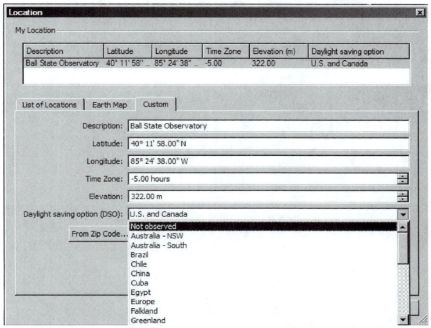

Figure 3-7 U.S. and Canada Time Zone for Daylight Savings Time

26. **Figure 3-8** displays the Location window after changing the time from Daylight Savings time to Standard time. You can see that the Daylight saving option has been changed from "U.S. and Canada" to "Not observed."

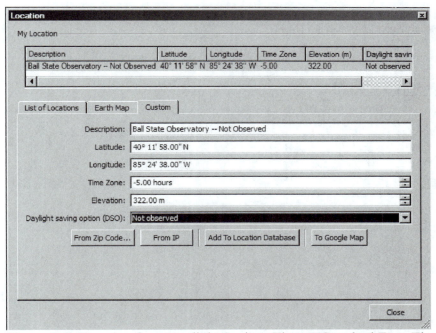

Figure 3-8 Changing from Daylight Savings Time to Standard Zone Time

27. To return the desktop window back to the current date and time click the "Computer Clock" (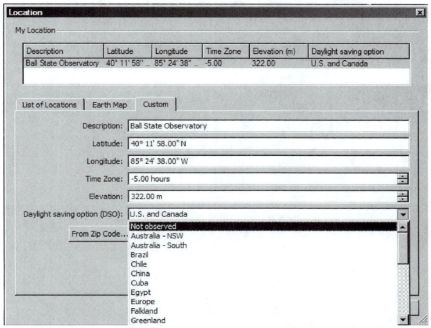) button.

© 2011 Cengage Learning. All Rights Reserved. May not be scanned, copied or duplicated, or posted to a publicly accessible website, in whole or in part.

Exercise B: Let's Find an Object in *TheSkyX*

1. Click "Edit" on the Standard toolbar at the top of the sky window, then "Find." The Find window opens and appears like the one illustrated in **Figure 3-9**.

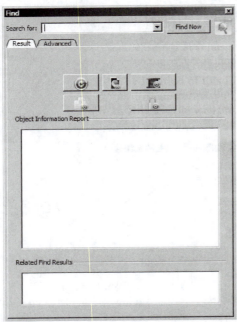

Figure 3-9 The Find Window

2. Find Messier Object Number 42 (M42).

3. To do this, type "M 42" in the "Search for" box as shown in **Figure 3-10**.

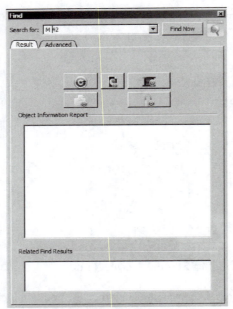

Figure 3-10 Searching for M 42

© 2011 Cengage Learning. All Rights Reserved. May not be scanned, copied or duplicated, or posted to a publicly accessible website, in whole or in part.

4. Now, click "Find Now." In the Find window "Result" tab detailed information about M 42 appears in the "Object Information Report" also shown in **Figure 3-11**.

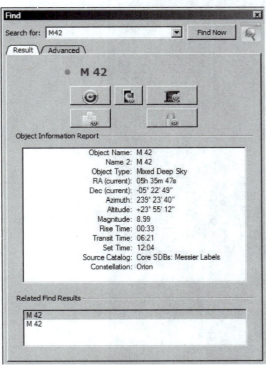

Figure 3-11 Find Window Displaying M 42

5. Information about M 42 may be found in the Object Information Report displayed in **Figure 3-11**.

6. Click the "Center" button (Center).

7. Write down the azimuth and altitude of M42 from the Object Information window as shown in **Figure 3-11** for this location.

Azimuth = _____ °   Altitude = _____ °

8. Close the Find window by clicking [X] in the upper right-hand corner of the window.

9. *TheSkyX* window looks exactly like the one illustrated in **Figure 3-12**. M42 is the circled object in the figure.

© 2011 Cengage Learning. All Rights Reserved. May not be scanned, copied or duplicated, or posted to a publicly accessible website, in whole or in part.

Figure 3-12 Viewing M42 in *TheSkyX* from the Ball State Observatory

There is another way to find M42 or *any other* celestial object in the sky.

Exercise C: Let's Find Another Object in *TheSkyX*

1. Instead of clicking on the Edit menu on the Standard toolbar at the top of the sky window, right-mouse click anywhere on the desktop. A menu appears on the desktop like the one shown in **Figure 3-13**.

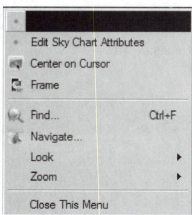

Figure 3-13 The Right-Mouse Click Menu Window

2. Now, click "Find." The Find window reappears on the desktop window as before.

© 2011 Cengage Learning. All Rights Reserved. May not be scanned, copied or duplicated, or posted to a publicly accessible website, in whole or in part.

3. Now, type "M 45" in the "Search for" line. The Find window appears, like the one illustrated in **Figure 3-14**.

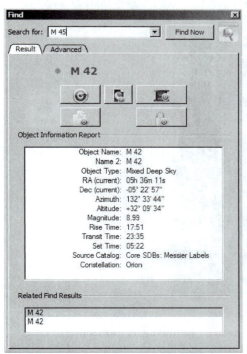

Figure 3-14 Find Window for M45

4. Now, click the "Find Now" (Find Now) button. The Find Result window opens with detailed information about M 45 displayed in the Object Information Report window for M45. It is displayed in **Figure 3-15**.

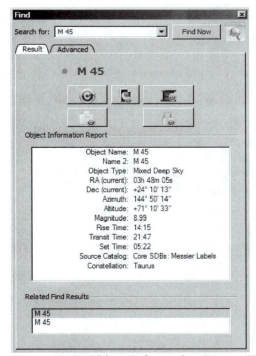

Figure 3-15 Object Information Report Window for M45

© 2011 Cengage Learning. All Rights Reserved. May not be scanned, copied or duplicated, or posted to a publicly accessible website, in whole or in part.

5. Click [Center] to center the Pleiades or "Seven Sisters" star cluster on the computer screen. The sky window looks exactly as shown in **Figure 3-16**.

Figure 3-16 Star Cluster M45 in Taurus

The user can find any object in *TheSkyX* database by following the previous exercises. The second method saves you time instead of clicking the "Common Names" and scrolling through the indexed objects.

The fastest way to find astronomical objects in *TheSkyX* database is to right click your mouse anywhere in the sky window. A small menu appears, allowing users to access the Find window and locate objects quickly and easily. This menu enables the user to perform a variety of other functions in *TheSkyX* such as changing the preferences and filters in your sky window.

Exercise D: Let's find M42 from Miami, Florida

1. Click "Input" on the Standard toolbar at the top of the sky window.

2. Click on "Location." This opens a menu window like the one displayed in **Figure 3-19**.

3. Click the ⊞ in the "List of Locations" tab next to the United States to expand the list of U.S. cities.

4. Scroll down the list until you find Miami, Florida. Now, select it. The Location Information window now looks like the one displayed in **Figure 3-20**.

5. Now, click the [Set As My Location] button.

© 2011 Cengage Learning. All Rights Reserved. May not be scanned, copied or duplicated, or posted to a publicly accessible website, in whole or in part.

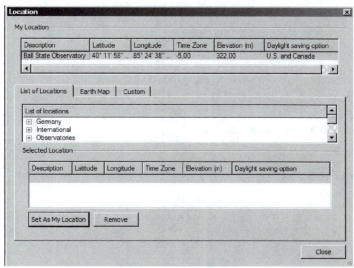

Figure 3-19 Site Information Window

6. Click [Close]. (Do *not* change the Date and Time.)

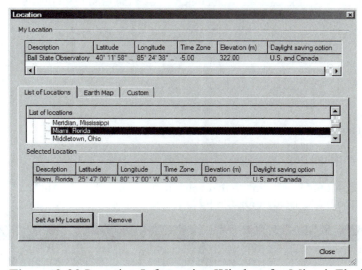

Figure 3-20 Location Information Window for Miami, Florida

7. Now, find M42 (Review Exercise A or B).

8. Write down its azimuth and altitude from this location.

   Azimuth = _____°   Altitude = _____°

Exercise E: Let's find M42 from Golden, Colorado

1. Click "Input" on the Standard toolbar at the top of the sky window.

2. Click on "Location." This reopens the menu window like the one displayed in **Figure 3-21**.

© 2011 Cengage Learning. All Rights Reserved. May not be scanned, copied or duplicated, or posted to a publicly accessible website, in whole or in part.

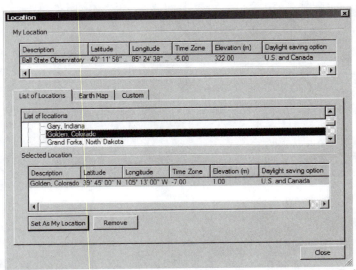

Figure 3-21 Location Information Window for Golden, Colorado

3. Since the "List of Locations" of the cities in the United States is already expanded scroll to
    Golden, Colorado. Click ⸢Set As My Location⸥. Remember to reset the time to 9:00PM STD.

4. The Location Information window appears like the one shown in **Figure 3-21**.

5. Click ⸢Close⸥.

6. Find M42.

7. Write down its azimuth and altitude from this location.

    Azimuth = _____° Altitude = _____°

8. Compare the values of the altitude and azimuth that you wrote down earlier for M42.
    They should be the same as those listed in **Figure 3-22**.

Location	Azimuth	Altitude
Ball State Observatory	144° 50′ 14″	+71° 10′ 33″
Miami, Florida	103° 49′ 10″	+83° 51′ 20″
Golden, Colorado	174° 59′ 45″	+74° 22′ 14″

Figure 3-22 Azimuths and Altitudes of M42 from Three Different Locations

9. The reason for the differences in **Figure 3-22** is discussed in **Chapter 5**. After you add

    your location to the list of cities, click the Save button (⸢Save⸥) located in the upper left
    of the sky window on the Standard toolbar. This saves all of the information you just
    entered into the file called *Normal.sky*. You can copy this file to a floppy disk or to
    another folder on your PC, to be used later, in case the information is lost or becomes
    corrupted.
    If you must re-install *TheSkyX*, the file that the program creates will not have your
    location or preferences in it. If you wish to copy this file to *TheSkyX's* documents
    folder, it should be renamed as a backup file and then copied into the folder such as the
    file:

C:\Program Files\Software Bisque\TheSkyX Student Edition\User\Documents\Normal.bak.

© 2011 Cengage Learning. All Rights Reserved. May not be scanned, copied or duplicated, or posted to a publicly accessible website, in whole or in part.

It is probably a good idea to keep the number of astronomical objects to a bare minimum until you become proficient in using *TheSkyX*. To accomplish this, you must first specify what type of astronomical objects you want to display on your computer screen. ***Exercise F***, which follows below, is designed to show how to display some basic celestial information in *TheSkyX*.

Exercise F: Displaying Celestial Information in *TheSkyX*

1. Click on "Display" on the Standard toolbar at the top of the sky window. This opens the Display menu window that is shown in **Figure 3-23**.

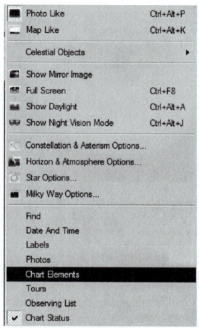

Figure 3-23 The Display Menu Window

2. Now, scroll down to "Chart Elements" as highlighted in **Figure 3-23**. This menu window determines what objects one may display on the computer screen. The Chart Elements menu window is displayed in **Figure 3-24**.

© 2011 Cengage Learning. All Rights Reserved. May not be scanned, copied or duplicated, or posted to a publicly accessible website, in whole or in part.

Figure 3-24 The Chart Elements Menu Window

3. By clicking the ⊞ next to any of the Celestial Objects will expand the list of objects as shown in **Figure 3-25**. This expanded menu allows the user to display all or to hide celestial objects displayed in the sky window.

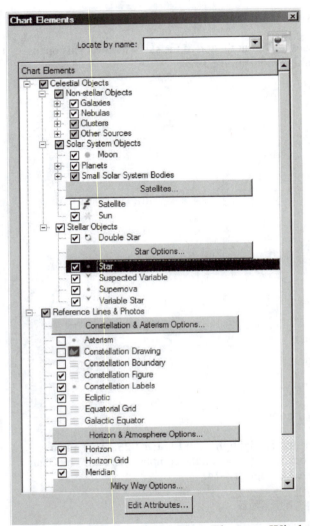

Figure 3-25 The Expanded Chart Elements Window Menu

4. The things that should be displayed on the desktop window are the Constellation Figure and Labels. Next, reference lines such as the Ecliptic, Meridian, and Horizon should also be displayed. These items appear to be checked in **Figure 3-25** as well as a sundry of other items.

5. The next thing that should perhaps be changed is the magnitude limits. To change the magnitude limits highlight "Stars" as shown in **Figure 3-25** then click the button labeled "Edit Attributes" located at the bottom of the Chart Elements window. A window appears such as the one shown in **Figure 3-26** appears.

© 2011 Cengage Learning. All Rights Reserved. May not be scanned, copied or duplicated, or posted to a publicly accessible website, in whole or in part.

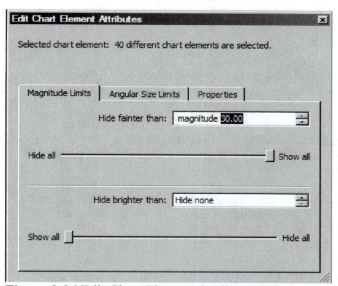

Figure 3-26 Edit Chart Element Attributes Window – Magnitude Limits Tab

6. The brightest magnitude should probably be set to about -3.0, while the faintest magnitude to probably +10.0 or less. If one lives in a medium-sized city (such as Muncie), then +4.5 is just about right for the faintest magnitude. Magnitudes are discussed in **Appendix F**.

    Notice that in the Magnitude Limits tab that one has the option to choose the magnitude limit in the "Hide fainter than:" description window. One may either "Hide all" or "Show all" the stars in the desktop window. The current setting is to "Show all" magnitudes of celestial objects fainter than 30.0. This number ranges from +30.0 to -6.0.

    Typically, the magnitude setting is done by sliding the bar to about half-way. This will display most celestial objects that are brighter than 8th magnitude in the desktop window as shown in **Figure 3-27**. This is typically the magnitude range for binoculars. Remember that the faintest star that can be seen from mountain tops is a 6th magnitude star. Make sure that the slide bar at the bottom of the window is set to the "Show all" setting.

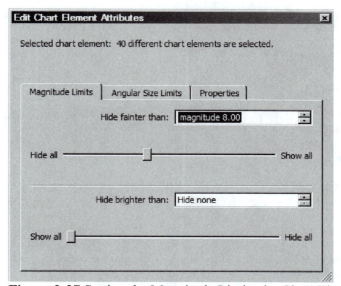

Figure 3-27 Setting the Magnitude Limits the Chart Elements Menu

© 2011 Cengage Learning. All Rights Reserved. May not be scanned, copied or duplicated, or posted to a publicly accessible website, in whole or in part.

7. Close the menu window.

8. While "Looking South," click 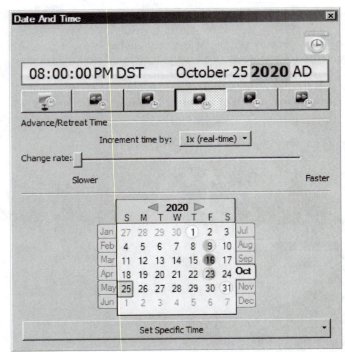 Look South if necessary, a red-solid line appears perpendicular to the horizon; this is the *local celestial meridian*.

   There is also a blue-colored line in the window. If not seen, then use **Move Up** to move the sky view upward on the desktop. This line represents the Sun's apparent path in the sky, the *ecliptic*.

9. One can change these colors at any time by using the "Tools" menu on the Standard toolbar at the top of the sky window. **Appendix E** describes how to change the preferences and attributes in *TheSkyX*.

Exercise G: Let's Find a Planet in *TheSkyX*

1. Run *TheSkyX*.

2. Set the location for Ball State Observatory.

3. Set Date: October 25, 2020
   Set Time to 8:00 PM DST

   These settings are displayed in **Figure 3-28**.

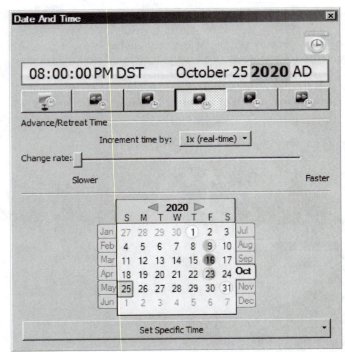

Figure 3-28 Date and Time Menu Window – October 25, 2020

4. Click "Edit" on the Standard toolbar at the top of the sky window and then select "Find." Type "Neptune" in the "Search for" description line.

© 2011 Cengage Learning. All Rights Reserved. May not be scanned, copied or duplicated, or posted to a publicly accessible website, in whole or in part.

5.   Click "Find Now." The Find Result window should appear like the one displayed in
     Figure 3-29.

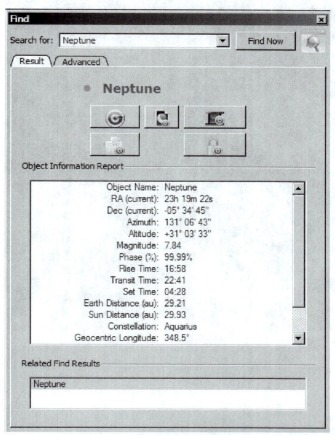

Figure 3-29 The Find Result Window Displaying the Object Information Report

6.   Look at the azimuth and altitude of Neptune. Make a note of these values. **Figure 3-30**
     displays the view of the southern sky on October 25 at 8:00 PM, facing south, from the
     Ball State Observatory. Neptune is the circled object in the figure.

© 2011 Cengage Learning. All Rights Reserved. May not be scanned, copied or duplicated, or posted to a publicly accessible website, in whole or in part.

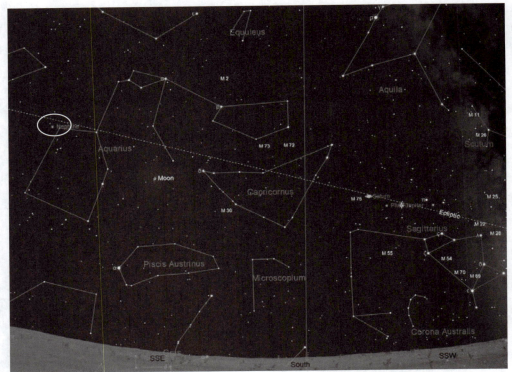

Figure 3-30 Southern Sky as Seen From the Ball State Observatory

7. **Figure 3-31** displays the azimuth and altitude information for Neptune.

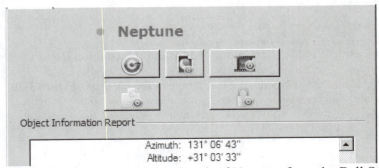

Figure 3-31 Azimuth and Azimuth of Neptune from the Ball State Observatory

Exercise H: Let's Find Neptune from Mumbai, India

1. Run *TheSkyX*.

2. Set Location to Mumbai, India. Go to "Input" on the Standard toolbar at the top of the sky window and click "Location" to open the Location menu window as displayed in **Figure 3-32**.

© 2011 Cengage Learning. All Rights Reserved. May not be scanned, copied or duplicated, or posted to a publicly accessible website, in whole or in part.

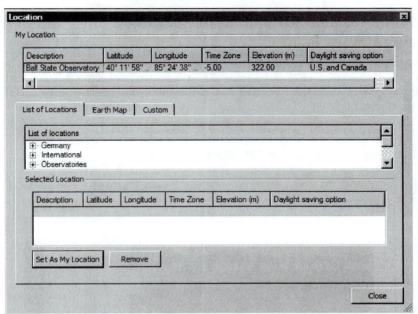

Figure 3-32 The Location Window Displaying Ball State Observatory Location

3. Click the ⊞ next to the "International" List of Locations to expand the list. Scroll down the list and highlight "Bombay," India as illustrated in **Figure 3-33**.

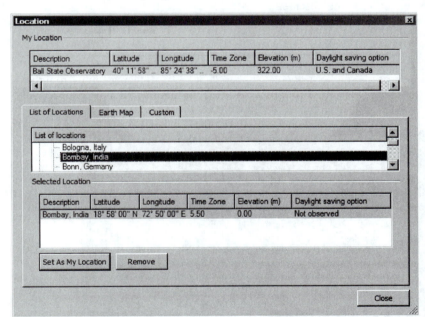

Figure 3-33 The Location Menu Window Displaying Mumbai, India

4. Click the button labeled [Set As My Location]. Now, click [Close].

5. The Location window now appears like the one displayed in **Figure 3-34**.

© 2011 Cengage Learning. All Rights Reserved. May not be scanned, copied or duplicated, or posted to a publicly accessible website, in whole or in part.

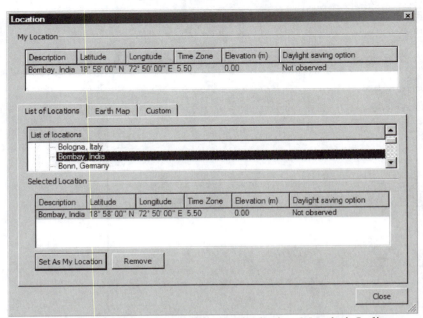

Figure 3-34 Location Menu Window Displaying Mumbai, India

6. Reset Date: October 25, 2020
    Reset Time: 8:00 PM STD
    Daylight saving option: "Not Observed"

7. Now, find Neptune as was done in **Exercise G**.

8. Look at the azimuth and altitude of Neptune. Make a note of these values. **Figure 3-35** displays the view of the southern sky on October 25 at 8:00 PM, facing south, from the Mumbai, India. Neptune is the circled object in the figure. Compare this figure to that of **Figure 3-30**. Note the change in Neptune' position in the sky.

Figure 3-35 Southern Sky as Viewed from Bombay, India

© 2011 Cengage Learning. All Rights Reserved. May not be scanned, copied or duplicated, or posted to a publicly accessible website, in whole or in part.

9.   Click on Neptune, and display the Object Information Report window.

10.   Note the altitude and azimuth values of Neptune. The Object Information Report window displayed looks exactly like the one shown in **Figure 3-36**.

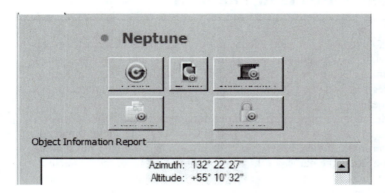

Figure 3-36 Azimuth and Azimuth of Neptune from Mumbai, India

11. **Figure 3-37** lists a comparison of the azimuths and altitudes of Neptune as seen from the Ball State Observatory and Mumbai, India. Note the difference in the values.

Location	Azimuth	Altitude
Ball State Observatory	131° 06′ 43″	+31° 03′ 33″
Mumbai, India	132° 22′ 27″	+55° 10′ 32″

Figure 3-37 Altitude and Azimuth of Neptune at Two Different Locations on Earth

It should be apparent now on how to set locations and display objects in *TheSkyX*. It is a simple task if one follows the suggestions given in the previous exercises.

If trouble develops later on, simply refer back to these exercises. The procedures outlined here are much easier than they seem at a glance. It is best for users to explore different possibilities by changing their location on Earth or *attributes* in the sky window. Remember: the *Normal.sky* file can be changed at any time!

Once becoming proficient in using *TheSkyX*, the settings can be changed and resaved at anytime. In the worst-case scenario, all one would have to copy the backup file, *Normal.bak.skyx* to the appropriate folder as *Normal.skyx* or re-install *TheSkyX*.

Have fun and enjoy *TheSkyX*!

© 2011 Cengage Learning. All Rights Reserved. May not be scanned, copied or duplicated, or posted to a publicly accessible website, in whole or in part.

Chapter 4

Naming Objects in *TheSkyX*

Astronomers need to be able to assimilate and exchange information about specific objects in the sky. Many systems have been devised over the last few thousand years to identify and name the most conspicuous ones. As our technology has advanced, studies of astronomical objects have become more precise. Today it is common for astronomers to assign numerous designations to a single celestial body.

Humans have observed the stars for millennia. Our ancestors named the bright stars as well as larger groups of stars called constellations. As was stated in Chapter 1, the ancients named many of the constellations after mythological beasts, gods, demigods, and ordinary household objects. Astronomers continue to use the names of the constellations first recorded by ancient astronomers thousands of years ago. It is here that we may begin to learn about where things are located in the sky and how they are named.

Astronomers today officially recognize 88 distinct constellations. *TheSkyX* displays all of them for you. From the mid-northern latitudes, you can see over half of them. Most are visible every night from your location at some time during the night. *TheSkyX* helps you to find them accurately. But, *you* must go outside on any clear night throughout the year and look for them yourself in order to appreciate them fully.

About six or so constellations are visible every night from the 40° North latitude circle all year round. These are the *circumpolar* constellations. They are all located in the northern sky near the North Star, Polaris. Using *TheSkyX* will definitely help you locate all these constellations easily during any season of the year. **Exercise A** is designed to help you locate some of the constellations visible from mid-northern latitudes.

Exercise A: Viewing Some Constellations

1. Run **TheSkyX** and open the file **Normal.skyx**.

2. Click **Input** on the toolbar at the top of the sky window, then **Location**.

3. Set your location to 40° North latitude. The city or longitude is not important here!
    Figure 4-1 displays the appropriate information that should be entered into the Location Site Information window.

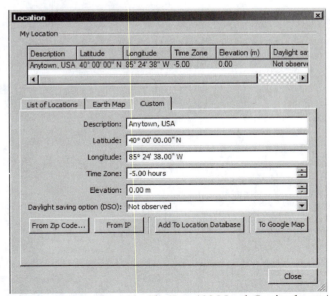

Figure 4-1 Setting *TheSkyX* to 40° North Latitude at Anytown, USA

© 2011 Cengage Learning. All Rights Reserved. May not be scanned, copied or duplicated, or posted to a publicly accessible website, in whole or in part.

4. Click "Close."

5. Set Date October 25, 2015 and Time to: 8:45PM
   Set Daylight saving option: "U.S. and Canada"

6. Did you remember to click on "Stop Clock?"

7. On the toolbar, click "Look North" (Look North).

8. Zoom to 125° x 125° by clicking on the "Zoom Out" button Zoom Out 125° x 125°.

9. If the constellation figures are not displayed on the desktop. Click "Display" on the tool bar, then "Chart Elements." **Figure 4-2** displays the "Chart Elements" window.

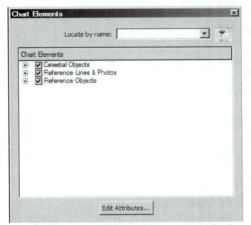

Figure 4-2 The Chart Elements Window

10. Click the " ⊞ " next to "Reference Lines & Photos" to expand the list of items.

11. **Figure 4-3** displays an expanded list of items. Click the box labeled "Constellation Figure".

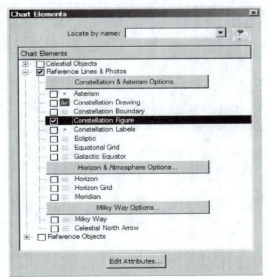

Figure 4-3 Displaying the Constellation Figures in *TheSkyX*

© 2011 Cengage Learning. All Rights Reserved. May not be scanned, copied or duplicated, or posted to a publicly accessible website, in whole or in part.

12. Now, let's display the Common` Names for the constellations, if they are not already displayed.

13. Click "Display" on the Standard toolbar at the top of the desktop window then click "Chart Elements" again. Now, click "Reference Lines & Photos" once again. Click the "Constellation Labels" box as shown in **Figure 4-4**.

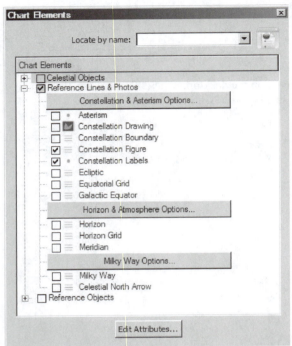

Figure 4-4 Displaying the Constellation Names *TheSkyX*

TheSkyX desktop should appear as the one shown in **Figure 4-5**.

Figure 4-5 *TheSkyX* Desktop Window After Toggling On the Constellation Figures and Names

© 2011 Cengage Learning. All Rights Reserved. May not be scanned, copied or duplicated, or posted to a publicly accessible website, in whole or in part.

14. Now, on *TheSkyX* toolbar click the "Go Forward" (Go Forward) button and observe the motion. If there is no motion, then the time skip feature must be set to a faster rate.

15. The Time Skip default setting is 1x (1x). Scroll down the time skip menu and click 1minute as shown in **Figure 4-6**.

Custom...

1x
10x
100x
1000x
10000x
1 second
1 minute
1 hour
1 day
1 lunar month
1 year
sunrise
sunset
start twilight
end twilight

Figure 4-6 Setting the Time Skip to 1minute

16. Now, click the Go Forward button (Go Forward) and list the names of the constellations that do not go below the northern horizon.

_____

_____

_____

_____

_____

_____

_____

These constellations are visible every night throughout the year and are known as the *circumpolar constellations*. **Exercise B** is designed to find a particular constellation in *TheSkyX*.

Exercise B: Find the Constellation of Orion

1. Use the Date and Time in **Exercise A**.

2. There are three ways in which you can find objects in *TheSkyX* software. You may click "Edit" on the Standard toolbar at the top of *TheSkyX* window, and then Find. Right click

© 2011 Cengage Learning. All Rights Reserved. May not be scanned, copied or duplicated, or posted to a publicly accessible website, in whole or in part.

the mouse anywhere in *TheSkyX* desktop window, and then "Find." Or you may click on "Display," at the top of *TheSkyX* desktop window, and then "Find."

3. Once you have opened the "Find" window, type in Orion and click on "Find Now." Click

    Center (Center). The Find "Result" window is shown in **Figure 4-7**.

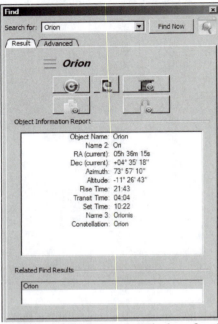

Figure 4-7 The Result Window for the Constellation of Orion

4. There is another way find a constellation in the Find window. After opening the Find window click on the "Advance" tab as shown in **Figure 4-8**.

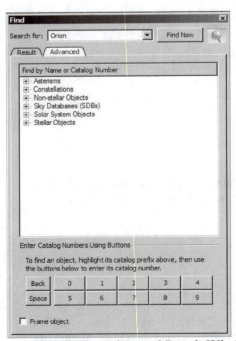

Figure 4-8 The Advanced Result Window in the Find Window for Orion

© 2011 Cengage Learning. All Rights Reserved. May not be scanned, copied or duplicated, or posted to a publicly accessible website, in whole or in part.

5. Click on the "⊞" beside "Constellations" to expand the list of constellations Scroll down the "Find by Name or Catalog Number" list to name Orion as shown in **Figure 4-9**.

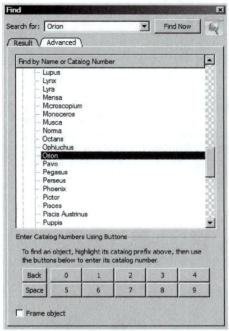

Figure 4-9 The Expanded Result Window in the Find Window for Orion

6. Double-click on "Orion."

7. After clicking on "Orion," an Object Information window appears, like the one displayed in **Figure 4-10**. This box indicates where the constellation is located in the sky and other pertinent information about the constellation, such as whether or not that it's visible.

    The time of day that Orion appears to rise, cross your local celestial meridian (transit), and set at your location is provided in this window. The time of meridian transit is the best time to view this constellation.

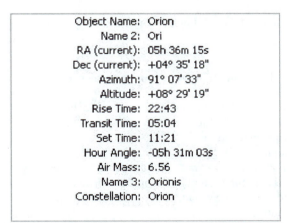

Figure 4-10 Object Information Window for the Constellation Orion

8. With that in mind, answer the following questions:

    At what time does Orion appear to rise? _____:_____

© 2011 Cengage Learning. All Rights Reserved. May not be scanned, copied or duplicated, or posted to a publicly accessible website, in whole or in part.

At what time does Orion appear to set? ____:____

When is the best time to view Orion?   ____:____

9.   Next click the Center Object button (Center) at the lower left-hand corner of the Command Center window. This will center Orion in your sky desktop window.

10.   Click on some of the objects in the constellation. Did you find M42?

11.   Now, let's display non-stellar objects that might be located in Orion. On *TheSkyX* Standard toolbar click "Display" then "Labels." Click the box labeled "Messier Objects." Your window should look like the one displayed in **Figure 4-11**. Be sure that the "Show Labels" box is checked at the top of the menu window.

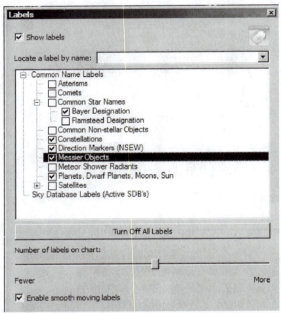

Figure 4-11 Displaying Labels of Non-Stellar Objects in *TheSkyX*

12. Did you find M42? If not, use the "Find" command. Then, click on the "Frame" button (   ) to center M42 and to view it.

13. A list of all of the Messier Objects is provided in **Appendix D** for your convenience.

That's all there is to it! Try finding another constellation—perhaps one you like better than Orion. You can use this exercise to find any astronomical object in *TheSkyX's* database. **Appendix B** contains a list of the 88 recognized constellations.

Now let's turn our attention to how astronomical objects are named or designated. Astronomical objects fall into two groups, *stellar* and *non-stellar*. The first group includes *only* the stars. The second group has a more diversified population of astronomical objects. The later group contains objects that lie well beyond our Solar System. The non-stellar group contains objects such as nebulae, star clusters, and galaxies.

The planets in contrast are not included in either of these groups. Because they are Solar System objects, astronomers usually designate them by their given name.

Let's first discuss how stars are designated. Over the millennia, people have used several systems to categorize the stars. The numerous cultures that have inhabited our planet have assigned many names to the stars over the millennia.

© 2011 Cengage Learning. All Rights Reserved. May not be scanned, copied or duplicated, or posted to a publicly accessible website, in whole or in part.

Proper Names

Astronomers usually refer to stars as either *prominent* stars or *representative* stars. Prominent stars are the bright stars. Stars, such as those displayed in **Figure 4-12,** are typically the bright stars seen in the winter months. They are located in the constellations of Taurus, Orion, Canis Major, and Canis Minor.

The dot sizes displayed in the desktop window, representing stars, indicate the stars' relative brightnesses, not their true physical sizes. In other words, the bigger the dot, the brighter the star; the smaller the dot, the fainter the star. The magnitude system that astronomers use in observational astronomy is discussed in **Appendix F**.

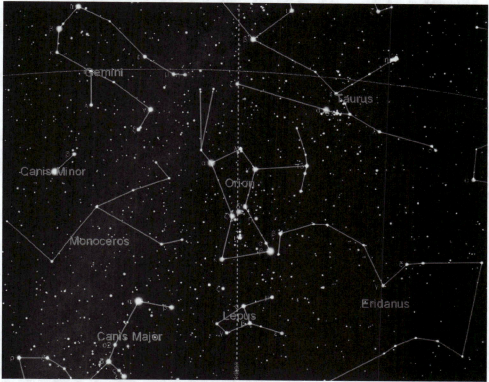

Figure 4-12 Prominent or Bright Stars of Winter

The bright stars of summer are typically those that are displayed in **Figure 4-13**. These stars are located in the constellations of Lyra, Cygnus, and Aquila and are usually observed nearly overhead about midnight at 40° North latitude.

© 2011 Cengage Learning. All Rights Reserved. May not be scanned, copied or duplicated, or posted to a publicly accessible website, in whole or in part.

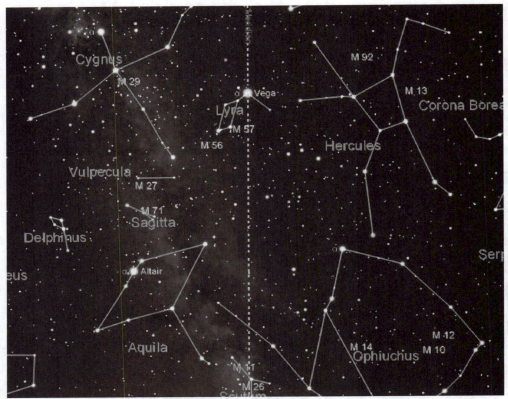

Figure 4-13 Prominent or Bright Stars of Summer

Figure 4-14 displays the summer constellations that are observed when looking south at 40°
North latitude. It is there that you will observe the stars located in the constellations of Scorpius
and Sagittarius.

Figure 4-14 Summer Constellations Looking South at 40° North Latitude

The brightest stars are usually designated with proper names. Only a few dozen stars are
frequently referred to by their proper names. Several hundred stars have been named this way,
but only a few are easily recognized by name. The bright stars are easily seen after the Sun goes
down and twilight falls.

The representative stars, in contrast, are those that are for the most part in the solar
neighborhood. They are close by, so to speak. Their distances are within 4 to 8 parsecs of

© 2011 Cengage Learning. All Rights Reserved. May not be scanned, copied or duplicated, or posted to a publicly accessible website, in whole or in part.

Earth. This is about 13 - 26 light years. Not many of the representative stars have proper names. The representative stars in the Milky Way Galaxy are usually fainter than the Sun's intrinsic brightness. These stars are usually small, cool, and very faint.

On any clear night in spring, you can find the bright star Arcturus in the evening sky. In summer, it might be Antares (heart of the scorpion), or perhaps Vega, the brightest star in the constellation Lyra. In late autumn, you might find the star Capella in Auriga, the charioteer. In the winter, it might be Betelgeuse or Rigel in the constellation of Orion. A list of some of the brightest stars is provided in **Appendix A**. This list includes the stars' designations, proper names, coordinates, apparent brightnesses, spectral types, intrinsic brightnesses, and their approximate distances. **Exercises C** and **D** are designed for you to locate some bright stars.

Exercise C: Find the Bright Star Aldebaran

    Set Location: Cleveland, Ohio
    Set Date: October 19, 2015.
    Set Time: 9:00PM.
    Set Daylight saving option: "U.S. and Canada"

The Location window looks like that in **Figure 4-15**.

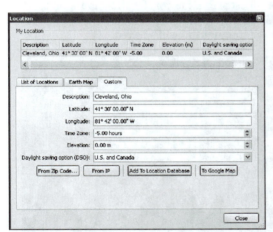 

Figure 4-15 Location, Date and Time Windows for **Exercise C**

1. Find Aldebaran.

2. In what constellation is Aldebaran located? _____

3. At what time does Aldebaran appear to rise? ____:____

4. At what time does Aldebaran appear to set? ____:____

5. When is the best time to view Aldebaran? ____:____

Exercise D: Find the Bright Star Altair

    Set Location: Golden, Colorado
    Set Date: November 1, 2015
    Set Time: 9:00PM
    Set Daylight saving option: "U.S. and Canada"

© 2011 Cengage Learning. All Rights Reserved. May not be scanned, copied or duplicated, or posted to a publicly accessible website, in whole or in part.

1. Find Altair.

2. In what constellation is Altair located? _____

3. At what time does Altair appear to rise? ____:____

4. At what time does Altair appear to set? ____:____

5. When is the best time to view Altair? ____:____

Bayer Letters

Because it is difficult for most of us to memorize the names of hundreds of stars, Johann Bayer developed a more convenient system in 1603. His system used Greek letters. The stars in the constellations observed in the northern sky were assigned letters according to their relative brightnesses.

Bayer used the Greek alphabet to designate the stars by their order of decreasing brightness in each constellation. Each naked-eye star is labeled with a Greek letter followed by the possessive case (genitive – second person singular) of the Latin name of the constellation in which it is found. This system started with the brightest star in the constellation and went to the faintest. In 1757, Nicholas Lacaille extended this system to include the southern constellations. **Figure 4-16** shows several stars with their Bayer letter designations displayed.

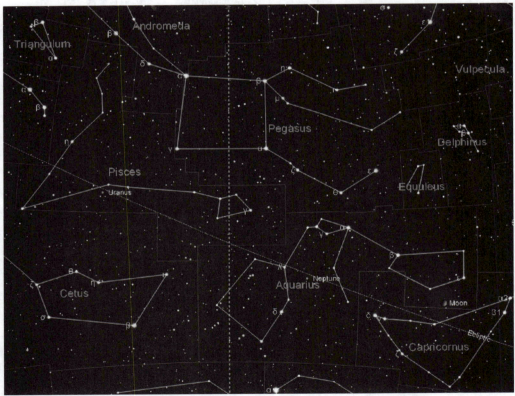

Figure 4-16 Stars with Bayer Letter Designations

For example, the brightest star in the constellation Lyra is shown in **Figure 4-17**. Its proper name is *Vega*. Its Bayer designation is α Lyrae – Alpha Lyrae. This is like saying "Alpha Lyree." The brightest star in the constellation of Cygnus is Deneb, which is Arabic for the "Tail" of the Swan. Its Bayer designation is α Cygni – "Cygnee". The second brightest star in

© 2011 Cengage Learning. All Rights Reserved. May not be scanned, copied or duplicated, or posted to a publicly accessible website, in whole or in part.

Cygnus is named Albireo and has a Bayer designation of β Cygni – Beta "Cygnee". Over the years, astronomers have abbreviated the constellation names to just the first three letters of the constellation, for example, β Cyg. **Appendix B** lists the constellation's Latin names, possessive forms and abbreviations of the constellations. The Greek alphabet is provided in **Appendix C**.

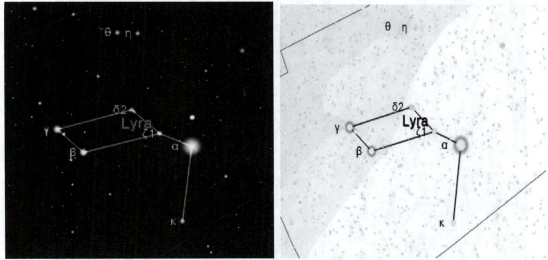

Figure 4-17 Stars in Lyra, the Lyre (Photo View Left, Chart View Right)

After the Greek letters have been used, the labeling continues with the lowercase letters of the Roman alphabet. If there are enough stars in a particular constellation, then the uppercase Roman alphabet letters are used up to and including the letter **Q**.

In Latin, some words (nouns and adjectives) are considered either masculine or feminine. Masculine words, as a rule, usually end with the letters "us," whereas feminine words end with the letter "a." Masculine constellation names ending in "us" are changed to the possessive (second person singular) case by replacing the letters "us" with the letter "i." Feminine constellation names ending in the letter "a" are changed to the possessive case by adding the letter "e" to the end of the constellation name.

Note that these are general rules to be followed and by no means are valid in all cases. There are several exceptions. For example Leo, the Lion, is a masculine word but it doesn't end in the letters "us." The possessive case for this constellation is Leonis. So, Denebola, the second brightest star in Leo, is designated as β Leonis. The same is true for the constellation Virgo, the Virgin. The possessive case for this constellation is Virginis. Therefore, the brightest star in Virgo is α Virginis. Its proper name is Spica.

In addition, if all of this were not confusing enough, a few constellations contain stars that are of about equal brightness. In these rare instances, the letters are assigned sequentially as one traces the pattern of the constellation. An example is the constellation Ursa Major, the Great Bear, which is shown in **Figure 4-18**. The "Big Dipper" is actually an asterism in Ursa Major. The stars that make up the "Dipper" are labeled sequentially as α, β, γ, δ, ε, ζ, and η as is displayed in **Figure 4-18**. **Appendix B** will assist one with the constellation names and **Appendix C** provides help with the Greek Alphabet.

© 2011 Cengage Learning. All Rights Reserved. May not be scanned, copied or duplicated, or posted to a publicly accessible website, in whole or in part.

Figure 4-18 Stars in Ursa Major

Exercise E: Find the Bright Star Wasat

Set Location: Atlanta, Georgia
Set Date: December 19, 2015
Set Time: 9:00PM
Set Daylight saving option: "Not Observed"

1. Find Wasat.

2. In what constellation is Wasat located? _____

3. What is its Bayer designation? _____

4. What is its magnitude? _____

5. Is Wasat the brightest star in the constellation? _____

6. At what time does Wasat appear to rise? ____:____

7. What time does Wasat appear to set? ____:____

8. What is the best time to view Wasat? ____:____

Exercise F: Find the Brightest Star in the Constellation of Gemini

Set Location: Portland, Maine
Set Date: January 15, 2020
Set Time: 9:00PM
Set Daylight saving option: "Not Observed."

© 2011 Cengage Learning. All Rights Reserved. May not be scanned, copied or duplicated, or posted to a publicly accessible website, in whole or in part.

1. Find the constellation Gemini.

2. What is the brightest star's proper name? _____

3. What is its Bayer designation? _____

4. What is its magnitude? _____

5. At What time does it appear to rise? ____:____

6. At what time does it appear to set? ____:____

7. When is the best time to view it? ____:____

Flamsteed Numbers

In 1712 astronomer John Flamsteed devised a simple system to designate stars. His system identified many more stars within a given constellation than the letters that Bayer had introduced a century earlier.

Flamsteed assigned numbers to the various stars in each constellation. The major difference in his system is that he ignored stars' apparent brightnesses in each constellation. Every star is designated in each constellation based on its position within the constellation. **Figure 4-19** illustrates several constellations and stars with their Flamsteed designations displayed.

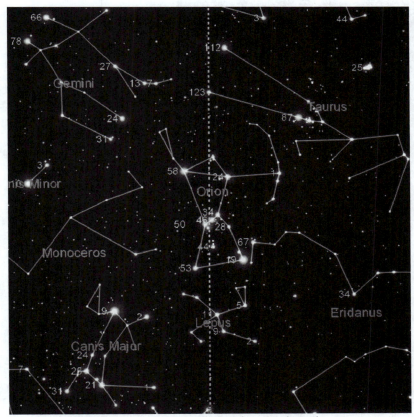

Figure 4-19 Stars with Flamsteed Number Designations

© 2011 Cengage Learning. All Rights Reserved. May not be scanned, copied or duplicated, or posted to a publicly accessible website, in whole or in part.

The numbers were assigned in order of their location in the constellation. They were labeled from *west to east* in each constellation; that is, in the *order* by which they *appear to cross* the local celestial meridian. The possessive form of the constellation name follows the number, as in the Bayer system. 6 Cygni is the Flamsteed designation for the second brightest star in the constellation Cygnus and is also known as β Cygni. Its proper name is Albireo and is shown in **Figure 4-20**.

Figure 4-20 6 Cygni, β Cygni, or Albireo

Flamsteed numbers are frequently used to identify stars that are much fainter than the naked-eye stars in each constellation. It is common practice to use the Bayer letters for the brighter stars in the constellations, because they provide some useful information about the apparent brightness of each star. *TheSkyX* displays both proper names and Flamsteed numbers for the stars visible in the nighttime sky.

Exercise G: Find the Brightest Star in the Constellation of Taurus

Set Location: San Diego, California
Set Date: February 3, 2025
Set Time: 9:00PM
Set Daylight savings option: to "U.S. and Canada"

1. Find the constellation of Taurus. Now, find the brightest star.

2. What is its proper name? _____

3. What is its Bayer designation? _____

4. What is its Flamsteed number? _____

5. What is its magnitude? _____

6. At what time does it appear to rise? _____:_____

© 2011 Cengage Learning. All Rights Reserved. May not be scanned, copied or duplicated, or posted to a publicly accessible website, in whole or in part.

7. At what time does it appear to set? ____:____

8. When is the best time to view it? ____:____

Exercise H: Find the Third Brightest Star in the Constellation of Orion

Set Location: Indianapolis, Indiana
Set Date: December 19, 2025
Set Time: 9:00PM
Set Daylight savings option: to "U.S. and Canada"

1. Find the constellation of Orion. Find the third brightest star.

2. What is its proper name? _____

3. What is its Bayer designation? _____

4. What is its Flamsteed number? _____

5. What is its magnitude? _____

6. At what time does it appear to rise? ____:____

7. At what time does it appear to set? ____:____

8. When is the best time to view it? ____:____

Binary and Multiple Stars

With the increased use of the telescope by astronomers in the 17th and 18th centuries, another interesting situation arose relating to stellar designation. Astronomers found that many individual stars were composed of two or more stars that were very close together. They appear as a single star to the naked-eye but as two or perhaps three separate stars through the telescope.

Binary stars are most often referred to as "double stars" but this is a misnomer! Today, astronomers make a distinction between a *double* star and a *binary* star. A double star is two stars that appear very close to each other in the sky. They share the *same direction* in space, but are at *different distances* from us. In other words, they are not gravitationally attached to each other. Binary stars, in contrast, not only share the *same direction* in space but both stars are at the *same distance* from us and orbit each other.

Since it would be a futile task to completely revise all the systems in use today, it is sometimes more suitable to just use the original designation. However, when a star is found to have more than one component, letters are assigned to the various components, in order of their decreasing brightness, after their Bayer designation.

In a binary system, the Roman letter A is assigned to the primary (brighter) component and the Roman letter B is assigned to the secondary (fainter) component. Again, the possessive form of the constellation name follows the letters, as in the Bayer and Flamsteed designations. If there are more than two stars, then subsequent letters are used (C, D, E, etc.) and the possessive form of the constellation name. There are exceptions to this general rule as well. For example, ζ Ursae Majoris (Mizar) is a well-known binary system. Mizar is the middle star in the handle of the "Big Dipper." The stars in this system are often referred to as Mizar A and Mizar B as well as ζ Ursae Majoris A and ζ Ursae Majoris B.

The closest star system to Earth is the α Centauri system. This star system is actually a tertiary system. The primary component of this system is designated as α Centauri A. The secondary and fainter component is designated as α Centauri B. The tertiary or faintest

© 2011 Cengage Learning. All Rights Reserved. May not be scanned, copied or duplicated, or posted to a publicly accessible website, in whole or in part.

component of this system is denoted α Centauri C. The brightest member, A, in this system is very much like our Sun in its size and temperature.

Another famous visual binary system is in the constellation of Cygnus the Swan. It is designated as β Cygni or Albireo. The two stars in this system are separated by approximately 54 seconds of arc and are easily resolved in a small telescope. They are also two stars that have very different surface temperatures and thus very different colors. In small to moderate-sized telescopes they appear as a blue and a yellow star. **Figure 4-21** depicts Albireo as viewed in a moderate-size (14-inch) telescope and rendered in *TheSkyX*.

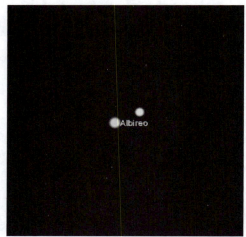

Figure 4-21 Albireo in the Constellation Cygnus

Figure 4-22 is a CCD (Charge-Coupled Device) image of Albireo taken at the Ball State Observatory through its 14-inch Celestron telescope. It is an electronic, digitized image of Albireo taken in three colors. The images were taken with a Photometrics Star 1 camera. Three images were taken through a Red, Green, and Blue filter and later combined to produce the tricolor image displayed in **Figure 4-22**.

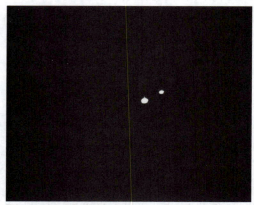

Figure 4-22 CCD Image of Albireo Taken at the Ball State Observatory
(Image provided by Author)

Variable Stars

Another situation arose about the same time that astronomers found many stars to be multiple star systems. It was discovered that many of these stars appeared to vary in brightness. This change in brightness may be due to a variety of reasons. It was thought that it may have been something intrinsic to the star, such as the very nature of the star itself. Or, that there may

© 2011 Cengage Learning. All Rights Reserved. May not be scanned, copied or duplicated, or posted to a publicly accessible website, in whole or in part.

have been an extrinsic cause, such as the star being a member of an eclipsing binary system. For whatever reason, the cause of the brightness variation in stars is still designated in a special way.

The last Roman letter used in Bayer's notation was the letter Q. Hence, variable stars begin their designations with the Roman letter R. They are designated in the *order of* their *discovery* within the constellation. Thus the first variable star in any constellation is designated with the Roman letter R, then S, T, U, V, W, X, Y, and Z followed by the possessive form of the constellation name. After the letter Z, the letters are doubled RR, RS, RT, to RZ; then SS, ST, SU, to SZ; then TT, TU, TV, to TZ; then UU, UV, UW, to UZ through ZZ and all followed by the possessive form of the constellation name. After the letters ZZ, additional variable stars may be designated with the letters AA, AB, AC, to AZ; then BB, BC, BD, to BZ; then CC, CD, CE, to CZ all the way through QZ each followed by the possessive form of the constellation name in which they are discovered. Note: because there is no letter J in the Roman alphabet, it does not appear in any of the designations. The designations R through QZ take care of the first 334 variable stars discovered in any constellation. If subsequent variable stars are discovered in a constellation, they are designated with a number preceded by the capital letter **V** – for *variable*. Thus, the 335th variable star in any constellation is designated as V335 followed by the possessive form of the constellation name in which it is found. In other words, the 335th variable star discovered in the constellation of Capricornus is designated as V335 Capricorni.

TheSkyX displays variable stars with the blue symbol shown in **Figure 4-23**. You can use the find command to locate a variety of variable stars in *TheSkyX*.

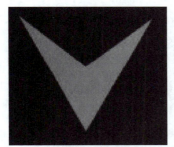

Figure 4-23 Symbol used for a Variable Star in *TheSkyX* Software.

Although difficult to see, **Figure 4-24** displays a field of variable stars in the constellation of Taurus the Bull near the Pleiades cluster, M45. Information on V352 Tauri is also displayed in right hand image of **Figure 4-24**.

Figure 4-24 Field of Variable Stars Near the Star Cluster M45 in Taurus the Bull

If you are interested in observing variable stars you may contact the American Association of Variable Stars Observers (AAVSO) at 25 Birch Street Cambridge, Massachusetts 02138-1205. This organization is a nonprofit scientific and educational international organization of

© 2011 Cengage Learning. All Rights Reserved. May not be scanned, copied or duplicated, or posted to a publicly accessible website, in whole or in part.

amateur and professional astronomers who study and catalog variable stars. You may also visit their Website at www.aavso.org.

Non-Stellar Objects

Most or all of the non-stellar objects lie well beyond our Solar System and are a mixed bag of astronomical objects. There are three basic categories of non-stellar objects. They are star clusters, nebulae, and galaxies. Usually these objects are designated with a letter and a number signifying the catalog in which they are found.

Star Clusters

There are two types of star clusters. The type of cluster is indicative of its location in the Milky Way. *Open clusters* are located *in* the spiral arms, and g*lobular clusters* are *centered on* and *around* the nucleus of our Galaxy. NGC 4755 is an excellent example of an open cluster. It is displayed as the left-hand image in **Figure 4-25**. Its designation is 4755 is its number in a catalog. It is also known as "The Jewel Box" cluster. Unfortunately, it is not visible from midnorthern latitudes. The second image displayed in **Figure 4-25** is one of several hundred globular clusters surrounding the nucleus of the Milky Way. It is displayed as the right-hand image in **Figure 4-25** and is designated NGC 6205 or M13.

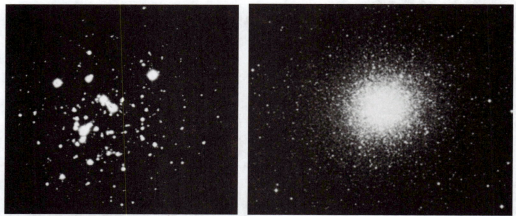

Figure 4-25 NGC 4755 (left) and NGC 6205/M13 (right)

Nebulae

Nebulae are more diversified in nature. They range from objects that are associated with early stages of star formation to those of stellar death.

Dark nebulae are cold regions of gas and dust that loom between the stars in the Milky Way and may harbor *protostars* (infrared stars). This type of nebula is only seen when silhouetted against a bright background of stars or a glowing nebula. One of the best examples of this type of nebula is located in Orion, the Hunter. It is also called the "Horsehead Nebula," because of its appearance, and is designated as IC 434. An image of it taken from *TheSkyX's* database is shown in **Figure 4-26**.

© 2011 Cengage Learning. All Rights Reserved. May not be scanned, copied or duplicated, or posted to a publicly accessible website, in whole or in part.

Figure 4-26 Horsehead Nebula in Orion, IC 434

Figure 4-27 displays a beautiful digitized image showing IC 434 and a more extensive view of the region around and near the Horsehead Nebula.

Figure 4-27 Regions Around and Near IC 434
(Image taken from *TheSkySix* Serious Astronomer)

Emission nebulae are associated with newborn stars and sometimes clusters of stars. They glow because the gases in the nebulae are excited by the ultraviolet radiation produced by the

© 2011 Cengage Learning. All Rights Reserved. May not be scanned, copied or duplicated, or posted to a publicly accessible website, in whole or in part.

surrounding stars. The nebula glows faintly with a pinkish-red color due to the hydrogen in the cloud. A good example of this type of nebula is displayed in **Figure 4-28**. It is called the "Eagle Nebula" because of its appearance. This image was taken by the author while at Kitt Peak using the SARA 0.9m telescope and is designated as M16 and NGC 6611 and is located in the constellation Serpens.

Figure 4-28 Eagle Nebula, M16, or NGC 6611 in Serpens
(Taken by Author with the SARA 0.9m Telescope at Kitt Peak)

Reflection nebulae, in contrast, appear bluish in color. In addition to gas, the cloud contains dust grains that reflect visible light from newly formed stars near the cloud. The Pleiades cluster (Seven Sisters) in the constellation of Taurus is perhaps the best known example of this type of nebulosity. This cluster contains a few dozen stars and is surrounded by bluish nebulosity. **Figure 4-29** shows a picture of the Pleiades taken by Dr. Jordan. It is designation as M45.

Figure 4-29 Reflection Nebula and Cluster in Taurus
(Photo courtesy of the author)

Other types of nebulae are associated with the death of elderly stars. When stars like our Sun reach the end of their life cycle, they go through a series of expansions and contractions. Eventually, they shed the outer layers of their atmosphere into space. These nebulae become planetary nebulae.

© 2011 Cengage Learning. All Rights Reserved. May not be scanned, copied or duplicated, or posted to a publicly accessible website, in whole or in part.

The most extreme expulsion of stellar material occurs when massive stars reach the end of their lives. They have a tendency to detonate and blow themselves apart. This event causes a sudden brightening of the star, and as a result, a supernova appears in the sky. These nebulae become supernova remnants. The last visible supernova occurred in the Large Magellanic Cloud in 1987. In any case, it is quite a traumatic event for a star.

Planetary nebulae result from stars that were once like our Sun who have gone through their entire life cycle. These nebulae are spherical shells of gas moving outward from the central part of the star. They usually appear as ring-shaped objects with a faint star in the center of the nebula. The planetary nebula in **Figure 4-30** is known as the "Ring Nebula." It is located in the constellation of Lyra and is usually referred to as M57. It is also designated NGC 6720. The image is a CCD image taken at the Ball State Observatory.

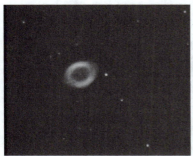

Figure 4-30 Ring Nebula in Lyra
(CCD image provided by the author)

Supernova remnants, on the other hand, appear as twisted knots of gas and serve as evidence of violent explosions—to put it mildly. They are more catastrophic than violent!
Figure 4-31 displays the supernova remnant in Taurus (the Bull) known as the *Crab Supernova Remnant*. This explosion was observed and recorded by Chinese astronomers in 1054 C.E. It was bright enough, according to historical accounts, to be seen in broad daylight. It is designated as NGC 1952 or as the first object in the Messier Catalogue (M1).

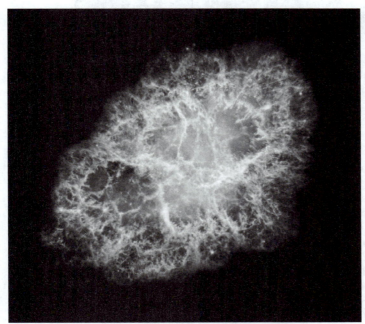

Figure 4-31 Crab Supernova Remnant, M1, in Taurus

Of the types of nebulae discussed here, only the brightest ones are seen either with the naked-eye or in binoculars and small telescopes. They appear as smudges, or more often as diffuse faint

© 2011 Cengage Learning. All Rights Reserved. May not be scanned, copied or duplicated, or posted to a publicly accessible website, in whole or in part.

blurs in telescopes. However, they are among some of the most interesting objects in our Galaxy. The brightest nebulae that can be observed and photographed with a small or moderate-size telescope are listed in some of the catalogs that are discussed hereafter.

The Messier Catalogue

In 1781 Charles Messier compiled a list (cataloged) of 110 faint, diffuse, non-stellar objects. He did so to assist those amateur astronomers who were specifically searching for comets. He published the positions of these objects so that other comet hunters would not waste their time making additional observations of them. Messier stated that objects in his catalog did *not* move and therefore were *not* comets! Many Messier objects are shown in **Figure 4-32** in and near the constellation of Sagittarius.

Figure 4-32 Messier Objects In and Near Sagittarius

Objects in Messier's Catalogue are numbered from 1 to 110 preceded by the capital letter **M**. A list of Messier objects is provided in **Appendix D**.

The New General Catalogue

William Herschel made several sky surveys near the end of the 18th century and into the beginning of the 19th century. He used larger and larger telescopes in his observations to do star gauging or counting measurements of stars in the Milky Way. He also recorded the positions and descriptions of thousands of faint non-stellar objects. In 1888 Herschel's observations were combined with the observations of many other astronomers into what became known as the New General Catalogue of Non-Stellar Objects. Several thousand of these non-stellar objects have been compiled for this catalog and are designated with the letters **NGC** and a number. The "Sombrero Galaxy" gets its name from its appearance and is designated as Object Number 4594 in the New General Catalogue or NGC 4594. This particular object is also listed as Object 104 in Messier's Catalogue or M104. The Find Result window for NGC 4594 is displayed in **Figure 4-33** along with an image of it. **Figure 4-34** displays an image of NGC 4594 taken from *TheSkyX* database.

© 2011 Cengage Learning. All Rights Reserved. May not be scanned, copied or duplicated, or posted to a publicly accessible website, in whole or in part.

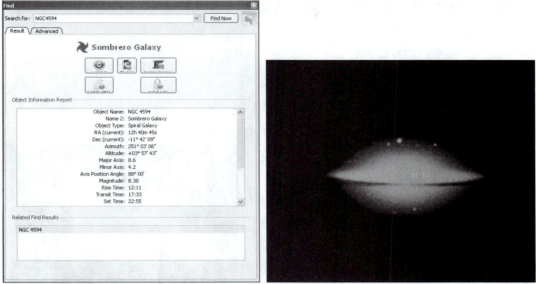

Figure 4-33 The Find Result Window for NGC 4594, the Sombrero Galaxy

Figure 4-34 NGC 4594, the Sombrero Galaxy, from *TheSkyX* Database

The NGC list of nonstellar objects is too large to include in this workbook. You can easily generate a complete observing list of NGC objects by visiting one of the following Web-sites: *www.ngcic.org* or *www.ngic.org/oblstgen.htm*. It is possible to create a complete list of NGC objects in any constellation at this website and download it to your PC or Mac.

A few years after the publication of the New General Catalogue, two supplemental catalogs were published. They contained information on more recently discovered objects (other than NGC objects). They were called the First and the Second Index Catalogues, respectively. The objects contained in both of these catalogs were denoted with a number preceded by the letters "IC." The astronomical objects listed in the IC catalogs are, as a rule, fainter than many of those listed in the NGC catalog. The IC lists contain several thousand more non-stellar objects. One can also create an observer's list of the IC objects within any constellation by accessing the same website as for the NGC objects. The Object Information window in the Find window displays IC 5152 along with its image in **Figure 4-35**.

© 2011 Cengage Learning. All Rights Reserved. May not be scanned, copied or duplicated, or posted to a publicly accessible website, in whole or in part.

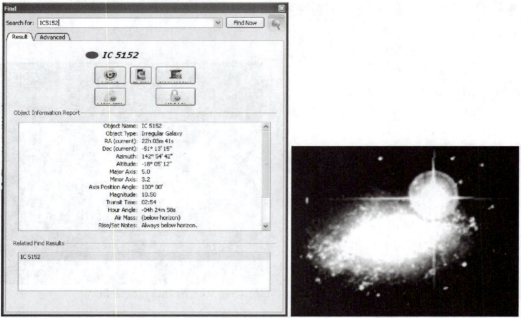

Figure 4-35 The Find Result Window for IC 5152

Hubble Space Telescope

Since its launch in April 1990, the Hubble Space Telescope has been quietly orbiting Earth and returning remarkable images and data on a variety of astronomical objects. It has compiled and assimilated more information in the last decade than all earth-based optical telescopes have historically, combined! The Find Result window is displayed for GSC 5847:2333 in **Figure 4-36**. This star has the proper name Diphda and also has the designations of β Ceti and 16 Ceti.

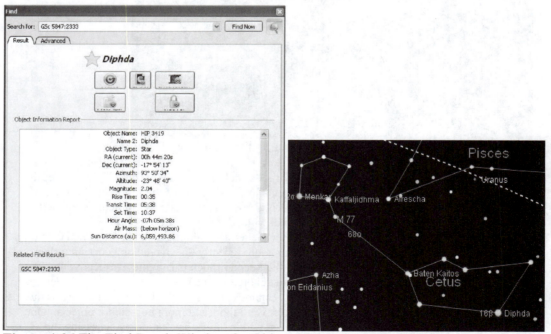

Figure 4-36 The Find Result Window for Object GSC 5847:2333, Diphda

© 2011 Cengage Learning. All Rights Reserved. May not be scanned, copied or duplicated, or posted to a publicly accessible website, in whole or in part.

Hubble has compiled data on tens of millions of stars and non-stellar objects as well. The information about these stars is stored on about 100+ CD-ROMs. Each star is cataloged with a number preceded by the letters **GSC,** for Guide Star Catalog.

Other Catalogs

Other catalogs that list information about stars are used in *TheSkyX*. The Harvard Smithsonian Astrophysical Observatory published a catalog that is often used by astronomers. The Observatory compiled data on about 250,000 stars. Stars in this catalog are labeled with a number and preceded by the letters **SAO**. **Figure 4-37** displays the Object Information window for the star SAO 147420. This is still the star β Ceti.

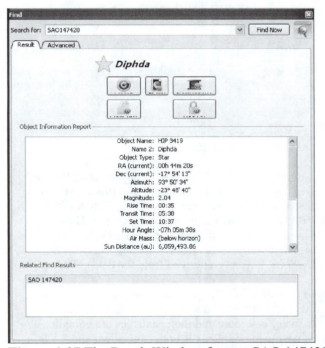

Figure 4-37 The Result Window for star SAO 147420, β Ceti, Diphda

Another catalog is the Henry Draper Catalog. This catalog, compiled between 1918 and 1924, contains spectral information on over 225,000 stars. Stars in this catalog are designated with a number preceded by the letters **HD**. The Object Information window for the star HD 4128 is displayed in **Figure 4-38**. Notice that this star is still β Ceti.

© 2011 Cengage Learning. All Rights Reserved. May not be scanned, copied or duplicated, or posted to a publicly accessible website, in whole or in part.

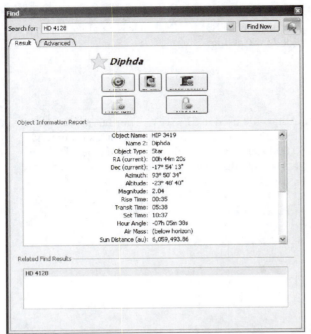

Figure 4-38 The Result Window for star HD 4128, Diphda, β Ceti

There are finally two other catalogs available in *TheSkyX* database. They are the **PPM** and **HIPPARCOS** catalogs. **PPM** is an acronym that stands for the **P**ositions and **P**roper **M**otion catalog. The stars in this catalog are designated with a number preceded by the letters **PPM**. Diphda is designated as PPM209214. **HIPPARCOS** is an acronym that stands for "**H**igh **P**recision **P**arallax **CO**llecting **S**atellite." The stars in this catalog are designated with a number preceded by the letters **HIP**. Diphda is designated in the HIPPARCOS catalog as HIP3419. The distances listed, for many of stars in *TheSkyX*, have been derived from the data collected by the HIPPARCOS satellite.

Today, we know of many diverse types of astronomical objects, and their designations can be confusing. Only computers have the ability and capacity to store, cross-reference, and assimilate information about these objects. Every day more and more statistics are compiled and stored in databases. This is one reason why using *TheSkyX* is both convenient and resourceful.

Numerous catalogs are accessible to those of us interested in observational astronomy. Each is compiled for a specific reason or to detail specific information about certain groups of celestial objects. This information has been cross-referenced in a worldwide database such as the one at the *Centre de Donnees Astronomiques de Strasbourg*. It's internet address is located at the following URL: http://cdsweb.u-strasbg.fr/CDS.html. Astronomers around the world can access this database via the World Wide Web each day. Most of the designations previously discussed and used in *TheSkyX* software can be found at this website.

© 2011 Cengage Learning. All Rights Reserved. May not be scanned, copied or duplicated, or posted to a publicly accessible website, in whole or in part.

Chapter 4

TheSkyX Review Exercises

Use *TheSkyX* to complete the following exercises. The first one has been done for you.

TheSkyX Exercise 1: Let's Find Object NGC 7000

1. Set Date: ***September 15, 2017***
    Set Time: to 8:45 PM DST
    Set Daylight saving option: "U.S. and Canada"

    Set Location: Ball State Observatory
    Longitude: 85°24′ 38″ West   Latitude: 40° 11′ 58″ North
    Set Time Zone to -5.0 hours
    Set Elevation to 322 meters

    When finished, the settings should look like those displayed in **Figure 4-39**.

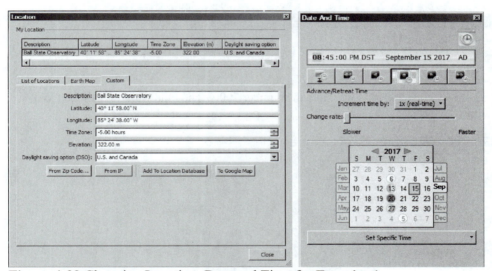

Figure 4-39 Changing Location, Date and Time for **Exercise 1**

2. Click "Edit" on the Standard toolbar at the top of the sky window then click ⟨ Find... Ctrl+F ⟩.

3. A shortcut to this command is to right-click mouse anywhere in the skyx desktop window. Either way, the Find window appears, like the one shown in **Figure 4-40**. You are now ready to search *TheSkyX* databases.

© 2011 Cengage Learning. All Rights Reserved. May not be scanned, copied or duplicated, or posted to a publicly accessible website, in whole or in part.

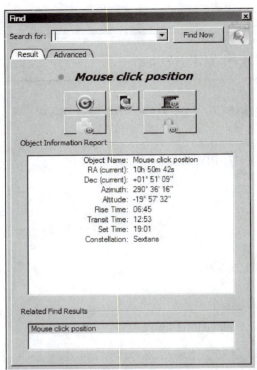

Figure 4-40 The Find Window

4. Let's find the NGC Object 7000.

5. In the Find window, type *"*NGC, *"* and the number 7000. The Find window will then look like **Figure 4-41**.

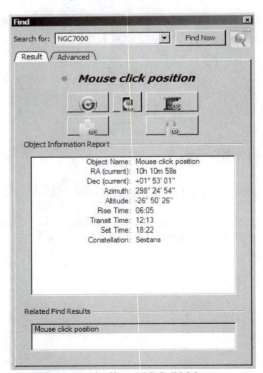

Figure 4-41 Finding NGC 7000

© 2011 Cengage Learning. All Rights Reserved. May not be scanned, copied or duplicated, or posted to a publicly accessible website, in whole or in part.

6. Click the Find Now button.

7. After clicking "Find Now," information about NGC 7000 appears as is shown in **Figure 4-42**.

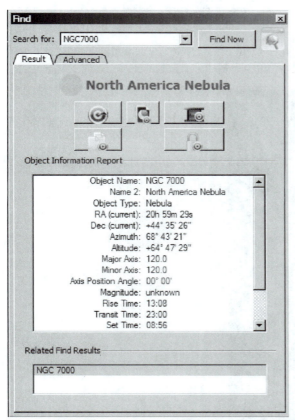

Figure 4-42 The Find Result Window for NGC 7000

8. Use the scroll bar to peruse the data provided in the area where the Object Information Report is located in regard to NGC 7000. The information that is not displayed in **Figure 4-42** is displayed in **Figure 4-43**. This information is user defined! You may display more or less information about your objects of interest.

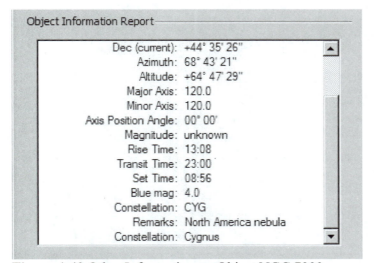

Figure 4-43 Other Information on Object NGC 7000

© 2011 Cengage Learning. All Rights Reserved. May not be scanned, copied or duplicated, or posted to a publicly accessible website, in whole or in part.

9.   Click the [Center] button to center NGC 7000 on your desktop.

10.   At this point you have two options to view NGC 7000. The first is the "Frame Option." If

you click the "Frame Option," [icon], then you will center NGC 7000 on the desktop with a
Field of View of 5-degrees as displayed in **Figure 4-44**.

Figure 4-44 The 5-degree Option of the North American Nebula, NGC 7000

The second option is the "Show Photo" option. If you click the "Show Photo option,"

[icon], then NGC 7000 will be displayed in "Photo Viewer" mode as displayed in
Figure 4-45.

© 2011 Cengage Learning. All Rights Reserved. May not be scanned, copied or duplicated, or posted to a publicly accessible website, in whole or in part.

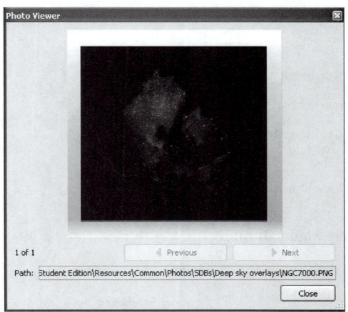

Figure 4-45 The Photo Viewer Option of the North American Nebula, NGC 7000

Now, it's your turn!

Complete the Following Exercises Using *TheSkyX*.

If you wish, you may leave the location set to the Ball State Observatory. Otherwise, set it to your location. The important thing is *not* to change the date and time.

TheSkyX Exercise 2: Find Object M30

1. What is the NGC number associated with M30? _____

2. In what constellation is M30 located? _____

3. What is its apparent magnitude? _____

4. What type of non-stellar object is this? _____

5. Does it have a common name? _____ If so, what is it? _____

6. Click the [Center] button in order to view an image of M30.

TheSkyX Exercise 3: Find the Planet Uranus

1. What time does Uranus appear to Rise? ____:____ Transit? ____:____ Set? ____:____.

2. What is the apparent magnitude and Phase of Uranus? _____; _____.

3. What are the Right Ascension and Declination of Uranus?

   R.A. _____H _____M _____S   Decl. = ___ _____° _____' _____"

© 2011 Cengage Learning. All Rights Reserved. May not be scanned, copied or duplicated, or posted to a publicly accessible website, in whole or in part.

4.   What are the Azimuth and Altitude of Uranus?

   Azi. _____H _____M _____S   Alt. = ___ _____° _____' _____"

5.   What is Uranus' distance from the Sun? _____ AUs.

6.   What is Uranus' distance from Earth? _____ AUs.

7,   In what constellation is Uranus currently located? _____.

8.   Click the [button icon] button to view Uranus.

TheSkyX Exercise 4: Find the Planet Jupiter

1.   What time does Jupiter appear to Rise? ____:____ Transit? ____:____ Set? ____:____.

2.   What is the apparent magnitude and Phase of Jupiter? _____; _____.

3.   What are the Right Ascension and Declination of Jupiter?

   R.A. _____H _____M _____S   Decl. = ___ _____° _____' _____"

4.   What are the Azimuth and Altitude of Jupiter?

   Azi. _____H _____M _____S   Alt. = ___ _____° _____' _____"

5.   What is Jupiter's distance from the Sun? _____ AUs.

6.   What is Jupiter's distance from Earth? _____ AUs.

7.   In what constellation is Jupiter currently located? _____.

8.   Click the [button icon] button to view Jupiter.

TheSkyX Exercise 5: Find Object M42

1.   What time does M42 appear to rise? ____:____

2.   What is its apparent magnitude? _____

3.   What is its NGC number? _____

4.   In what constellation is it located? _____

5.   What type of object is it? _____

6.   Click the [Center button icon] or [button icon] button to view this object.

© 2011 Cengage Learning. All Rights Reserved. May not be scanned, copied or duplicated, or posted to a publicly accessible website, in whole or in part.

TheSkyX Exercise 6: Find Object NGC 224

1. What time does NGC 224 appear to rise? ____:____

2. What is its apparent magnitude? _____

3. What Messier number does it have? _____

4. In what constellation is it located? _____

5. What type of object is it? _____

6. Does it have a common name? _____

7. Click the [Center] or [image] button to view this object.

TheSkyX Exercise 7: Find Object IC 446

1. What time does IC 446 appear to rise? _____:_____

2. When does this object transit your local meridian? _____:_____

3. In what constellation is it located? _____

4. What type of object is it? _____

5. What is its apparent magnitude? _____

6. Click the [Center] or [image] button to view this object.

TheSkyX Exercise 8: Find Object NGC 2244

1. What time does NGC 2244 appear to rise? _____:_____

2. When does this object transit your local meridian? _____:_____

3. In what constellation is it located? _____

4. What type of object is it? _____

5. What is its apparent magnitude? _____

6. Click the [Center] or [image] button to view this object.

© 2011 Cengage Learning. All Rights Reserved. May not be scanned, copied or duplicated, or posted to a publicly accessible website, in whole or in part.

TheSkyX Exercise 9: Find Object M79

1.   What time does M79 appear to rise? _____:_____

2.   When does this object transit your local meridian? _____:_____

3.   In what constellation is it located? _____

4.   Does it have an NGC number? _____ If so, what is it? _____

5.   What type of object is it? _____

6.   What is its apparent magnitude? _____

7.   Click the [Center] or [ ] button to view this object.

TheSkyX Exercise 10: Find Object M13

1.   What time does M13 appear to rise? _____:_____

2.   When does this object transit your local meridian? _____:_____

3.   In what constellation is it located? _____

4.   Does it have an NGC number? _____ If so, what is it? _____

5.   What type of object is it? _____

6.   What is its apparent magnitude? _____

7.   What is the object's angular size? _____

8.   Click [Center] or [ ] to view this object.

TheSkyX Exercise 11: Locate and Identify an Object

1.   On the Standard toolbar at the top of the sky window click "Orientation," then "Navigate."
     Or right-click the mouse button and click "Navigate." Now, click the "Sky Chart Center" tab and
     enter the following information into the "Equatorial Coordinates" box:

     Right Ascension: $13^H 30^M 00^S$ Declination: 48° 00′ 00″ North   Epoch: 2017.0

     Click the [Center on RA/Dec] bar.

     Now, click "Orientation" on the Standard toolbar again and click "Zoom To" Binocular or 50°.

2.   What object is located at the following coordinates? Right Ascension =$13^H 30^M 35^S$ and
     Declination = +47° 06′ 32″ _____

3.   What is its Messier Number? _____

© 2011 Cengage Learning. All Rights Reserved. May not be scanned, copied or duplicated, or posted to a publicly accessible website, in whole or in part.

4. Does it have a Common Name? _____

5. What is its NGC Number? _____

6. What time does this object appear to rise? _____:_____

7. When does this object transit your local meridian? _____:_____

8. In what constellation is it located? _____

9. What type of object is it? _____

10. What is its apparent magnitude? _____

11. Click the button to view this object.

TheSkyX Exercise 12: Locate and Identify an Object

1. Set Date: July 4, 2020
   Set Time: 9:45 PM DST
   Set Daylight saving option:"U.S. and Canada"

2. On the Standard toolbar at the top of the sky window click "Orientation," then click "Navigate."
   Or right-click the mouse button and click "Navigate." Now, click the "Sky Chart Center" tab and
   enter the following information into the "Equatorial Coordinates" box:

   Right Ascension: $20^H 45^M 5.0^S$ Declination: $40° 15' 15.0''$ North
   Epoch: 2020.0

   Click the [Center on RA/Dec] bar.

   Now, click on the Orientation on the standard toolbar and click Zoom To Binocular or 50°.

3. What constellation is located in the center of the field of view? _____

4. Are there any Messier Objects in this constellation? _____

5. List some of them: _____, _____, _____, _____, _____, _____, _____, _____.

6. Do any of them have Common Names? _____

7. What are the coordinates of M 29?

       R.A. = ____H ____M _____S   Declination = ____ _____° _____' _____''

8. What time does this object appear to rise? _____:_____

9. When does this object transit your local meridian? _____:_____

10. What type of object is it? _____

11. What is its apparent magnitude? _____

© 2011 Cengage Learning. All Rights Reserved. May not be scanned, copied or duplicated, or posted to a publicly accessible website, in whole or in part.

12. Does it have an NGC Number? _____. If so, what is it? _____

13. Click the button to view this object.

© 2011 Cengage Learning. All Rights Reserved. May not be scanned, copied or duplicated, or posted to a publicly accessible website, in whole or in part.

Chapter 4

TheSkyX Review Questions

Answer the following review questions:

1. In what order are the Bayer letters assigned to stars? _____

2. Is δ-Capricorni brighter than α-Capricorni? _____
   How do you know? _____

3. In what order were Flamsteed numbers assigned to stars? _____

4. How many nonstellar objects did Messier include in his catalogue? _____

5. Why did Messier compile his list, and how were these objects selected?
   _____

   _____

6. What nomenclature is used to identify variable stars? _____

7. Is V 143 Cyg a valid variable star designation? _____ If not, why not?
   _____. What is its correct designation?
   _____.

8. Is V469 Lyrae a valid variable star designation? _____

9. If a star is found to be multiple, how are its various components designated?

   _____

10. What is the order of discovery of TZ Orionis? _____

© 2011 Cengage Learning. All Rights Reserved. May not be scanned, copied or duplicated, or posted to a publicly accessible website, in whole or in part.

Chapter 5

Locating Celestial Objects in *TheSkyX*

Have you ever searched for a friend's house in a different city and found it next to impossible to find? Did you get lost? Perhaps there were only a few street signs, or perhaps that your friend didn't give you very clear instructions to follow. You probably stopped at a gas station and asked for directions or used your cell phone and called your friend or perhaps texted them to ask for more detailed directions.

Even though street signs may offer little assistance, think of what it would have been like to not have had any signs at all! We continually encounter problems in finding our way around or in getting from one place to another when it involves directions. Today, everyone seems to be obsessed with using GPS devices to find their way around…even to find garage sales! In observational astronomy, however, general procedures have been adopted to help avoid these problematic situations. In other words, astronomers have developed a system of coordinates to locate the positions of celestial objects very precisely.

Angles and Coordinate Systems

This chapter contains exercises that assist in learning how to use coordinate systems in observational astronomy. To understand an astronomer's method of locating objects in the sky, it helps to have a working knowledge of angles and a three-dimensional view of the sky.

Coordinate systems are usually set up in such a way that we must first choose a *frame of reference*, that is, a place from which our detailed measurements of the sky are made. In this instance, Earth is the *only* frame of reference we have! We begin by envisioning the sky above us, as ancient astronomers did, to be a large sphere. In doing so, this will amount to determining just two angles to locate objects on this sphere.

Many people have difficulty measuring angles. To avoid the most common mistakes made, we will take a very simplistic approach. We begin our discussion of angles by reviewing how angles are measured and what units are used to express angular measurement.

When it comes to angles, astronomers view things in the sky in two ways. The first is *angular separation*, or *how far apart* two objects appear to be in the sky. The second is *angular size*, or *how large* objects appear to be in the sky. Either way, angles are used to describe the separation and size of astronomical objects.

Angles are measured in units called *degrees* (°). There are 360 of them in a full circle. Therefore, one degree represents 1/360th of a full circle. Degrees are usually used for making large angular measurements. For example, there are two stars in the Ursa Major known as the *pole pointers* (α and β Ursae Majoris) because they always point to Polaris. They have an *angular separation* of about 5°. The stars α and β Ursae Majoris, the pole pointers, in the Ursa Major as seen in *TheSkyX* are illustrated in **Figure 5-1**.

Figure 5-1 α and β Ursae Majoris, Pole Pointers in Ursa Major

© 2011 Cengage Learning. All Rights Reserved. May not be scanned, copied or duplicated, or posted to a publicly accessible website, in whole or in part.

The Sun and (Full) Moon, in contrast, have an angular size of about ½°. **Figure 5-2** displays images of the Sun and Full Moon rendered from *TheSkyX's* image database. Even though the Sun and Full Moon have about the same angular size, the Sun is over 400 times larger than the Moon and over 400 times farther away than the Moon is from Earth. Despite this, the Sun appears to be about the same size as that of the Full Moon in Earth's sky.

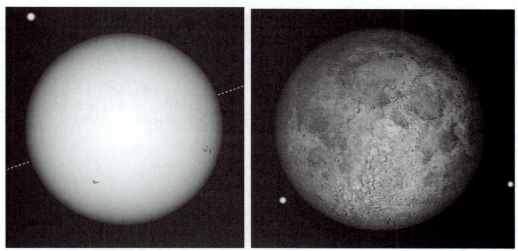

Figure 5-2 Moon and Sun

To measure an angle smaller than a degree, you need additional units. Therefore, 1° can be further divided into 60 equal smaller divisions called *minutes of arc* or *arcminutes*. Arcminutes are usually designated with an apostrophe mark ('). Don't confuse this mark with linear measurements of feet!

An average human eye (with 20/20 vision) can resolve two objects that have an angular separation of about 12 arcminutes. This happens to be the angular separation between the two stars that make up the "middle star" in the handle of the *Big Dipper*. Their proper names are Alcor and Mizar. They are also designated as 80 and ζ Ursae Majoris, respectively. They are shown in **Figure 5-3**. Try to find Mizar and Alcor some clear night. They are easy to locate in the sky, and you can consider them an eye test.

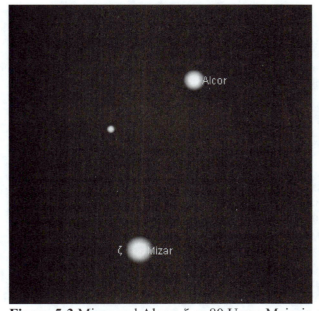

Figure 5-3 Mizar and Alcor: ζ or 80 Ursae Majoris, respectively

© 2011 Cengage Learning. All Rights Reserved. May not be scanned, copied or duplicated, or posted to a publicly accessible website, in whole or in part.

   If you look at Mizar and Alcor through a small telescope, which has a greater resolution than the eye, you will immediately notice that the angular separation between Alcor and Mizar is much greater and that Mizar has a very close companion. It is, indeed, a binary star system.

   To measure the angular separation between ζ Ursae Majoris A and B, one needs smaller angular units. To accommodate very small angles, such as between binary stars, we further divide 1 arcminute into 60 equal smaller divisions called *seconds of arc* or *arcseconds*. Arcseconds are usually designated with a quotation mark (″). Don't confuse this mark with linear measurements of inches! An arcsecond represents the smallest angular unit used in measuring angles. There are 1,296,000 of them in a full circle. ζUrsae Majoris A and B are separated by only 14 arcseconds in angle. To see both stars easily you need, at least, a pair of binoculars or a small telescope. *TheSkyX* cannot display ζ Ursae Majoris A and B because its minimum zoom is 30″.

   A realistic example of how small an arcsecond really is would be for a friend to hold a penny with their index finger and thumb. Now, observe that penny they are holding while they are standing at a distance of 2½ miles away! This is how large an arcsecond appears in the sky. The angle equivalencies are as follows:

$$360° \text{ (degrees)} = \text{full circle}$$

$$1° \text{ (degree)} = 60' \text{ (arcminutes)}$$

$$1' \text{ (arcminute)} = 60'' \text{ (arcseconds}$$

This is the most common procedure that astronomers use to measure angles in astronomy.

   However, another system in observational astronomy uses *time units* to measure angle instead of the conventional degrees, minutes, and seconds of arc. This system is based on a rotating Earth. Because Earth rotates once on its axis (full circle) every 24 hours, it is convenient to divide the sky into 24 equal parts instead of 360. This system employs a sphere divided into 24 equal intervals called *hours*. Each 1-hour interval represents 15° in angle. These hour intervals can be further divided into *60 equal* smaller intervals called *minutes*, and each minute interval can be further divided into *60 equal* smaller intervals called *seconds*. This is the system that you will eventually learn and the one that astronomers use to locate celestial objects in the sky.

   It is convenient to use equivalents to convert angular units to time units and vice versa. The following equivalents are very helpful:

$$15° = 1^{H}$$
$$15' = 1^{M}$$
$$15'' = 1^{S}$$

   Other equivalent expressions are used in the conversion of angular units to time units, too. Because $1^{H} = 60^{M}$, which equals 15° of angle, then

$$4^{M} = 1°$$

and, because $1^{M} = 60^{S}$, which equals 15′ of angle, then

$$4^{S} = 1'$$

   With information concerning angles and time at hand, one should now be ready to find things in the sky. First, however, some knowledge of coordinate systems is helpful. There are two coordinate systems used in the sky. Before discussing them, it is essential to review the coordinate system used on Earth.

© 2011 Cengage Learning. All Rights Reserved. May not be scanned, copied or duplicated, or posted to a publicly accessible website, in whole or in part.

The Geographic Coordinate System

The coordinate system that is most familiar to everyone is the system that is used on Earth's surface. It is the *Geographic Coordinate System*. In the geographic coordinate system, the surface of Earth is the frame of reference, and the *geographic equator* is the basic plane of origin from which and along which the coordinates of *latitude* and *longitude* are measured.

Not only are angular measurements used in this system but *directions* are also specified. In the geographic coordinate system, the *geographic equator* is a fundamental great circle. A *fundamental great circle* is a circle whose plane passes through the center of a sphere and has a diameter equal to that of the sphere. It divides the sphere into two equal halves. **Figure 5-4** illustrates the location of the north and south geographic poles as well as the geographic equator. By virtue of the way it is drawn, the geographic equator divides Earth into two equal halves, north and south.

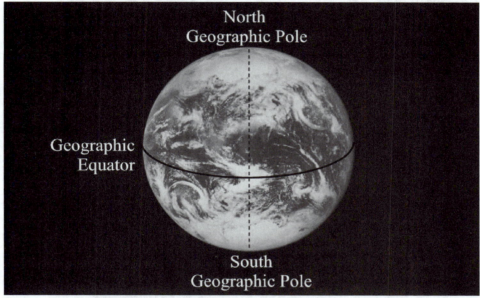

Figure 5-4 Geographic Equator

The north and south geographic poles are the two points about which Earth rotates and therefore defines its axis of rotation. The geographic equator is actually defined as the circle drawn *exactly* halfway between (90° from) the north and south geographic poles.

As stated earlier, the geographic equator is a great circle and the primary reference circle in the geographic coordinate system. Circles that are formed by intersections of the sphere with planes not passing through the center of the sphere have diameters smaller than that of the sphere and are consequently called *small circles*. Therefore, circles parallel to the geographic equator are defined as small circles (or latitude circles).

Secondary reference circles are used to divide Earth in half in an east-west direction. These secondary reference circles are also great circles and are shown in **Figure 5-5**.

© 2011 Cengage Learning. All Rights Reserved. May not be scanned, copied or duplicated, or posted to a publicly accessible website, in whole or in part.

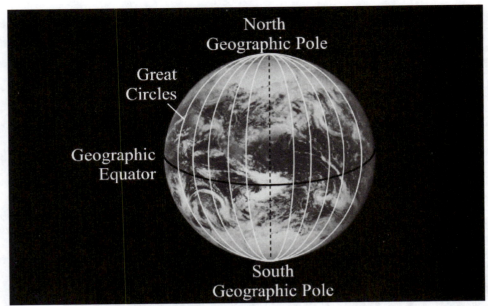

Figure 5-5 Great Circles on Earth

These great circles are drawn on the Earth in a north-south direction through the north and south geographic poles. The great circle passing through Greenwich, England (agreed on internationally since 1884) is called the *reference meridian* or the *prime meridian*. All other great circles that pass through the poles and are parallel to the prime meridian are also known as *meridians* (or longitude circles) on Earth.

To specify locations on the surface of Earth, geographers use the coordinates of latitude and longitude. *Latitude* is an angle that is measured north or south of the geographic equator as viewed from Earth's center, along a *meridian* or *longitude circle*. It is measured from the geographic equator to the latitude (small) circle on which the location is situated. The measurement of the latitude of Muncie, Indiana, is displayed in **Figure 5-6**.

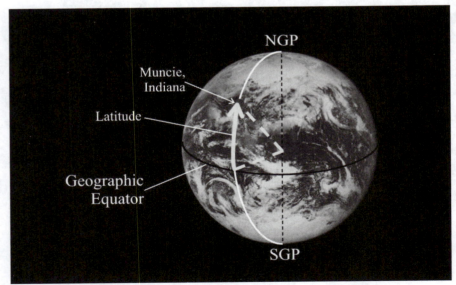

Figure 5-6 Latitude of Muncie, Indiana

Muncie, Indiana is located 40° 11′ 35″ north of the geographic equator. The *values* of latitude range from 0° to 90° (0°at equator and 90°at the NGP or SGP). Directions are specified as being either north or south of the geographic equator.

© 2011 Cengage Learning. All Rights Reserved. May not be scanned, copied or duplicated, or posted to a publicly accessible website, in whole or in part.

Longitude is an angle measured around Earth's equator, as seen from its center. The *values* of longitude range from 0° to 180°. Directions are designated as being either east or west of Greenwich, England (prime meridian). The measurement of the longitude of Muncie, Indiana, is displayed in **Figure 5-7**.

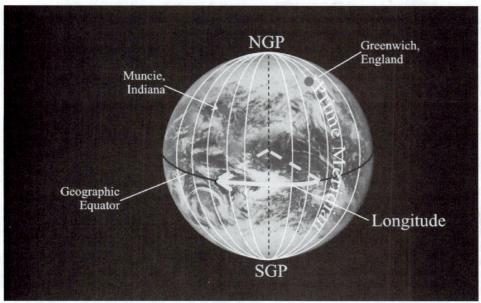

Figure 5-7 Longitude of Muncie, Indiana

Muncie, Indiana is located 85° 23′ 11″ west of Greenwich, England. The geographic coordinate system is an easy system to use to find the location of any city, observatory, or missile silo on Earth. The location of the Ball State Observatory is 40°11′ 58″ N and 85°24′ 38″ W.

In *TheSkyX* one can change these coordinates to any location on Earth. Usually the latitude and longitude of the place where the observer lives is used. *TheSkyX* lets the observer move to any location on the surface of Earth (0°- 90° N–S in latitude and 0° - 180° E–W in longitude).

When the Student Version of *TheSkyX* is installed it searches for the user's location from the IP Address of your PC or Mac. If one is not connected to the Internet, then the program may default to the site of Golden, Colorado in the United States. Otherwise, the observer's location may have to be keyed manually into the location database as described in Chapter 1 or Chapter 3.

To observe from other sites around the U.S. or the world, you must access the "Location" window. To do this, click "Input" on the Standard toolbar at the top of the sky window then "Location." A list of sites is displayed in the *TheSkyX* location window. **Figure 5-8** displays the Location menu window with the "List of Locations" in *TheSkyX* database. One should notice immediately that the list of locations is compressed. That is, there is a plus sign (⊞) next to the list of locations that one may wish to find. To expand any of the "Lists of Locations" click the ⊞ next to one of the location lists. This opens up the list. One then scrolls down the list to find the location of interest. Click the plus sign next to the "United States." A partial list of U.S. cities is shown is **Figure 5-9**.

© 2011 Cengage Learning. All Rights Reserved. May not be scanned, copied or duplicated, or posted to a publicly accessible website, in whole or in part.

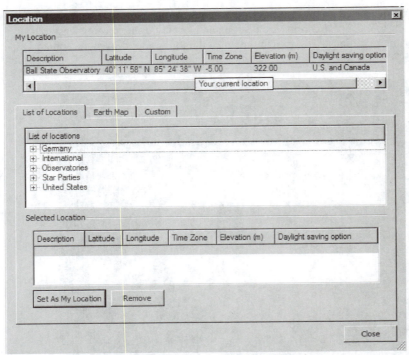

Figure 5-8 Lists of All Locations in *TheSkyX* Database

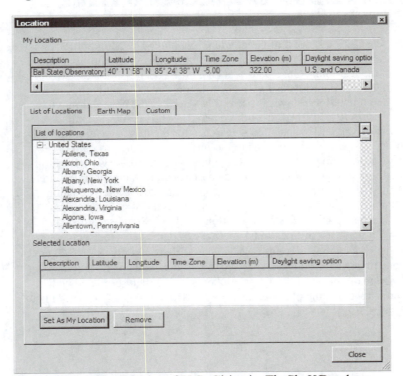

Figure 5-9 A Partial List of U.S. Cities in *TheSkyX* Database

Figure 5-10 displays Muncie, Indiana in *TheSkyX* database. Once you have opened the United States' list of locations you may now access any of the cities in the U.S. in *TheSkyX's* database. To set the location in *TheSkyX* database, scroll down the list and choose the appropriate site from which you wish to observe. You must then click the "Set As My Location" button near the bottom of the window.

© 2011 Cengage Learning. All Rights Reserved. May not be scanned, copied or duplicated, or posted to a publicly accessible website, in whole or in part.

If the desired location is not found in the list of indexed sites, then it must add it to the list. The site's latitude and longitude must be known before adding it to ***TheSkyX's*** database. There are two ways in which one may change their location. The first method is to use the "Earth Map."

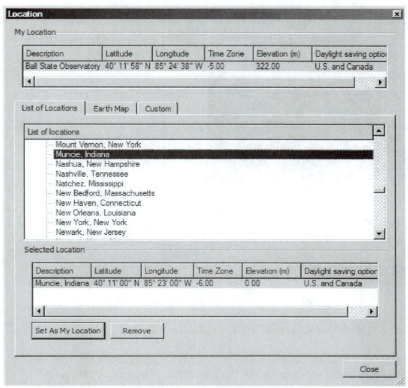

Figure 5-10 Muncie, Indiana in *TheSkyX's* List of U.S. Cities

If one clicks the tab labeled "Earth Map," a window appears like the one shown in **Figure 5-11** appears. Unless you have better than 20-20 vision it is suggested to zoom in close on the Earth map area. Each city in the United States has a red stick pin associated with it. One simply clicks on the appropriate pin to change their location on Earth. If one desires Buffalo, New York as the site for observing, then choose Buffalo and click "Close" afterwards.

© 2011 Cengage Learning. All Rights Reserved. May not be scanned, copied or duplicated, or posted to a publicly accessible website, in whole or in part.

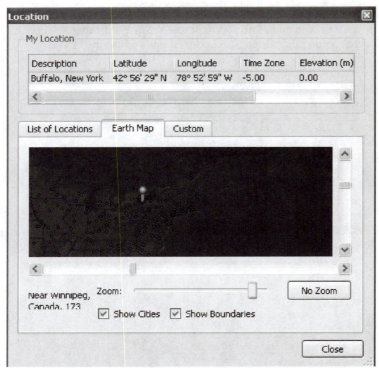

Figure 5-11 Earth Map Displaying Buffalo, New York as the Location

You *do not* have to click the "Set As My Location" button in this mode. The change is automatic.

    The second method involves clicking on the "Custom" tab. It is in this mode that you can enter *any* location on Earth. However, you must know the location's latitude and longitude precisely when using this mode of entry. One begins by typing in the location's "Description" (name of city or town) that you desire. Second, type the "Latitude" and "Longitude" which can be found from an Internet source or from a GPS device. Third, the location's "Time Zone" is entered next in hours from (- for West, + for East) of Greenwich, England and its elevation above sea level if known. And, finally the Daylight saving option (DSO). Most of the United States moves their clocks ahead in the spring months and back in the autumn months. The exception is Arizona and Hawaii. This will be discussed in detail in Chapter 7. Once all of this information is entered then it can be added to *TheSkyX* database by clicking the "Add to Location Database" button.

    Exercises A and **B** illustrates how you can view the sky from the north geographic pole and the geographic equator.

Exercise A: View *TheSkyX* From the North Geographic Pole

1. Run *TheSkyX* and open the file *Normal.skyx*.

2. Go to "Input" on the Standard toolbar at the top of the sky window.

3. Click on the "Custom" tab.

4. Type "North Pole" in the "Description" line.

5. Set your latitude to 90°N. The longitude is unimportant in this exercise because all the longitudes converge at the pole.

6. The "Location" window should appear like that in **Figure 5-12**.

© 2011 Cengage Learning. All Rights Reserved. May not be scanned, copied or duplicated, or posted to a publicly accessible website, in whole or in part.

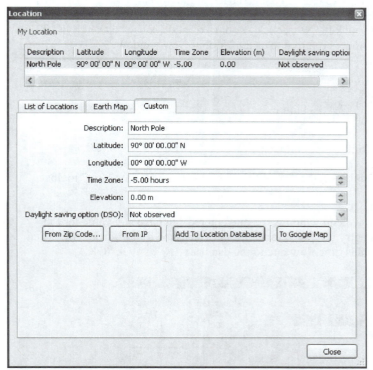

Figure 5-12 Setting Your Location to the North Geographical Pole

7. If you click the Look Up button (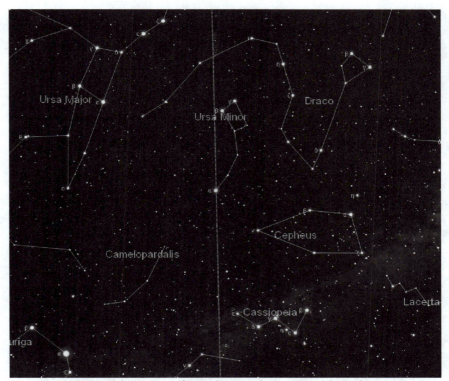Look Up), you will notice that the reasonably bright star Polaris in the constellation of Ursa Minor is directly overhead as seen in this zoomed out window in **Figure 5-13**.

Figure 5-13 Zenith as Seen from the North Geographic Pole

© 2011 Cengage Learning. All Rights Reserved. May not be scanned, copied or duplicated, or posted to a publicly accessible website, in whole or in part.

8. The view displayed in **Figure 5-13** can be changed back to the *Normal view*. In *TheSkyX*, the Normal view is the "Naked Eye" view of 100°. To accomplish this, go to "Orientation" on the Standard toolbar and click it. Next, click "Zoom To" and choose "Naked Eye 100°." This resets the sky back to what is referred to as the Normal view window.

9. Click the "Look North" () button.

10. Change the time flow rate from 1x to 5 minutes (5 minutes ▾).

11. This is accomplished by clicking on the "Time Flow" button on the "Time Skip" toolbar then "Custom."

12. After opening the Time Flow menu window, the rate entered can have any value for the Time Flow in this mode of entry. Type "5" and "minutes" as shown in **Figure 5-14**. Then click "Add." This will add this time flow rate to the database. Now, click "OK."

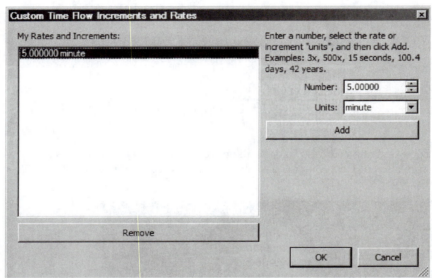

Figure 5-14 Adjusting the Time Flow Increments and Rates to 5 Minutes in *TheSkyX*

13. Zoom out to about 200°, then push the "Go Forward" (Go Forward) time step button and observe.

14. Stop the animation by pushing the Stop (Stop) button.

15. Describe the motion you observed. _____
    _____.

16. Do any objects go below the horizon? _____.

17. Click the Look East button (Look East) on the Standard Toolbar. What does the motion Look like near the eastern horizon? _____.

18. You may close the file and do not save it, or continue with **Exercise B**.

© 2011 Cengage Learning. All Rights Reserved. May not be scanned, copied or duplicated, or posted to a publicly accessible website, in whole or in part.

Exercise B: Viewing *TheSkyX* From the Geographic Equator

1. Run *TheSkyX* and open the file ***Normal.skyx***, or go to "Input" on the Standard toolbar at the top of the sky window.

2. Click on the "Custom" tab.

3. Type the words "Geographic Equator" in the "Description" line.

4. Set your latitude to 0°N.

5. Set your longitude for 0°W. Your Location window should appear like the one in **Figure 5-15**.

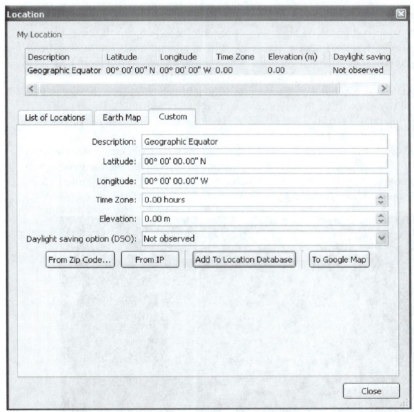

Figure 5-15 Setting Your Location to the Geographical Equator

6. Click the Look North button (Look North) to look North.

7. Enter a time flow of 5 minutes again (5 minutes ▾).

8. Push the Go Forward time flow button (Go Forward).

9. Stop the time flow by pushing Stop .

© 2011 Cengage Learning. All Rights Reserved. May not be scanned, copied or duplicated, or posted to a publicly accessible website, in whole or in part.

10. Describe what happened. _____.

11. Do any objects go below the horizon? _____.

12. Click the Look East button (Look East) to look East.

13. Describe the motion. _____.

14. Close the file and do not save it.

The Celestial Sphere

Having reviewed Earth coordinates, one can now apply this knowledge to the sky. For an astronomer, the sky now becomes the frame of reference. Since ancient times the sky has been known as the *celestial sphere*. The celestial sphere is an "apparent sphere" with its radius centered on the observer, standing on Earth, extending outward to infinity.

If Earth's axis of rotation is extended outward to infinity, it appears to touch the sky at two points. It is around these two points that the celestial sphere *appears* to rotate. These two points are defined as the *north* and *south celestial poles*. Their location in the sky is directly above the north and south geographic poles on Earth. If a line is drawn *exactly* halfway between or 90° from the north and south celestial poles, it defines a *fundamental great circle* in the sky known as the *celestial equator*. The celestial equator lies directly above the geographic equator in the sky. Moreover, like the geographic equator on Earth, divides the sky into two equal halves, north and south. The celestial sphere is illustrated in **Figure 5-16**.

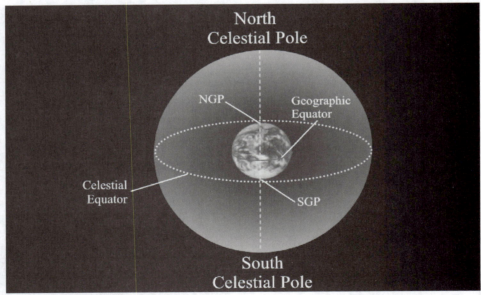

Figure 5-16 Celestial Sphere

It is on this sphere that astronomers put their coordinates to locate celestial objects. There are two coordinate systems that are used in observational astronomy.

© 2011 Cengage Learning. All Rights Reserved. May not be scanned, copied or duplicated, or posted to a publicly accessible website, in whole or in part.

The Horizon Coordinate System

The easiest sky coordinate to use in observational astronomy is based on the position of the observer, and uses the *horizon* as the *fundamental great circle*. This system, which is similar in design to the geographic coordinate system, is based on the *astronomical horizon* and is known as the *Horizon Coordinate System*.

Before setting up this coordinate system, two points in the sky need to be defined. These points are first located in order to draw the fundamental great circle that is used in this system. The first point to locate is defined as the *zenith point*. It is the point in the sky that is *directly above* your head. The second point to locate is known as the *nadir*. It is a point in the sky that is exactly 180° from your zenith. That is, it is located at the point in the sky directly below the observer – on the other side of the Earth. These two points in the sky represent the "poles" in this system.

The astronomical horizon is determined as a result of the direction of gravity for an observer's position on Earth. It is a plane that is perpendicular (90°) to the direction of gravity, and coincides with the observer's line of sight. That is, where land and sky appear to meet. This plane extends outward to infinity and divides the celestial sphere into two equal halves again; that is, half the sky is above the astronomical horizon and half the sky is below it.

The great circle that is used as the fundamental reference circle in this coordinate system is the astronomical horizon. It is then defined as a circle drawn on the celestial sphere that is *exactly* halfway between (90° from) the zenith and the nadir. **Figure 5-17** illustrates the location of the zenith, nadir, and astronomical horizon projected onto the celestial sphere.

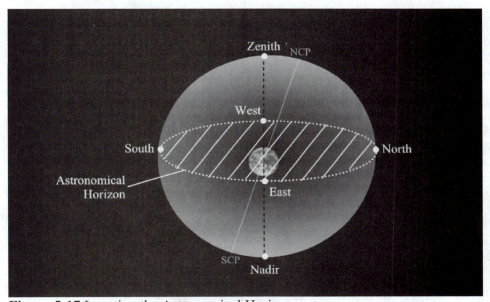

Figure 5-17 Locating the Astronomical Horizon

Using the horizon coordinate system to find celestial objects once again involves measuring just two angles. The first, is an angular measurement made from the astronomical horizon along a secondary great circle known as a *vertical circle* (remember: all circles passing through poles are great circles) to the object. The object's position is somewhere on this vertical circle with respect to the astronomical horizon.

The first angle is measured *above* the astronomical horizon and is defined as an object's *altitude*. The values for altitude range from 0° to 90°. The altitude of any object on the horizon is 0° and that at the zenith is 90°. The measurement of altitude on the celestial sphere is depicted in **Figure 5-18**.

© 2011 Cengage Learning. All Rights Reserved. May not be scanned, copied or duplicated, or posted to a publicly accessible website, in whole or in part.

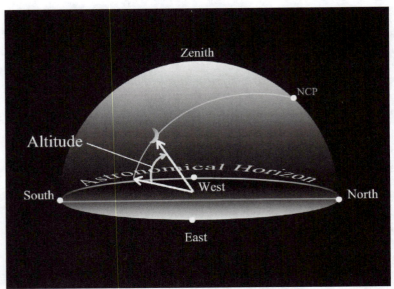

Figure 5-18 Measuring Altitude on the Celestial Sphere

The second angle is measured *around* the astronomical horizon. Like longitude on Earth, astronomers measure this angle from a well-defined point on the astronomical horizon. This well-defined point is the *North Point* on the horizon. It is the reference point from which to measure this angle. The angle is defined as an object's *azimuth*.

Azimuth is the angle measured from the North Point on and along the astronomical horizon toward east, and clockwise about the zenith point, to the vertical circle passing through the object. The values for azimuth range from 0° to 360°.

The major compass points on the horizon are commonly referred to as the *Cardinal Points* (N, E, S, and W). Their azimuths are 0°, 90°, 180°, and 270°, respectively. The North Point actually has two values, 0° and 360°. Therefore, the azimuth of any object that appears to be rising due East is 90°. The measurement of azimuth on the celestial sphere is illustrated in **Figure 5-19**. Once one has mastered these coordinates, it is an easy and useful coordinate system for locating celestial objects from any locale on Earth.

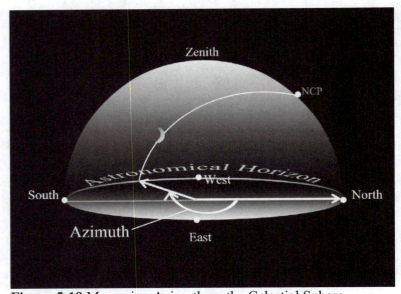

Figure 5-19 Measuring Azimuth on the Celestial Sphere

In *TheSkyX* there is a *horizontal grid* that displays approximate altitudes and azimuths of celestial objects. This grid can be displayed on the desktop window. In order to place the

© 2011 Cengage Learning. All Rights Reserved. May not be scanned, copied or duplicated, or posted to a publicly accessible website, in whole or in part.

horizon grid on the desktop click the "Show Horizon Grid" (Show Horizon Grid) button located on the Display toolbar. When this button is clicked, the horizontal grid is activated. Upon activation, lines appear on the desktop window that represent the altitude and azimuth in the sky. The angular separation between each line of altitude is 10° of angle, and that of azimuth is 15° of angle.

When you are looking north, Polaris is located above the North Point on the horizon. For any observer at 40° N latitude, Polaris is located about four (4) lines above the northern horizon, or at an altitude of 40°, as shown in **Figure 5-20**. Polaris is located on the fourth circle above the North Point on the horizon and is the circled object in the figure.

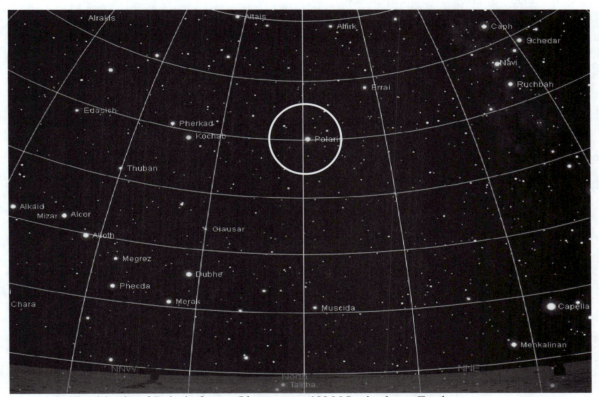

Figure 5-20 Altitude of Polaris for an Observer at 40° N Latitude on Earth

Displaying the horizon coordinates for any object in the sky is simple. Point your cursor on any object desired, and information about the object is displayed on the desktop. For example, when you put the cursor on Polaris with the mouse, a window appears as the one displayed in **Figure 5-21**. The azimuth and altitude appear on the fourth and fifth lines, respectively in this window.

Name 2: Polaris
RA (current): 02h 55m 29s
Dec (current): +89° 20' 12"
Azimuth: 00° 43' 42"
Altitude: +40° 33' 47"
Magnitude: 1.97
Transit Time: 03:44

Figure 5-21 Azimuth and Altitude for Polaris at 40° North Latitude

© 2011 Cengage Learning. All Rights Reserved. May not be scanned, copied or duplicated, or posted to a publicly accessible website, in whole or in part.

As one can see in **Figure 5-21**, information about the object that is displayed is the object's proper name, its position in the sky (RA and Dec), its Azimuth and Altitude, its magnitude or brightness, and its the meridian crossing (transit) time. After clicking the "Show Horizon Grid" button on the Display toolbar one should be able to estimate an object's altitude above the horizon using this grid. The azimuth is displayed across the top of the desktop. That too can be estimated from this grid.

Figure 5-22 shows a variety of information in the Find Result window regarding Polaris. Actually, what is displayed in **Figure 5-22** is the "Object Information Report." This appears in the desktop window after one left-mouse clicks on any object. It clearly displays information about any object that has been chosen on *TheSkyX* desktop. In fact, the user may choose what information they want to display. That is, one may choose to display more or less information about any celestial object of interest.

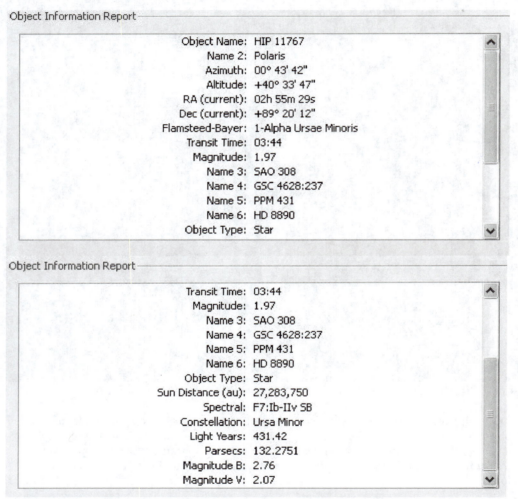

Figure 5-22 The Object Information Report in the Find Result Window

For the current Object Information Report window, in **Figure 5-22**, the Object name is the HIPPARCOS Catalog designation then the Proper Name of the object. Under that, are Polaris' Azimuth and Altitude and its position in the sky (RA and Dec). In addition the Flamsteed and Bayer designations are displayed. The meridian crossing (transit) time is displayed as well as its magnitude or brightness and other alternate designations for Polaris.

Any object's position, altitude and azimuth, meridian crossing (transit) time can be displayed for any time of day, any day of the year, from any location on Earth. There are two major disadvantages of the horizon coordinate system. As Earth rotates on its axis, objects in the sky appear to move westward 15 degrees every hour. Consequently, their altitudes and

© 2011 Cengage Learning. All Rights Reserved. May not be scanned, copied or duplicated, or posted to a publicly accessible website, in whole or in part.

azimuths constantly change with time. Second, because Earth is spherical your location on Earth makes objects appear to be in different places in the sky.

The problem is most noticeable if one observes a star or constellation from two locations that are separated by several hundred miles along a north-south direction on Earth. A case in point: remember that when you viewed Polaris from 40° N latitude? Its altitude above the northern horizon was about 40°. Then, when you viewed Polaris from the north geographic pole (90°N), as you did in **Exercise A**, it was located at your zenith, an altitude of 90°. The fact that the altitudes of stars and constellations change when viewed from different locations on Earth led ancient astronomers thousands of years ago to conclude that Earth was round.

Observing objects in the sky works best with the horizon coordinate system if you and a friend with whom you might want to share this information live in the same town. It is a simple system to learn, and convenient to use. *TheSkyX* provides this information so that anyone can go outside and locate celestial objects easily. The main thing to remember is the shortcomings of this system. That is, the horizon coordinates *constantly change with time* and *depend on the observer's location* on Earth.

Most amateur astronomers can see several hundred celestial objects without the aid of a telescope. With a telescope, however, thousands–even millions–of additional objects come into view. Of these, the stars are the most numerous. Because distances to stars are enormous, they are considered infinitely far away. Once that this assumption is understood their positions can be marked with coordinates on the celestial sphere.

The Equatorial Coordinate System

One might be wondering by now if there is a coordinate system that remains fixed in time and is independent of an observer's location on Earth. Would such a system be difficult to devise because of a rotating Earth? In fact, there is such a system, and it is really quite simple to use and master. The frame of reference is once again the celestial sphere. It is most appropriate for this system to be fixed to it.

The *fundamental great circle* in this system, however, is the *celestial equator* from which this coordinate system derives its name. It is known as the *Equatorial Coordinate System*. The equatorial coordinate system measures angles from and around the great circle that lies halfway between the projections of Earth's geographic poles onto the celestial sphere. The celestial equator is precisely halfway between (90° from) these two points in the sky and is often considered as a projection of Earth's equator onto the sky. These projected poles (north and south celestial poles) are the two points in the sky that are "fixed" and about which the entire celestial sphere *appears* to rotate. This coordinate system is affixed to the celestial sphere and moves with it.

At first it may seem that finding one's way around the sky using this system is at best difficult. But, it is soon realized that once again we end up measuring just two angles that are *exactly* like the ones in the geographic coordinate system used on Earth.

The first angle is measured from the celestial equator along a vertical circle perpendicular to (90°) the celestial equator that passes through an object in the sky. Vertical circles on the celestial sphere are also known as *hour circles*. The reason, of course, is that the sky has been divided into 24 equal parts along the celestial equator, with each part representing 15° of angle or 1 hour of time.

The angular measurement is also directional. That is, it is measured both north and south of the celestial equator. This angle is defined as an object's *declination*. It is measured exactly as latitude is on Earth.

However, so that this coordinate is not confused with the geographic coordinate of latitude the notation of $\pm$ is used instead of north and south. The + sign denotes angles measured north of the celestial equator while the – sign denotes angles measured south of the celestial equator. The measurement of declination on the celestial sphere is illustrated in **Figure 5-23**.

© 2011 Cengage Learning. All Rights Reserved. May not be scanned, copied or duplicated, or posted to a publicly accessible website, in whole or in part.

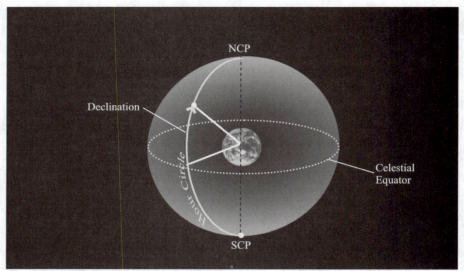

Figure 5-23 Measuring Declination on the Celestial Sphere

As stated earlier, the celestial equator is a great circle and the primary reference circle in the equatorial coordinate system. Circles that are formed by intersections of a sphere with planes not passing through the center of the sphere, have diameters smaller than that of the celestial sphere and are consequently called *small circles*. Therefore, the circles that are parallel to the celestial equator are defined as small circles. These small circles, above or below the celestial equator, are known as *declination circles*.

As in other coordinate systems, secondary reference circles are used to divide the sky in half in an east-west direction. These secondary reference circles are also *great circles*. The celestial equator is the fundamental great circle along which the second coordinate is measured. This second coordinate requires a reference point on the celestial equator from which to measure this angle–a Greenwich, England, so to speak!

Any point on the celestial equator could be chosen as the point of origin for the second coordinate. However, the point that is chosen on the celestial equator is the point where the celestial equator and the *ecliptic* intersect. The intersection point is near the constellation boundaries of *Pisces* and *Aquarius*.

The *ecliptic* is defined as the apparent path of the Sun. Over the course of a year the Sun *appears* to move eastward along this path. It appears to move through 13 constellations. That's right, 13! The 13 constellations through which the Sun appears to move each year are known as the *zodiacal constellations* or simply, the *zodiac*. In antiquity, as in *astrology* today, only 12 of these 13 constellations are recognized as constellations of the zodiac. The Sun's apparent motion along the ecliptic is actually cause by Earth's orbital motion, or its revolution around the Sun.

At two locations on the celestial sphere, the celestial equator and ecliptic intersect each other. These are defined as the *vernal* and *autumnal equinoxes*. Astronomers have chosen the *vernal equinox* as the reference point or more importantly as the *zero point* for the secondary fundamental plane, which *is* the vertical circle passing through this point and the point 180° from it. It so happens that the vernal equinox is also the point on the celestial sphere where the Sun appears to cross the celestial equator on or about March 21 each year. The vertical circle passing through the two equinoxes is known as the *equinoctial colure*.

An angle is then measured eastward along the celestial equator from the vernal equinox to an hour circle passing through any object on the celestial sphere. This angle is known as the coordinate of *right ascension*. Right ascension is the *only* angle in observational astronomy that is measured in hours, minutes, and seconds of time rather than degrees, minutes, and seconds of arc (angle). Its value ranges from 0^H to 24^H, with subdivisions of minutes and seconds of time. The measurement of right ascension on the celestial sphere is illustrated in **Figure 5-24**.

© 2011 Cengage Learning. All Rights Reserved. May not be scanned, copied or duplicated, or posted to a publicly accessible website, in whole or in part.

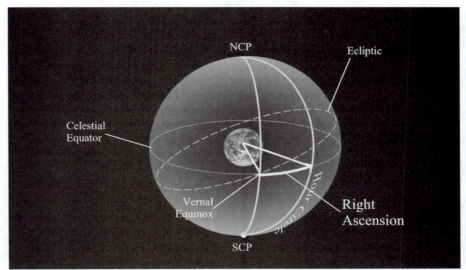

Figure 5-24 Measuring Right Ascension on the Celestial Sphere

At first this system seems confusing and almost impossible to use. Why use hours instead of degrees? It becomes clearer when one thinks about this angle being based on a rotating Earth. As Earth rotates on its axis, the *celestial sphere appears* to rotate counterclockwise about the north celestial pole. Thus, if an observer goes outside and faces north, after a while they will realize that the sky's motion about the NCP (Polaris) is counterclockwise and looking south, on the other hand, the motion of the sky is from east to west. Perhaps **Exercise C** will help visualize these motions.

Exercise C: Viewing the Motion About the North Celestial Pole

1. Run *TheSkyX* and open the file *Normal.skyx*.

2. The location doesn't matter.

3. Click the "Look North" (![Look North])button on the Orientation toolbar.

4. Set the Time Flow rate to 1 minute on the Time Skip toolbar (![1 minute]).

5. Click the "Go Forward" (![Go Forward]) button.

6. Notice the motion about the NCP (Polaris) is counterclockwise.

7. Click the "Stop" button (![Stop]).

8. Now, click the "Look South" (![Look South]) button.

9. Click the "Go Forward" (![Go Forward]) button again.

© 2011 Cengage Learning. All Rights Reserved. May not be scanned, copied or duplicated, or posted to a publicly accessible website, in whole or in part.

10. Notice that the motion while looking south is from east to west.

11. Click the "Show Horizon Grid" ( Show Horizon Grid) button. Right away one sees that the grid lines representing azimuth and altitude *do not* move as the sky appears to move from east to west.

12. Turn off the Horizon Grid. Now, click the "Show Equatorial Grid" ( Show Equatorial Grid) button. Notice in this window that the blue lines representing right ascension and declination *do* appear to move from east to west with the sky.

13. Close the file and do not save it.

As a result of Earth's rotation, each vertical (hour) circle passing through the celestial poles appears to move 15° westward in the sky as seen in **Exercise C**. The measurement of this motion is with respect to an imaginary reference circle drawn through your zenith.

The reference circle, passing through your zenith, is a north–south line drawn from the south point on the horizon to the north point on the horizon. This north–south line is defined as the *local celestial meridian*. The local celestial meridian in the previous exercise is found in the middle of the sky window while facing due South. In **Exercise C** it should notice that *one* of these blue vertical circles crosses the local celestial meridian every hour. In *TheSkyX*, the default setting does not display the local celestial meridian. However, **Exercise D** instructs one on how to display it.

Right ascension may be thought of as a measurement of celestial longitude of astronomical objects. It is very similar to longitude measured on Earth. The basic difference between geographic longitude and right ascension is that *longitude* on Earth *is* an angle *measured* both *east and west* of Greenwich, England, whereas *right ascension* is an angle *measured only eastward* from the vernal equinox. Both angles, of course, are measured around fundamental great circles (equators).

TheSkyX displays equatorial coordinates for all the objects in its data files. Right ascension and declination of the cursor's position can be displayed in *TheSkyX* at any time. They are usually displayed in the "Status" window on the desktop.

In *TheSkyX* there is an Equatorial Grid that displays approximate declinations and right ascensions of celestial objects. This grid can be displayed on the desktop window. In order to place the equatorial grid on the desktop simply click on the "Show Equatorial Grid"

(Show Equatorial Grid) button located on the Display toolbar. When this button is clicked, the equatorial grid is activated. Upon activation, blue lines appear on the desktop window that represents declination and right ascension in the sky. The angular separation between each line of declination is 10° of angle, and that of right ascension is 1^H of time.

Displaying equatorial coordinates for any object in the sky is a simple task. Point your *cursor* to any object desired on *TheSkyX* desktop and the equatorial and horizon coordinates will appear on the desktop. For example, when you put the cursor on Polaris with the mouse, a window appears as the one displayed in **Figure 5-25**. The right ascension (RA) and declination (Dec) appear on the second and third lines, respectively in this window.

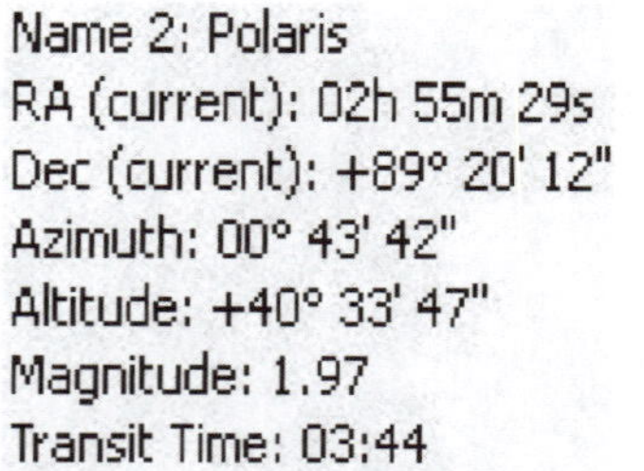

Name 2: Polaris
RA (current): 02h 55m 29s
Dec (current): +89° 20' 12"
Azimuth: 00° 43' 42"
Altitude: +40° 33' 47"
Magnitude: 1.97
Transit Time: 03:44

Figure 5-25 Right Ascension and Declination of Polaris

© 2011 Cengage Learning. All Rights Reserved. May not be scanned, copied or duplicated, or posted to a publicly accessible website, in whole or in part.

In any event, the object's right ascension and declination are displayed for any time of day, any day of the year, from any location on Earth. Unlike horizon coordinates, right ascension and declination of celestial objects are the *same* for every observer on Earth. They are independent of time and your location. Anytime one searches for an astronomical object, the "Find" window opens and displays the Object Information Report about the object of interest. Along with the Object Name, the equatorial coordinates and horizon coordinates are always displayed as they were in **Figure 5-22** for the star Polaris.

At this point, it would be helpful to visualize the orientation of the sky. Imagine standing outside and facing North as was done in **Exercise C**. Looking up into the sky, draw a line (in your mind's eye) from the north point on the horizon, through your zenith, to the south point on the horizon. This is the *local celestial meridian*. In *TheSkyX*, this reference line is not displayed in the default *Skyx* window. However, it is an easy task for one to display this reference line. **Exercise D** illustrates how to display this reference line and the orientation of the celestial sphere.

Exercise D: Orientation of the Celestial Sphere

1. Run *TheSkyX* and open the file ***Normal.skyx***.

2. Open the file *Normal.skyx*. Click "Display" on the Standard toolbar at the top of the skyx desktop window.

3. The Display menu window opens and appears like the one shown in **Figure 5-26**.

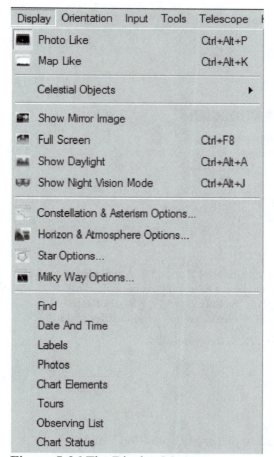

Figure 5-26 The Display Menu

© 2011 Cengage Learning. All Rights Reserved. May not be scanned, copied or duplicated, or posted to a publicly accessible website, in whole or in part.

4. Click on "Chart Elements" in the Display menu.

5. Click the "Plus Sign" next to "Reference Lines & Photos" which opens the options that are available in this menu.

6. The "Reference Lines and Photos" menu is displayed in **Figure 5-27**. It is in this menu that one can change the entire appearance of *TheSkyX* desktop window. For example, one may either display the "Constellation Drawing" or the "Constellation Figure" (this is the connect-the-dot appearance). Now, one can choose the option to display the "Constellation Boundary" in the sky or not. Reference lines such as the "Ecliptic," the "Equatorial Grid," or the "Galactic Equator" can also be displayed using this window. The main thing here is to explore the possibilities and create a desktop that you will enjoy…but remember it can be changed at any time!

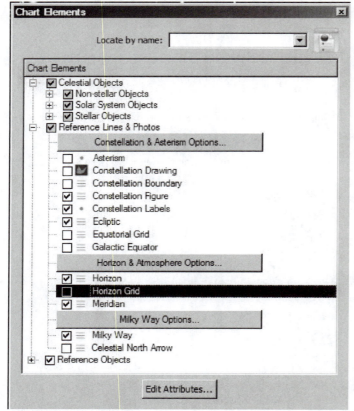

Figure 5-27 Reference Lines menu

7. Scroll down the "Chart Elements" menu until you find the "Horizon & Atmosphere Options" menu. Click "Horizon" and "Meridian" boxes to display them in *TheSkyX* desktop window. Choosing these options displays the horizon and the local celestial meridian in the *Normal.skyx* window. **Figure 5-28** illustrates this window after making these selections. Once the change is made close (⊠) the window.

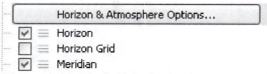

Figure 5-28 Displaying the Horizon and Meridian in *TheSkyX* Desktop Window

© 2011 Cengage Learning. All Rights Reserved. May not be scanned, copied or duplicated, or posted to a publicly accessible website, in whole or in part.

8. After clicking the Meridian option, *TheSkyX* displays it in the *Normal.skyx* window. The default color setting in *TheSkyX* is a *vertical red line*. **Figure 5-29** displays the local celestial meridian looking north with the constellation boundaries displayed. One may change the color of this line at any time. The instructions for changing the colors and other Preferences in *TheSkyX* are described in **Appendix E**.

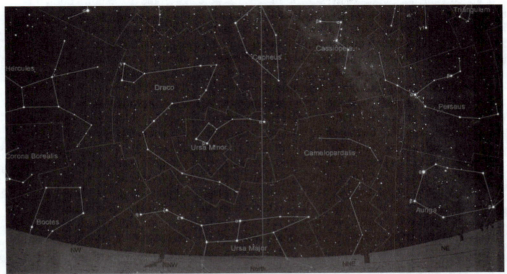

Figure 5-29 Local Celestial Meridian Facing North at 40° N Latitude

While facing north, it is easy to locate Polaris or more correctly the north celestial pole. They are located approximately 40°above the north point on the horizon for an observer at 40°N latitude as shown in **Figure 5-29**. The north celestial pole is situated on the local celestial meridian, whereas Polaris is not! In **Figure 5-30** *TheSkyX* displays the local celestial meridian facing south from the 40°N latitude location also with the constellation boundaries displayed.

Figure 5-30 Local Celestial Meridian Facing South at 40° N Latitude

It is obvious, from looking at **Figure 5-30**, the local celestial meridian passes through the south point on the horizon. What is *not* so obvious, however, is the location of the celestial equator in this figure. Recall that the celestial equator is 90° from the north celestial pole; therefore looking south along the local celestial meridian, the celestial equator should be found

© 2011 Cengage Learning. All Rights Reserved. May not be scanned, copied or duplicated, or posted to a publicly accessible website, in whole or in part.

at an altitude of 50° above the south point. To see this, click "Show Horizon Grid" on the Display toolbar.

This can be easily illustrated for an observer located at 40° N latitude. First imagine standing outside and drawing a circle north-south in the sky to represent the local celestial meridian. Draw it from the North Point on the horizon, through the zenith, to the South Point on the horizon. Next, mark the location of the north celestial pole at 40° above the north point on the horizon. Then, measure an angle of 90° from the north celestial pole and mark this location on this circle. This point represents the location of the celestial equator.

Now, draw a line from where you are standing on Earth to the north celestial pole mark and another line to the celestial equator mark and put in the angles. The angle left over is the angle between the South Point on the horizon and the celestial equator. This will be 50°! This is illustrated schematically for you in **Figure 5-31**. The large half-circle, in **Figure 5-31**, represents your local celestial meridian looking from the east.

In reality, the celestial equator is projected against the background stars along an east–west direction through the sky. It eventually passes directly through the East and West Points on the horizon and disappears below your horizon. This can be seen in *TheSkyX* by clicking on either of the two buttons "Look East" or "Look West" on the "Orientation" toolbar.

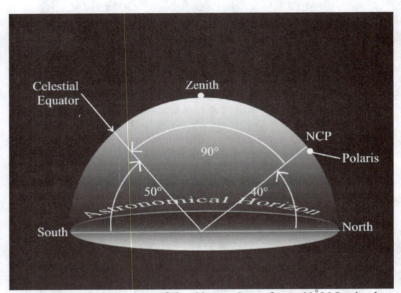

Figure 5-31 Orientation of the Sky as Seen from 40° N Latitude

There are two places on Earth where this is not the case. They are the north and south geographic poles. At these two locations on Earth the celestial equator coincides with or lies on the astronomical horizon.

Figure 5-32, displays the equatorial grid. The circle labeled +00° at the right edge in this figure represents the celestial equator in *TheSkyX*. In this figure, the Sun is located just to east of the local celestial meridian and just below the celestial equator. The Sun is *on* the celestial equator *twice* a year, once on the vernal equinox and again on the autumnal equinox.

Perpendicular to the celestial equator, and passing through the celestial poles, are vertical circles of right ascension spaced 1^H (15°) apart. Although not shown in **Figure 5-32**, the hours of right ascension are displayed on the screen in *TheSkyX* at the bottom of the desktop window.

© 2011 Cengage Learning. All Rights Reserved. May not be scanned, copied or duplicated, or posted to a publicly accessible website, in whole or in part.

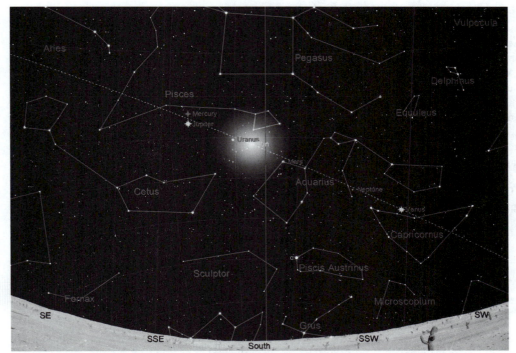

Figure 5-32 Celestial Equator as Seen from 40° N Latitude Looking South

Since the surface of Earth is spherical, it is interesting to note that the orientation of the sky changes the farther north or south one travels. Viewing the sky from different locations on Earth allows the observer to see how the orientation and positions of astronomical objects change.

Traveling north on Earth, the celestial equator appears lower in the sky while the north celestial pole appears higher. Traveling south on Earth, the celestial equator appears higher in the sky while the north celestial pole appears lower.

Knowing one's location on Earth makes viewing the sky much more easier and fun. One can determine the sky's orientation for any particular location by estimating the altitude of the celestial pole and the celestial equator (Remember **Exercises A** and **B**?). If you are in the Northern Hemisphere, the *altitude* of the north celestial *pole* is the *same* as your *latitude*, whereas the *altitude* of the *celestial equator* is equal to *90° minus* your *latitude*. The same is true for anyone living in the Southern Hemisphere, except there it is the south celestial pole that is used.

Many objects in the sky appear to rise and set throughout the day. Some, however, do not. These objects are situated around the celestial poles. The motion of objects in the sky is also affected by your location on Earth. If you are located at the north or south geographic pole, objects in the sky appear to move parallel to the horizon. That is, nothing appears to rise or set at either of the geographic poles. Objects that never rise or set are known as *circumpolar objects*. However, if you are located at the geographic equator, then *every* object in the sky appears to rise and set.

Even though the sky appears to change its orientation as one travels to different locations on Earth, the coordinates of right ascension and declination in the sky remain the same for observers in both hemispheres. This is true no matter where one goes on Earth. These coordinates are unaffected by the short-term motions of Earth. However, moving ahead or backward in time hundreds or thousands of years one will observe that these coordinates are affected! This long-term effect on the equatorial coordinates is due to the Sun and Moon gravitationally tugging on Earth. The Earth responds to this tugging by *precessing*. Like a top, Earth's axis of rotation changes its orientation in space. One result of this motion is that Polaris *will not* always be our North Star. In fact, in 2500 B.C.E. the star α Draconis (Thuban) was then the current North Star. This star, by the way, is the one that the Egyptians used as their North Star and to build their pyramids.

© 2011 Cengage Learning. All Rights Reserved. May not be scanned, copied or duplicated, or posted to a publicly accessible website, in whole or in part.

Finally, many people usually think of the vernal equinox, summer solstice, autumnal equinox, and the winter solstice as being points in time. That is, they represent the beginning of the Seasons on Earth. These are associated with the approximate dates of March 21, June 21, September 21, and December 21, respectively. The seasons are discussed in detail in **Chapter 8**.

Astronomers, however, think of these as points in space. At the time of the vernal equinox, the Sun appears to be crossing the celestial equator from the southern hemisphere into the northern hemisphere of the sky. On the summer solstice, the Sun appears to be at its greatest declination north of the celestial equator. At the time of the autumnal equinox, the Sun once again appears to be crossing the celestial equator, only this time it appears to be moving from the northern hemisphere into the southern hemisphere of the sky. On the winter solstice, the Sun appears to be at its greatest declination south of the celestial equator. And so it goes, year after year after year!

Thinking about the beginning of each season as a point in space allows astronomers to assign them equatorial coordinates. They are as follows:

Vernal equinox	R.A. = $0^H 0^M$	Declination =	$0° 0'$
Summer solstice	R.A. = $6^H 0^M$	Declination =	$+23° 26'$
Autumnal equinox	R.A. = $12^H 0^M$	Declination =	$0° 0'$
Winter solstice	R.A. = $18^H 0^M$	Declination =	$-23° 26'$

The equatorial coordinates of these points in space are illustrated in **Figure 5-33**.

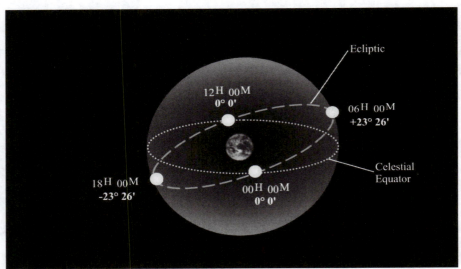

Figure 5-33 Equatorial Coordinates of the Equinox and Solstice Points

For telescopic observations the coordinates of right ascension and declination are readily used, and astronomical objects can be easily found. Observers need to determine the orientation of the equatorial system according to their location on Earth and the time of observation. *TheSkyX* is very proficient in keeping track of sky coordinates and time. It has an extensive database that lists the right ascension and declination of thousands of astronomical objects. It calculates their changes in altitude and azimuth caused by Earth's rotation. In addition, it constantly updates the coordinates of right ascension and declination due to Earth's precession. This makes going outside and looking for things simple and more pleasant.

© 2011 Cengage Learning. All Rights Reserved. May not be scanned, copied or duplicated, or posted to a publicly accessible website, in whole or in part.

Chapter 5

TheSkyX Review Exercises

Before doing the following exercises, changes in *TheSkyX* are necessary.

TheSkyX Exercise 1: Observing *TheSkyX* at Ball State Observatory

Run *TheSkyX* and open the file *Normal.skyx*.

Set the location, date, and time as follows:

Location: Ball State Observatory
Date: November 25, 2018
Time: 9:00 PM EST
Time Zone: -5 hours

1. What bright star lies near the north celestial pole? _____

2. Go ahead in time to the year 9,999 C.E.; what bright star would be located close to the north celestial pole and would be our North Star? _____

3. Go back in time to the year 2,800 B.C.E.; what bright star would be located close to the north celestial pole and would be our North Star? _____

Return to the date and time described at the start.

4. What are the altitude and azimuth of Polaris?

    Altitude = _____° _____'   Azimuth = _____° _____'

5. What are the right ascension and declination of Polaris?

    R.A. = _____ H _____ M _____ S   Declination = __ ___° ____'____"

6. What constellation is located in the south on your local celestial meridian?

    _____

7. What are the altitude and azimuth of Alpheratz?

    Altitude = _____° _____'   Azimuth = _____°____'

8. What are the right ascension and declination of Alpheratz?

    R.A. = _____ H _____ M _____ S   Declination = __ ___° ____' ____"

9. In what constellation is Jupiter located? _____

10. What are the altitude and azimuth of Jupiter?

    Altitude = _____° _____'   Azimuth = _____° _____'

11. What are the right ascension and declination of Jupiter?

    R.A. = ___ H ___ M ___ S   Declination = __ ___° ___' ___"

© 2011 Cengage Learning. All Rights Reserved. May not be scanned, copied or duplicated, or posted to a publicly accessible website, in whole or in part.

12. What are the right ascension and declination of M42?

    R.A. = ___^{H} ___^{M} ___^{S} Declination = __ ___° ___′ ___″

13. Is M42 currently visible? ____. If so, where would you have to look to see it?
    _____

14. What are the altitude and azimuth of M42?

    Altitude = ____° ____′   Azimuth = ____° ____′

15. Is M45 currently visible? _____

16. What type of object is M45? _____

17. What are the right ascension and declination of M45?

    R.A. = ___^{H} ___^{M} ___^{S} Declination = __ ___° ___′ ___″

18. What are the altitude and azimuth of M45?

    Altitude = ____° ____′   Azimuth = ____° ____′

19. What astronomical object is located at R.A. = 00^H 42^M 45.2^S and Declination = +41° 6′ 19″?
    _____

20. What type of object is the object in Question 19? _____

21. In what constellation is the bright star Sirius located? _____

22. How far away is Sirius? _____

23. What are the equatorial coordinates of IC 2118?

    R.A. = ___^{H} ___^{M} ___^{S} Declination = __ ___° ___′ ___″

24. What type of object is IC 2118? _____

25. In what constellation is the minor planet Vesta located? _____

26. How far is Vesta from the Sun? _____

27. How far is Vesta from Earth? _____

28. Find the Sun.

    a. What are the right ascension and declination of the Sun?

    R.A. = ___^{H} ___^{M} ___^{S} Declination = __ ___° ___′ ___″

    b. In what constellation is the Sun located on this date? _____

29. What is the classification of the object located at the coordinates of:

    R.A. = 13^H 29^M 15^S Declination = +47° 11′ 54″?

    _____

© 2011 Cengage Learning. All Rights Reserved. May not be scanned, copied or duplicated, or posted to a publicly accessible website, in whole or in part.

30. What are the altitude and azimuth of the object in Question 29?

Altitude = _____° _____'   Azimuth = _____° _____'

TheSkyX Exercise 2: Observing *TheSkyX* in Little Rock, Arkansas

Run *TheSkyX* and open the file ***Normal.skyx***, or make the following changes in *TheSkyX*.

Set the location, date, and time as follows:

Location: Little Rock, Arkansas.
Date: November 25, 2018.
Time: 9:00 PM CST.
Time Zone: -6 hours

1. What bright star lies near the north celestial pole? _____

2. Go ahead in time to the year 9,999 C.E.; what bright star would be located close to the north celestial pole and would be our North Star? _____

3. Go back in time to the year 2,800 B.C.E.; what bright star would be located close to the north celestial pole and would be our North Star? _____

Return to the date and time described at the start.

4. What are the altitude and azimuth of Polaris?

Altitude = _____° _____'   Azimuth = _____° _____'

5. What are the right ascension and declination of Polaris?

R.A. = _____H _____M _____S   Declination = __ ___° ____' ____"

6. What constellation is located to the South on your local celestial meridian?
_____

7. What are the altitude and azimuth of Alpheratz?

Altitude = _____° _____'   Azimuth = _____° _____'

8. What are the right ascension and declination of Alpheratz?

R.A. = _____H _____M _____S   Declination = __ ___° ____'____"

9. In what constellation is Jupiter located? _____

10. What are the altitude and azimuth of Jupiter?

Altitude = _____° _____'   Azimuth = _____° _____'

11. What are the right ascension and declination of Jupiter?

R.A. = ___H ___M ___S   Declination = __ ___° ___' ___"

12. What are the right ascension and declination of M42?

R.A. = ___H ___M ___S   Declination = __ ___° ___'___"

© 2011 Cengage Learning. All Rights Reserved. May not be scanned, copied or duplicated, or posted to a publicly accessible website, in whole or in part.

13. Is M42 currently visible? _____. If so, where would you have to look to see it?
_____

14. What are the altitude and azimuth of M42?

Altitude = _____° _____'   Azimuth = _____° _____'

15. Is M45 currently visible? _____

16. What type of object is M45? _____

17. What are the right ascension and declination of M45?

R.A. = ___^{H} ___^{M} ___^{S} Declination = __ ___° ___' ___"

18. What are the altitude and azimuth of M45?

Altitude = _____° _____'   Azimuth = _____°_____'

19. What astronomical object is located at R.A. = 00^H 42^M 45.2^S and Declination = +41° 6' 19"?
_____

20. What type of object is the object in Question 19? _____

21. In what constellation is the bright star Sirius located? _____

22. How far away is Sirius? _____

23. What are the equatorial coordinates of IC 2118?

R.A. = ___^{H} ___^{M} ___^{S} Declination = __ ___° ___' ___"

24. What type of object is IC 2118? _____

25. In what constellation is the minor planet Vesta located? _____

26. How far is Vesta from the Sun? _____

27. How far is Vesta from Earth? _____

28. Find the Sun.

a. What are the right ascension and declination of the Sun?

R.A. = ___^{H} ___^{M} ___^{S} Declination = __ ___° ___' ___"

b. In what constellation is the Sun located on this date? _____

29. What is the classification of the object located at the coordinates of:

R.A. = 13^H 29^M 15^S Declination = +47° 11' 54"?

_____

30. What are the altitude and azimuth of the object in Question 29?

Altitude = _____° _____'   Azimuth = _____° _____'

© 2011 Cengage Learning. All Rights Reserved. May not be scanned, copied or duplicated, or posted to a publicly accessible website, in whole or in part.

31. What does the negative sign for the altitude mean? _____

32. Did you notice any differences in the answers in **Exercise 1** and **Exercise 2**? _____.

    Were some the same? _____ Were some different? _____

    Why were some of the answers different? _____
    _____

    Why were some of the answers the same? _____
    _____

TheSkyX Exercise 3: Observing *TheSkyX* at Orange, California

Run *TheSkyX* and open the file *Normal.skyx*.

Set the location, date, and time as follows:

Location: Orange California.   Longitude: 117° 51' 55"
Date: November 25, 2018.   Latitude: 33° 41' 4"
Time: 9:00 PM PST.   Elevation: 60 meters
Time Zone: -8 hours

1. What bright star lies near the north celestial pole? _____

2. Go ahead in time to the year 9,999 C.E.; what bright star would be located close to the north celestial pole and would be our North Star? _____

3. Go back in time to the year 2,800 B.C.E.; what bright star would be located close to the north celestial pole and would be our North Star? _____

Return to the date and time described at the start.

4. What are the altitude and azimuth of Polaris?

    Altitude = ____° ____'   Azimuth = ____° ____'

5. What are the right ascension and declination of Polaris?

    R.A. = ____^{H} ____^{M} ____^{S} Declination = __ ___° ____'____"

6. What constellation is located in the south on your local celestial meridian?

    _____

7. What are the altitude and azimuth of Alpheratz?

    Altitude = ____° ____'   Azimuth = ____°____'

8. What are the right ascension and declination of Alpheratz?

    R.A. = ____^{H} ____^{M} ____^{S} Declination = __ ___° ____' ____"

9. In what constellation is Jupiter located? _____

10. What are the altitude and azimuth of Jupiter?

    Altitude = ____° ____'   Azimuth = ____° ____'

© 2011 Cengage Learning. All Rights Reserved. May not be scanned, copied or duplicated, or posted to a publicly accessible website, in whole or in part.

11. What are the right ascension and declination of Jupiter?

R.A. = ___^{H} ___^{M} ___^{S} Declination = __ ___° ___′ ___″

12. What are the right ascension and declination of M42?

R.A. = ___^{H} ___^{M} ___^{S} Declination = __ ___° ___′ ___″

13. Is M42 currently visible? ____. If so, where would you have to look to see it?
    _____

14. What are the altitude and azimuth of M42?

Altitude = ____° ____′   Azimuth = ____° ____′

15. Is M45 currently visible? _____

16. What type of object is M45? _____

17. What are the right ascension and declination of M45?

R.A. = ___^{H} ___^{M} ___^{S} Declination = __ ___° ___′ ___″

18. What are the altitude and azimuth of M45?

Altitude = ____° ____′   Azimuth = ____° ____′

19. What astronomical object is located at R.A. = 00^H 42^M 45.2^S and Declination = +41° 6′ 19″?
    _____

20. What type of object is the object in Question 19? _____

21. In what constellation is the bright star Sirius located? _____

22. How far away is the star Sirius? _____

23. What are the equatorial coordinates of IC 2118?

R.A. = ___^{H} ___^{M} ___^{S} Declination = __ ___° ___′ ___″

24. What type of object is IC 2118? _____

25. In what constellation is the minor planet Vesta located? _____

26. How far is Vesta from the Sun? _____

27. How far is Vesta from Earth? _____

28. Find the Sun.

    a. What are the right ascension and declination of the Sun?

R.A. = ___^{H} ___^{M} ___^{S} Declination = __ ___° ___′ ___″

    b. In what constellation is the Sun located on this date? _____

© 2011 Cengage Learning. All Rights Reserved. May not be scanned, copied or duplicated, or posted to a publicly accessible website, in whole or in part.

29. What is the classification of the object located at the coordinates of:

$$R.A. = 13^H\ 29^M\ 15^S \qquad Declination = +47°\ 11'\ 54''?$$

_____

30. What are the altitude and azimuth of the object in Question 29?

Altitude = ____° ____' Azimuth = ____° ____'

31. Did you notice any differences in the answers in **Exercise 1** and **Exercise 2** and **Exercise 3**? _____.

Were some the same? _____ Were some different? _____

Why were some of the answers different? _____
_____

Why were some of the answers the same? _____
_____

TheSkyX Exercise 4: Observing *TheSkyX* at the Tropic of Cancer

Run *TheSkyX* and open the file *Normal.skyx*, or make the following changes in *TheSkyX*.

Set the location, date, and time as follows:

Location: Latitude: +23° 26′ 00″   Longitude: 85° 24′ 45″
Date: November 25, 2018.
Time: 9:00 PM EST.
Time Zone: -5 hours.

1. What bright star lies near the north celestial pole? _____

2. Go ahead in time to the year 9,999 C.E.; what bright star would be located close to the north celestial pole and would be our North Star? _____

3. Go back in time to the year 2,800 B.C.E.; what bright star would be located close to the north celestial pole and would be our North Star? _____

Return to the date and time described at the start.

4. What are the altitude and azimuth of Polaris?

Altitude = ____° ____' Azimuth = ____° ____'

5. What are the right ascension and declination of Polaris?

R.A. = ____^{H} ____^{M} ____^{S} Declination = __ ___° ____' ____"

6. What constellation is located to the South on your local celestial meridian?

_____

7. What are the altitude and azimuth of Alpheratz?

Altitude = ____° ____' Azimuth = ____° ____'

© 2011 Cengage Learning. All Rights Reserved. May not be scanned, copied or duplicated, or posted to a publicly accessible website, in whole or in part.

8. What are the right ascension and declination of Alpheratz?

    R.A. = ____^H ____^M ____^S   Declination = __ ___° ____′ ____″

9. In what constellation is Jupiter located? _____

10. What are the altitude and azimuth of Jupiter?

    Altitude = ____° ____′   Azimuth = ____° ____′

11. What are the right ascension and declination of Jupiter?

    R.A. = ___^H ___^M ___^S Declination = __ ___° ___′ ___″

12. What are the right ascension and declination of M42?

    R.A. = ___^H ___^M ___^S Declination = __ ___° ___′ ___″

13. Is M42 currently visible? ____. If so, where would you have to look to see it?
    _____

14. What are the altitude and azimuth of M42?

    Altitude = ____° ____′   Azimuth = ____° ____′

15. Is M45 currently visible? _____

16. What type of object is M45? _____

17. What are the right Ascension and declination of M45?

    R.A. = ___^H ___^M ___^S Declination = __ ___° ___′ ___″

18. What are the altitude and azimuth of M45?

    Altitude = ____° ____′   Azimuth = ____° ____′

19. What astronomical object is located at R.A. = 00^H 42^M 45.2^S and Declination = +41° 6′ 19″?
    _____

20. What type of object is the object in Question 19? _____

21. In what constellation is the bright star Sirius located? _____

22. How far away is Sirius? _____

23. What are the equatorial coordinates of IC 2118?

    R.A. = ___^H ___^M ___^S Declination = __ ___° ___′ ___″

24. What type of object is IC 2118? _____

25. In what constellation is the minor planet Vesta located? _____

26. How far is Vesta from the Sun? _____

27. How far is Vesta from Earth? _____

© 2011 Cengage Learning. All Rights Reserved. May not be scanned, copied or duplicated, or posted to a publicly accessible website, in whole or in part.

28. Find the Sun.

    a. What are the right ascension and declination of the Sun?

    R.A. = ___^{H} ___^{M} ___^{S} Declination = __ ___° ___′ ___″

    b. In what constellation is the Sun located on this date? _____

29. What is the classification of the object located at the coordinates of:

    R.A. = 13^H 29^M 15^S Declination = +47° 11′ 54″?

    _____

30. What are the altitude and azimuth of the object in Question 29?

    Altitude = ____° ____′   Azimuth = ____° ____′

31. What does the negative sign for the altitude mean? _____

32. Did you notice any differences in the answers in **Exercise 1** and **Exercise 4**? _____

    Were some the same? _____ Were some different? _____

    Why were some of the answers different? _____
    _____

    Why were some of the answers the same? _____
    _____

33. Did you notice any differences in the answers in **Exercise 2** and **Exercise 4**? _____

    Were some the same? _____ Were some different? _____

    Why were some of the answers different? _____
    _____

    Why were some of the answers the same? _____
    _____

34. Did you notice any differences in the answers in **Exercise 3** and **Exercise 4**? _____

    Were some the same? _____ Were some different? _____

    Why were some of the answers different? _____
    _____

    Why were some of the answers the same? _____
    _____

© 2011 Cengage Learning. All Rights Reserved. May not be scanned, copied or duplicated, or posted to a publicly accessible website, in whole or in part.

TheSkyX Exercise 5: Observing *TheSkyX* at 40° Latitude

Run *TheSkyX* and open the file ***Normal.skyx***, or make the following changes in *TheSkyX*.

Set the location, date, and time as follows:

Location: Latitude: 40° 00′ 00″   Longitude: 85° 24′ 38″
Date: December 25, 2018
Time: 5:00 PM EST
Time Zone: -5 hours

1. What are the altitude and azimuth of Polaris?

    Altitude = ____° ____′   Azimuth = ____°____′

2. What are the right ascension and declination of Polaris?

    R.A. = ____H ____M ____S   Declination = __ ___° ____′ ____″

3. What constellation is located to the south on your local celestial meridian?

    _____

4. What are the altitude and azimuth of Alpheratz?

    Altitude = ____° ____′   Azimuth = ____° ____′

5. What are the right ascension and declination of Alpheratz?

    R.A. = ____H ____M ____S   Declination = __ ___° ___′ ____″

6. In what constellation is Jupiter located? _____

7. What are the altitude and azimuth of Jupiter?

    Altitude = ____° ____′   Azimuth = ____° ____′

8. What are the right ascension and declination of Jupiter?

    R.A. = ___H ___M ___S   Declination = __ ___° ___′ ___″

9. In what constellation is Saturn located? _____

10. What are the altitude and azimuth of Saturn?

    Altitude = ____° ____′   Azimuth = ____° ____′

11. What are the right ascension and declination of Saturn?

    R.A. = ___H ___M ___S   Declination = __ ___° ___′ ___″

12. What are the right ascension and declination of M42?

    R.A. = ___H ___M ___S   Declination = __ ___° ___′ ___″

13. Is M42 currently visible? _____ If so, where would you have to look to see it?

    _____

© 2011 Cengage Learning. All Rights Reserved. May not be scanned, copied or duplicated, or posted to a publicly accessible website, in whole or in part.

14. What are the altitude and azimuth of M42?

              Altitude = ____° ____′   Azimuth = ____° ____′

15. Is M45 currently visible? _____

16. What are the right ascension and declination of M45?

              R.A. = ___^{H} ___^{M} ___^{S} Declination = __ ___° ___′ ___″

17. What are the altitude and azimuth of M45?

              Altitude = ____° ____′   Azimuth = ____° ____′

18. What is located at R.A. = 00^H 36^M 45.2^S Declination = +42° 16′ 19″?
    _____

19. What type of object is the object in Question 18? _____

20. In what constellation is the bright star Procyon located? _____

21. How far away is Procyon? _____

22. Find the Sun. Center (Center) it in the desktop window.

    a. What are the Sun's right ascension and declination?

            R.A. = ___^{H} ___^{M} ___^{S} Declination = __ ___° ___′ ___″

    b. What are the Sun's Altitude and Azimuth?

            Altitude = ____° ____′   Azimuth = ____° ____′

    c. What time does it appear to transit the local celestial meridian today? _____:_____

23. Find the Moon. Center (Center) it in the desktop window.

    a. What are the Moon's right ascension and declination?

            R.A. = ___^{H} ___^{M} ___^{S} Declination = __ ___° ___′ ___″

    b. What are the Moon's altitude and azimuth?

            Altitude = ____° ____′   Azimuth = ____° ____′

    c. At what time does it appear to transit the local celestial meridian today? _____:_____

24. Do you suppose there was a solar eclipse today? _____

25. If so, what time does it occur? _____:_____

© 2011 Cengage Learning. All Rights Reserved. May not be scanned, copied or duplicated, or posted to a publicly accessible website, in whole or in part.

TheSkyX Exercise 6: Observing TheSkyX at the Geographic Equator

Run *TheSkyX* and open the file ***Normal.skyx***, or make the following changes in *TheSkyX*.

Set the location, date, and time as follows:

Location: Latitude: 0° 0′ 0″   Longitude: 85° 24′ 38″
Date: December 25, 2018
Time: 5:00 PM EST
Time Zone: -5 hours

1. What are the altitude and azimuth of Polaris?

    Altitude = _____° _____′   Azimuth = _____° _____′

2. What are the right ascension and declination of Polaris?

    R.A. = _____H _____M _____S   Declination = __ ___° _____′ _____″

3. What constellation is located to the South on your local celestial meridian?

    _____

4. What are the altitude and azimuth of Alpheratz?

    Altitude = _____° _____′   Azimuth = _____° _____′

5. What are the right ascension and declination of Alpheratz?

    R.A. = _____H _____M _____S   Declination = __ ___° _____′ _____″

6. In what constellation is Jupiter located? _____

7. What are the altitude and azimuth of Jupiter?

    Altitude = _____° _____′   Azimuth = _____° _____′

8. What are the right ascension and declination of Jupiter?

    R.A. = ___H ___M ___S   Declination = __ ___° ___′ ___″

9. In what constellation is Saturn located? _____

10. What are the altitude and azimuth of Saturn?

    Altitude = _____° _____′   Azimuth = _____° _____′

11. What are the right ascension and declination of Saturn?

    R.A. = ___H ___M ___S   Declination = __ ___° ___′ ___″

12. What are the right ascension and declination of M42?

    R.A. = ___H ___M ___S   Declination = __ ___° ___′ ___″

13. Is M42 currently visible? _____ If so, where would you have to look to see it?

    _____

© 2011 Cengage Learning. All Rights Reserved. May not be scanned, copied or duplicated, or posted to a publicly accessible website, in whole or in part.

14. What are the altitude and azimuth of M42?

> Altitude = ____° ____'   Azimuth = ____° ____'

15. Is M45 currently visible? _____

16. What are the right ascension and declination of M45?

> R.A. = ___^{H} ___^{M} ___^{S} Declination = __ ___° ___' ___"

17. What are the altitude and azimuth of M45?

> Altitude = ____° ____'   Azimuth = ____° ____'

18. What astronomical object is located at R.A. = 22^H 30^M 02^S and Declination = –20°40' 39"?
    _____

19. What type of object is the object in Question 18? _____

20. In what constellation is the bright star Procyon located? _____

21. How far away is Procyon? _____

22. Find the Sun. Center ([Center]) it in the desktop window.

    a. What are the Sun's right ascension and declination?

    > R.A. = ___^{H} ___^{M} ___^{S} Declination = __ ___° ___' ___"

    b. What are the Sun's altitude and azimuth?

    > Altitude = ____° ____'   Azimuth = ____° ____'

    c. What time does it appear to transit the local celestial meridian today? _____:_____

23. Find the Moon. Center ([Center]) it in the desktop window.

    a. What are the Moon's right ascension and declination?

    > R.A. = ___^{H} ___^{M} ___^{S} Declination = _ ___° ___' ___"

    b. What are the Moon's altitude and azimuth?

    > Altitude = ____° ____'   Azimuth = ____° ____'

    c. What time does it appear to transit the local celestial meridian today? _____:_____

24. Do you suppose there was a solar eclipse today? _____

25. If so, what time did it occur? _____:_____

© 2011 Cengage Learning. All Rights Reserved. May not be scanned, copied or duplicated, or posted to a publicly accessible website, in whole or in part.

Chapter 5

TheSkyX Review Questions

Part A

Answer these basic questions about Earth and the geographic coordinates:

1. The two points on Earth intersected by its axis of rotation are known as the _____ and _____.

2. What is the name of the line drawn on Earth dividing it into two equal halves, north and south? _____.

3. Great circles are circles that pass through the North and South Poles on Earth. The Earth is divided into 24 equal parts east to west along its equator. The angle between any two of these circles is ___ hour or ____°.

4. Any two points on the surface of Earth that are located on *one* of these great circles and is located in the *same* hemisphere will have the same _____.

5. All points on the surface of Earth that have latitude circles that are not zero are known as _____ circles.

6. What is the latitude of the north geographic pole? _____°

7. What is the longitude of the north geographic pole? _____°

Part B

Answer these basic questions about the sky coordinates

1. Which coordinate in the equatorial coordinate system is like longitude in the geographic coordinate system? _____

2. Which coordinate in the equatorial coordinate system is like latitude in the geographic coordinate system? _____

3. Which coordinate in the horizon coordinate system is similar to longitude in the geographic coordinate system? _____

4. Which coordinate in the horizon coordinate system is similar to latitude in the geographic coordinate system? _____

5. What units are used to measure the coordinates of right ascension (R.A.)? _____

6. What direction in the sky is the coordinate of R.A. measured? _____

7. What is the range in values of right ascension? _____

8. What is the name of the point in the sky that defines the secondary circle in the equatorial coordinate system? _____

© 2011 Cengage Learning. All Rights Reserved. May not be scanned, copied or duplicated, or posted to a publicly accessible website, in whole or in part.

Chapter 6

Motions in *TheSkyX*

Observing the stars continues to be an essential activity for many cultures today. For millennia seafarers have used the stars to navigate by, something that continues even into modern times. Being familiar with how stars are designated and how to find their location in the sky is important in observational astronomy. Now it is time to focus attention on the motions observed in the sky. In ancient times, most astronomers believed that such motions were actually caused by the objects moving around us. This error led some of them to formulate intricate cosmologies to explain these motions. Today, we understand these motions and can explain them correctly.

There are basically two kinds of motions observed in the sky, *short-term* motion and *long-term* motion. The short-term motion is called *daily* or *diurnal* motion and is responsible for the daily rising and setting of the Sun, Moon, planets, stars, and everything else in the sky. A consequence of short-term motion is that most celestial objects move in a westerly direction across the sky during the course of a day. That is, they rise in the east and set in the west. Short-term motion has a period of time that is on the order of a day or less (hours, minutes, seconds).

As mentioned in the last chapter, the entire celestial sphere appears to rotate about two fixed points in the sky, the north and south celestial poles. This motion, however, is an apparent one. Daily motion is actually a result of Earth rotating on its axis. Diurnal motion is responsible for the time-dependent change of the horizon coordinates and the ever-changing positions of objects in the sky. The rate at which objects move in the sky is 15° per hour from east to west.

In this chapter and subsequent ones, there will be many Sky files that may be thought of as sky "tours." These files will have to be downloaded into a folder that was described in Chapter 1.

The folder path for **Windows 7** users is:
> C:\Users*Your Username*\Documents\Software Bisque\TheSkyX Student Edition\\ **Sky Files**

The folder path for **Windows Vista** users is:
> C:\Users*Your Username*\Documents\Software Bisque\TheSkyX Student Edition\\ **Sky Files**

The folder path for **Windows XP** users is:
> C:\Documents and Settings*Your Username*\My Documents\Software Bisque\\ **TheSkyX Student Edition\Sky Files**.

The folder path for **MacIntosh** users is:
> **User\Software Bisque\TheSkyX Student Edition\Documents\Sky Files**.

Each user must create "Chapter Folders" for *Chapter 6*, *Chapter 7*, *Chapter 8*, and *Chapter9*. Since this folder was not created during the installation procedure described in **Chapter 1** the user must create these folders. The URL for downloading these Sky files is: http://www.cengage.com/astronomy/jordan/skyx. These files are by chapter and labeled with a "*filename*" and an extension labeled "*skyx*." These Sky files have been created for this and other chapters to assist in visualizing and demonstrating what is seen in the sky.

The following exercises are designed to help open and utilize these Sky files. These exercises are designed to enable one to view both the short-term and long-term motions seen in the sky. The first few exercises give explicit details in how to open and view these Sky files. Later exercises will indicate which files to open and view.

The Sun's apparent rising on the eastern horizon is perhaps the most obvious example of diurnal motion. The Sky file named **SunRise.skyx** is a sky window that illustrates this motion. **Exercise A** illustrates how to open the Sky file **SunRise.skyx** and view the apparent sunrise in *TheSkyX*.

© 2011 Cengage Learning. All Rights Reserved. May not be scanned, copied or duplicated, or posted to a publicly accessible website, in whole or in part.

Exercise A: Observing the Diurnal Motion of the Sun

1.   Run *TheSkyX* and open the file ***Normal.skyx***.

2.   Click "File" on the Standard toolbar then the *Open* (Open) file.

3.   Look for the Sky files in Documents\Chapter 6. An Open w*indow*, like the one displayed in **Figure 6-1**, appears. The location of this file in your PC or Mac is *C:\...\Software Bisque\ TheSkyX Student Edition\Documents\Sky Files\Chapter6*.   All the Sky files in this chapter, once downloaded from Brook-Cole website, may be saved to this file folder. The Sky files in this folder may be used for other exercises at later times.

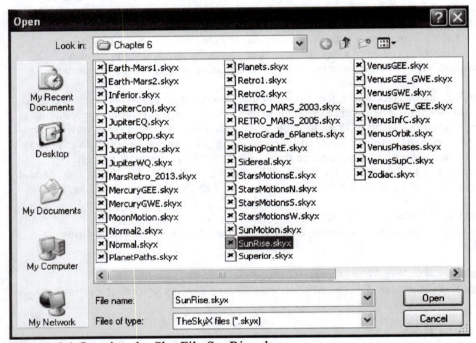

Figure 6-1 Opening the Sky File SunRise.skyx

4.   Select the proper demonstration filename–in this case ***SunRise.skyx***. The extension is listed on the "Files of type" line at the bottom of the window (*.skyx).

5.   Click the Open button to open the file.

6.   After opening the file, click the Go Forward button and watch. If the motion is too fast, then click the Stop button and then the Step Backward button to move the Sun back below the eastern horizon. Using the Step Forward button allows the observer to view the sky window at a much slower rate. Notice that the direction in which the observer is facing is east. If

© 2011 Cengage Learning. All Rights Reserved. May not be scanned, copied or duplicated, or posted to a publicly accessible website, in whole or in part.

you want to replay this Sky file, then re-open the file as was described in Step 2. It may be necessary to reset the Time Flow rate to 30 seconds or less for the best results.

7. To close this file, click "File" on the Standard toolbar at the top of the window then click

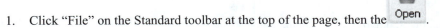

    Open again. At this point you can either choose to open a new file, or to continue with **Exercise B**, or to exit the program. It is not necessary to save this Sky file.

Exercise B: Observing the Diurnal Motion of the Stars, Part I

1. Click "File" on the Standard toolbar at the top of the page, then the Open .

2. An Open w*indow*, like the one displayed in **Figure 6-1**, appears again with all of the Sky files displayed. This time double-click on the Sky file named ***StarsMotionS.skyx***.

3.   After opening the file, click the Go Forward button and watch. Once again, if the motion is

    too fast, click the Stop button and then the Step Backward button to reposition the center of

    the constellation Leo back on the local celestial meridian. Click the Step Forward button to view the sky window at a much slower rate. As before, if you wish to replay this Sky file at a slower rate, you must again reset the Time Flow rate to 30 seconds or less for best results.

4. Notice the ***view*** is looking ***south*** in this sky window. The stars appear to move left to right or from east to west. If you look closely, the Moon is near the bright star Spica, and the planet Saturn is in the constellation of Leo, the Lion. The sky is not completely dark in this window but will darken as the time progresses.

5. To close this file, click "File" on the Standard toolbar at the top of the window and the

    Open (Open) button again. At this point you can either choose to open a new file, or to continue with **Exercise C**, or to exit the program. It is not necessary to save this Sky file.

Exercise C: Observing the Diurnal Motions of the Stars, Part II

1. Click "File" on the Standard toolbar again, then Open .

2. Double-click on the Sky file named ***StarsMotionE.skyx***.

3. After opening the file, click the Go Forward button and watch.

4. This Sky file is the essentially the same as the one in **Exercise B** except the view is looking east and the time of day is approximately two hours earlier. Notice in this sky window that all the objects appear to be rising at an angle to the eastern horizon. As before, as the sky continues to darken more stars become visible.

© 2011 Cengage Learning. All Rights Reserved. May not be scanned, copied or duplicated, or posted to a publicly accessible website, in whole or in part.

5. If the motion is too fast, repeat the procedure described in **Exercises A** and **B** to view this sky window at a slower rate.

6. To close this file, click "File" on the Standard toolbar at the top of the window then click again. At this point you can either choose to open a new file, or to continue with **Exercise D**, or to exit the program. It is not necessary to save this Sky file.

Exercise D: Observing the Diurnal Motions of the Stars, Part III

1. Click "File" on the Standard toolbar again, then .

2. Double-click on the Sky file named ***StarsMotionW.skyx***.

3. After opening the file, click the ⌐Go Forward⌐ button and watch.

4. This file is the same as in **Exercise C** except the view is looking west. Notice the stars appear to be setting below the western horizon at an angle. As the sky continues to darken more stars become visible. If you watch this window long enough, the Moon and Saturn will also appear to set.

5. If the motion is too fast, repeat the procedure described in **Exercises A** and **B** to view this sky window at a slower rate.

6. To close this file, click "File" on the Standard toolbar at the top of the window then click again. At this point you can either choose to open a new file, or to continue with **Exercise E**, or to exit the program. It is not necessary to save this Sky file.

Exercise E: Observing the Diurnal Motions of the Stars, Part IV

1. Click "File" on the Standard toolbar again, then .

2. Double-click on the Sky file named ***StarsMotionN.skyx***.

3. After opening the file, click the ⌐Go Forward⌐ button and watch.

4. This file is the same as the previous exercises' windows except that the view is looking north. In the middle of the sky window is the bright star Polaris, the North Star. Notice that stars in the sky window appear to move counterclockwise about Polaris, or more correctly about the north celestial pole. Notice, too, that the circumpolar stars never go below the horizon.

   As the sky continues to darken, more and more stars become visible. Their motions become very evident after you have observed for some time. Stars *above* Polaris appear to move from east to west (to the left). Stars *below* Polaris appear to move from west to east

© 2011 Cengage Learning. All Rights Reserved. May not be scanned, copied or duplicated, or posted to a publicly accessible website, in whole or in part.

(to the right). Stars to the *left* of Polaris appear to move *downward*. Stars to the *right* of Polaris appear to move *upward*. The overall motion about the north celestial pole is in a counterclockwise direction. The counterclockwise motion observed in this sky window is caused by Earth itself rotating counterclockwise (west to east) on its axis. The Time Flow rate has been set to 5-minute interval.

5. As before, if the motion is too fast, repeat the procedure described in the previous exercises once again in order to view this sky window at a slower rate.

6. To close this file, click "File" on the Standard toolbar at the top of the window then click

    Open again. At this point you can either choose to open a new file, or to continue with **Exercise F**, or to exit the program. It is not necessary to save this Sky file.

    Exercises B, C, D, and **E** could all have been done as one exercise. The more often you use this software, the easier this sequence will become.

    In ancient times people noted different types of objects in the sky. They called the constellations simply the "fixed" stars. These stars didn't move in the sky except from east to west. Their patterns remained the same year after year for millennia. Ancient astronomers took these things very seriously, because the stability probably had a calming effect on their thinking and their culture. There was no chaos in this universe, only order.

    Long-term motion is often referred to as *annual* motion. It corresponds to time periods longer than 24 hours (weeks, months, and even years). Observing the nighttime sky over longer periods of time, one notices, as the ancients did, that not all stars remain "fixed." Some of them move relative to the "fixed" stars or constellations.

    The ancient Greeks called these moving objects the *planetes* or "wandering stars." *Planetes* is the origin for the word planets. There are five naked-eye planets, which move along well-defined paths near the ecliptic. The Sky file named ***Planets.skyx*** in **Exercise F** illustrates how these objects appear to move in Earth's sky.

Exercise F: Observing the Planets' Long-Term Motion

1. Start *TheSkyX,* or if it is already running click "File" on the Standard toolbar, then **Open** .

2. Now, select the Sky file named ***Planets.skyx***.

3. Double-click on it to open the file.

4. After opening the file, click the **Go Forward** button. Once again, if the motion is too fast, repeat the procedure as described in the previous exercises.

5. To reset the window to the beginning, at any time, simply reset the time to the initial time (12-25-14 – 8:00 PM STD) in the Sky file or re-open the file.

6. To close this file, click "File" on the Standard toolbar at the top of the window then click

    Open again. At this point you can either choose to open a new file, or to continue with **Exercise G**, or to exit the program. It is not necessary to save this Sky file.

    The most obvious object in the night sky is the Moon. After a few nights of observing, one soon realizes that it too moves relative to the constellations. In ancient times, the Moon was considered a planet. Ancient Greek and Renaissance astronomers alike took the definition of *planetes* literally when it came to the objects in the sky: the planets "wandered." The Sky file

© 2011 Cengage Learning. All Rights Reserved. May not be scanned, copied or duplicated, or posted to a publicly accessible website, in whole or in part.

named ***MoonMotion.skyx*** illustrates the Moon's motion in the sky over a period of several days. Observe the motion in the next exercise carefully! Pay particular attention to the path that the Moon makes relative to the stars and compared it to that of the Sun's (blue line). **Exercise G** demonstrates the Moon's long-term motion in the sky.

Exercise G: Observing the Moon's Long-Term Motion

1. Start *TheSkyX,* or if it is already running click "File" on the Standard toolbar, then **Open**.

2. Now, select the Sky file named ***MoonMotion.skyx***.

3. Double-click on it to open the file.

4. The Moon is located in the center of this sky window in the constellation of Pisces just above the Sun.

    Click the **Step Forward** button to watch the motion in this file. Click the **Move Left** button to follow the Moon's motion over a period of a few days.

5. Close the file. Once again, it is not necessary to save this Sky file.

     If one observes the Sun over a period of several months, it too appears to move relative to the "fixed" stars. Its motion, however, is along a well-defined path known as the *ecliptic*. During the course of a year, the Sun appears to move through thirteen constellations in the sky. The Sky file named ***SunMotion.skyx*** demonstrates the Sun's eastward motion along the ecliptic. In *TheSkyX*, a blue-colored line in all skyx windows represents the ecliptic. This color is the default color setting in *TheSkyX* and may be changed at any time (see **Appendix E**).

     The Sun's motion is both north and south of the celestial equator. **Exercise H** illustrates the Sun's motion in the sky. Remember that this motion is an apparent motion. The Sun's apparent motion along the ecliptic is caused by Earth revolving around the Sun, not the Sun revolving around Earth as most ancients believed.

Exercise H: Observing the Sun's Long-Term Motion

1. Start *TheSkyX,* or if it is already running click "File" on the Standard toolbar, then **Open**.

2. Now, select the Sky file named ***SunMotion.skyx***.

3. Double-click on it to open the file.

4. Click the **Go Forward** button to watch this file. Click the **Stop** button when finished. If the motion is too fast, then reset the Time Flow rate as described in earlier exercises or reopen the file.

5. Close the file. It is not necessary to save this Sky file.

     The constellations that are centered on the ecliptic are known as the *zodiacal constellations* or the *zodiac*. **Exercise I** illustrates the motion of the Sun through these constellations.

© 2011 Cengage Learning. All Rights Reserved. May not be scanned, copied or duplicated, or posted to a publicly accessible website, in whole or in part.

Exercise I: Observing the Sun's Motion through the Zodiac

The constellation names in this exercise are very familiar (Virgo, Libra, Scorpius, Sagittarius, Capricornus, and so on).

1. Start *TheSkyX,* or if it is already running click "File" on the Standard toolbar, then 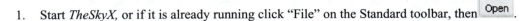.

2. Now, select the Sky file named ***Zodiac.skyx***.

3. Double-click on it to open the file.

4. Click the 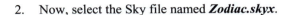 button to watch this file. Click the Stop button when finished.

5. If the motion is too fast, then reset the Time Flow rate as described in earlier exercises or reopen the file.

6. Use the directional button, Move Left, on the Orientation toolbar to follow the Sun around the sky. Make sure that you move through a full circle so that the Sun makes one complete circuit around the sky along its apparent path, the ecliptic.

7. Close the file when finished. It is not necessary to save this Sky file.
   The paths of the planets and the Moon are always found somewhere along and within ±8° of the Sun's apparent path. The Sky file named ***PlanetPaths.skyx*** illustrates the motion of all these objects near the ecliptic. This is very easy to understand when you think of our Solar System as being essentially a "flat" system. **Exercise J** illustrates the motion of Solar System objects in Earth's sky.

Exercise J: Observing the Motion of Sun, Moon, and Planets in the Sky

1. Start *TheSkyX,* or if it is already running click "File" on the Standard toolbar , then Open.

2. Now, select the Sky file named ***PlanetPaths.skyx***.

3. Double-click on it to open the file.

4. Click the Go Forward button to watch this file. Click the Stop button when finished.

5. If the motion is too fast, then reset the Time Flow rate as described in earlier exercises or

   reopen the file and use the Step Forward button.

6. It is not necessary to save this Sky file.

   Another short, but interesting, exercise that may be done while in this sky window, is to use *TheSkyX's* 3-Dimensional Solar System Mode option. This mode allows you to view the Solar

© 2011 Cengage Learning. All Rights Reserved. May not be scanned, copied or duplicated, or posted to a publicly accessible website, in whole or in part.

System in a three-dimensional (3-D) representation. It depicts all the planets in their respective orbits and their location with respect to the Sun and Earth at the time of one's observation. **Exercise K** shows how to view the Solar System in a 3-Dimensional mode.

Exercise K: Observing a 3-Dimensional Model of Our Solar System

1. Use the sky window in **Exercise J** and click the "3-Dimensional Solar System Mode,"

    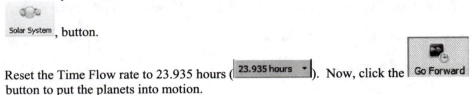, button.

2. Reset the Time Flow rate to 23.935 hours (23.935 hours ▾). Now, click the Go Forward button to put the planets into motion.

3. Notice that all of the objects in this sky window are essentially in a flat orientation. Their motion is counterclockwise about the Sun. It is helpful here to use the "View Distance" slide-bar at the bottom left edge of the window as shown in **Figure 6-2**. This slide bar allows the user to zoom in or out while watching this motion. One may zoom into a distance of 1 Astronomical Unit or out to 50 Astronomical Units. The current Time Flow rate is set at 1 sidereal day. If you wish to speed up the motion, then reset the Time Flow to another rate.

4. Click the button at the right-hand edge of the window and scroll up about half-way with the mouse. This allows the user to view the Solar System above the plane of Earth's orbit.

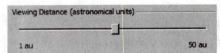

Figure 6-2 The Viewing Distance Slide Bar in the 3-D Mode

5. Notice that with the exception of Pluto, all the planetary orbits are very close to Earth's orbit.

6. Clicking on the "3-Dimensional Solar System Mode" button (Solar System) again returns the sky to the normal viewing mode. This mode can be activated at any time.

7. Close the file. It is not necessary to save this Sky file.

All the counterclockwise motion observed in the previous exercise translates to eastward motion as viewed from Earth's surface. In other words, long-term motion of the Sun, Moon, and planets is generally in an eastward direction as projected against the background stars in Earth's sky. The *eastward motion* of planets is often referred to as *direct motion*.

The Moon takes about a *month* to complete an entire circuit around the celestial sphere, whereas it takes the Sun a *year* to accomplish the same thing. Planets, in contrast, moving at different rates through the sky, have time periods that vary in duration. The duration to complete a circuit around the celestial sphere with respect to the stars depends upon their distance from the Sun. No matter which object is observed, each has its own time period with respect to the background stars. This period of time is known as the planet's *sidereal period*.

Sometimes planets are observed moving westward in the sky during this long-term cycle. This *westward* motion is called *retrograde motion*. Several Sky files have been rendered to display both direct and retrograde motions of the planets.

Figure 6-3 illustrates the retrograde motion of the planet Jupiter as projected against the background stars. In this figure, the trail was added to reproduce the motion that is observed over a period of several months from March 23 through June 10, 2023.

© 2011 Cengage Learning. All Rights Reserved. May not be scanned, copied or duplicated, or posted to a publicly accessible website, in whole or in part.

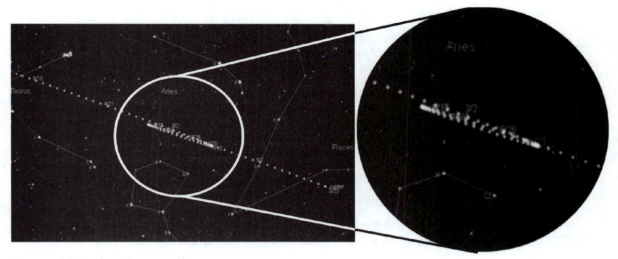

Figure 6-3 Jupiter Retrograding

The Sky file named ***JupiterRetro.skyx***, for example, illustrates Jupiter's retrograde motion in Ophiuchus from approximately January 28 to about November 1, 2019.
Exercises L and **M**, which follow, demonstrate motions of several planets as observed in Earth's sky projected against the background stars. Watch these motions carefully!

Exercise L displays planets moving eastward or in direct motion. The planets Mercury and Venus, however, do appear to exhibit retrograde motion in this sky window. **Exercise M** demonstrates the appearance of Jupiter's motion in Earth's sky when Jupiter is retrograding. By clicking on the "3-Dimensional Solar System Mode" button, you can view the positions of Earth and Jupiter in the Solar System during this period of time.

Exercise L: Observing Retrograde Motion, Part I

1. Open the Sky file named ***Retro1.skyx*** and click  .

2. Periodically clicking ⬛ Step Forward displays the daily displacement of Jupiter and other planetary positions. At the beginning of this sequence, Mercury's motion is eastward through the constellation of Libra. Eventually it begins to retrograde in the constellation of Ophiuchus through Scorpius and finally Libra again. During the same time, Jupiter's motion is also eastward through the constellation of Virgo. However, it too begins to retrograde in the constellation of Libra in early March of 2006. The time sequence begins on October 13, 2005 and ends on September 20, 2006. As a rule, it generally takes longer for a superior planet to retrograde than an inferior one.

3. Close the file, or continue with **Exercise M**. It is not necessary to save this Sky file.

Exercise M: Observing Retrograde Motion, Part II

1. Open the Sky file named ***Retro2.skyx***, and then clicks the ⬛ Go Forward button.

© 2011 Cengage Learning. All Rights Reserved. May not be scanned, copied or duplicated, or posted to a publicly accessible website, in whole or in part.

2. Periodically clicking displays the daily planetary positions.

3. Click the "3-Dimensional Solar System Mode" button, Solar System , and observe Jupiter's position in its orbit relative to Earth's.

4. What is the approximate elongation of Jupiter on March 1, 2006? _____°. This is the approximate date on which Jupiter begins to retrograde in Earth's sky.

5. What is the approximate elongation of Jupiter on May 13, 2006? _____°. This is the date on which Jupiter is approximately half-way through the retrograde period.

6. What is the approximate elongation of Jupiter on July 5, 2006? _____°. This is the approximate date on which Jupiter ends to retrograde in Earth's sky.

7. Close the file. It is not necessary to save either of these Sky files.

Of all these motions, the Sun's motion along the ecliptic and the retrograde motion of the planets are both apparent motions. The Sun's eastward motion along the ecliptic is actually caused by Earth revolving around it. The north–south motion of the Sun along this path is caused by the 23.5° tilt of Earth's axis. This motion can be viewed by reviewing the Sky file **SunMotion.skyx**.

The tilt of Earth is what makes the Sun appear to move north and south of the celestial equator. This, in turn, is responsible for the change of the apparent rising point of the Sun on the eastern horizon throughout the course of a year. It is also the *main cause* of the Seasons on Earth (See **Chapter 9**).

The Sun's apparent rising point moves both northward and southward along the eastern horizon. *TheSkyX* easily demonstrates this motion. The Sky file named **RisingPointE.skyx** was designed to do this. When one faces east, first locate where the ecliptic intersects the eastern horizon. This is easy, just look for the zodiacal constellations!

If one imagines now that the Sun is located at this point of intersection, as it appears to rise, then this point *is* the apparent rising point of the Sun on the eastern horizon. After watching over a period of a few months, it is soon realized that this intersection point moves both northern and southward along the eastern horizon. Ancient astronomers observed this phenomenon thousands of years ago and constructed accurate calendars based on this very observation. Stonehenge, on the Salisbury Plain, in England is a stone calendar that marks the location of the Sun's rising point on the eastern horizon at specific times of the year.

Exercise N was created and designed to illustrate this motion. When the Sun is located at the northernmost or southernmost point on the horizon, the Sun appears to rise on the summer solstice or winter solstice, respectively. The word *solstice* is a Greek word meaning "*sun standing still.*" on the horizon. That is, the Sun appears to stop its northerly and southerly motion along the eastern horizon at these two points. When the Sun's apparent rising point is due east, the Sun appears to rise due east on the vernal or autumnal equinox, respectively.

Exercise N: Observing the Sun's Rising Point on the Eastern Horizon

1. Open the Sky file named **RisingPointE.skyx**.

2. Click Look East , if not already facing that direction, and Zoom out to a 140° x 140° field of

    view (140° x 140°).

© 2011 Cengage Learning. All Rights Reserved. May not be scanned, copied or duplicated, or posted to a publicly accessible website, in whole or in part.

3. Find where the ecliptic intersects the eastern horizon. Click [Go Forward] and watch the intersection point (yellow-dotted line) move northward, then southward along the eastern horizon. To replay this file reset the time and date to March 20, 2013 at 7:30 AM. It is approximately 15 minutes before the apparent sunrise.

4. As a point of interest, one should click the [Look West] button and observe the apparent setting point of the Sun moving along the western horizon.

5. Close the file. It is not necessary to save this Sky file.

Since the retrograde motion of planets is an apparent motion, it may seem a bit more difficult to determine its cause. It is due to the combined motion of Earth and the planets that causes this motion. A similar phenomenon is observed while passing a slower moving vehicle on a four-lane highway. As one looks out their window at the car being passed, the other vehicle appears to be moving backward with respect to distant trees, houses, or billboards. The same is true for planets. Faster moving planets overtake and pass slower moving ones. When this happens, slower moving planets appear to back up, or move westward, in Earth's sky as seen projected against the background stars, as in **Figure 6-3**.

The Sky file named *EarthMars1.skyx* illustrates the orbits of Earth and Mars in a 3-Dimensional mode as they revolve around the Sun. There are two things to notice when viewing this sky window. The first is that both planets are moving in the same direction around the Sun. The second is that Earth is moving faster than Mars and will overtake it. In fact, this occurs quite regularly in our Solar System. Earth overtakes Mars approximately every 779 days.

This period of time is referred to as a Mars' *synodic period*. That is, every 2.1 years Earth catches up with and overtakes Mars in its orbit. When this event occurs, Mars is about 180° from the Sun or in a configuration that is known as *opposition*. During this time Mars appears to *retrograde* or *move westward* in Earth's sky. **Exercise O** illustrates the position and motion of both Earth and Mars in a 3-Dimensional-representation of our Solar System.

Exercise O: Observing the Motion of Earth and Mars about the Sun

1. Open the Sky file named *EarthMars1.skyx*.

2. After opening the file, click [Go Forward] and watch the motion just described.

   If the 3-D Solar system window is not open, click the [Solar System] button.

3. Close the file. It is not necessary to save this Sky file.

The Sky file named *EarthMars2.skyx* displays how the motion of Mars appears in Earth's sky projected against the background stars. The time interval is approximately the same as it was in the *EarthMars1.skyx* Sky file. **Exercise P** illustrates the motion of Mars in Earth's sky during the time when Mars is in opposition. The time interval, during which this retrograde motion occurs, is from approximately July 4, 2005, to March 5, 2006. At the time that this particular opposition occurred, Mars was at its closest distance to Earth in 60,000 years. The next opposition of Mars occurred in the middle of November, 2007. The next time that Mars appears to retrograde will be sometime during the middle of October, 2011 in the constellation of Leo and again in early December, 2013 in the constellation of Virgo.

© 2011 Cengage Learning. All Rights Reserved. May not be scanned, copied or duplicated, or posted to a publicly accessible website, in whole or in part.

Exercise P: Observing Retrograde Motion of Mars in Earth's Sky

1. Open the Sky file named ***EarthMars2.skyx***.

2. After opening the file, click [Go Forward] and watch the motion as previously described.

3. Close the file. It is not necessary to save this Sky file.

    In fact, all the outer planets appear to retrograde in this fashion but their synodic periods are different. Jupiter's retrograde motion is illustrated in the Sky file named ***JupiterRetro.skyx***. Viewing this Sky file will be left as an exercise for the student.
    Exercise Q assists in setting the sky window to view the retrograde motion of Mars in the year 2013.

Exercise Q: Observing Mars Retrograde in December, 2013

1. Re-open the Sky file used in **Exercise P**.

2. Click "Input" on the Standard toolbar located at the top of the sky window.

3. Click "Location." The location setting may be of your own choosing.

4. Now, click the "Input" again and "Date and Time."

5. Set Date: November 5, 2013
    Time: 9:30 PM STD
    Daylight saving option: "U.S. and Canada."

6. Click [X].

7. Set the Time Flow rate to [1 day ▼] or less.

8. Find Mars. Click "Edit" on the Standard toolbar located at the top of the sky window, then click "Find;" or right-click the mouse anywhere in the sky window desktop, then click

    "Find." After the Find window opens, type "Mars" and click "Find Now" then click .

9. Now, click the [Go Forward] button and watch.

10. Does Mars appear to retrograde? _____. On what date does Mars appear to begin to retrograde? _____. If you change your location on Earth, how will this affect what will be observed? _____.

11. Try it! Set your location to Moscow, CIS, and then Close.

12. Click the [Go Forward] button, and observe. Are there any changes? _____.

© 2011 Cengage Learning. All Rights Reserved. May not be scanned, copied or duplicated, or posted to a publicly accessible website, in whole or in part.

13. This file may be saved by clicking the button on the File toolbar. But, first, a word of caution!

    Caution: Before saving this file one *should change* the filename!

    By clicking the file button, the Sky window that was just viewed replaces the Sky file named ***EarthMars2.skyx***. To *avoid this problem*, one uses the 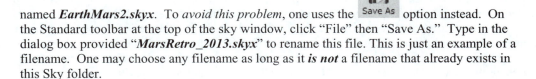 option instead. On the Standard toolbar at the top of the sky window, click "File" then "Save As." Type in the dialog box provided "***MarsRetro_2013.skyx***" to rename this file. This is just an example of a filename. One may choose any filename as long as it ***is not*** a filename that already exists in this Sky folder.

14. Now, close the file.

    Today we recognize the Sun as a star, the Moon as a satellite of Earth, and the planets as individual worlds orbiting the Sun. They are members of our Solar System environment.

    To observe them takes some time and patience but distinguishing planets in the sky is really not as difficult as you might think. Usually they appear as "bright" stars in the sky. After observing for a few days or weeks, one will notice if the object moves relative to the distant stars. If it does, then it's a planet! This is the most tried and true method.

    There is another easy way to ascertain whether an object is a planet or a bright star. First, find its location in the sky with respect to the ecliptic. A "bright" star along the ecliptic may be a planet. Once again observe it for a few days or weeks to see if it moves; if it does, then it is definitely a planet!

    The most reliable resource for locating planets is, of course, *TheSkyX*. It calculates the positions of all Solar System objects with amazing accuracy. One can find planets easily for any year or any time of the year using this software. Whether or not a planet is visible at the time of the observation remains to be "seen!"

Classifying Planets:

To make sense of planetary motions, it can be helpful to have an understanding of the way planets are classified, how they move, and where they are located in our Solar System. There are several ways in which we classify planets. One way is by their distance from the Sun.

Using Earth's distance as our "standard" distance in the Solar System, one finds that planets can be categorized into two types of planets. The first type is called an *inferior planet*. An inferior planet is a planet whose orbit lies *inside* Earth's orbit. That is, its semimajor axis or average distance from the Sun is *less than* 1 astronomical unit (AU), or about 93,000,000 miles.

There are two inferior planets, Mercury and Venus. The Sky file that is named ***Inferior.skyx*** shows the orbits and motions of Mercury and Venus with respect to Earth and the Sun. Inferior planets all have periods of revolution of less than one year. **Exercise R** shows the motion of Mercury, Venus, and Earth, as viewed from a location outside and above the Solar System. After you open this Sky file, press Go Forward to watch the motion of these planets. Using the scroll button at the right-hand side of the window allows one to change their viewing perspective from above to below the plane of the Solar System.

© 2011 Cengage Learning. All Rights Reserved. May not be scanned, copied or duplicated, or posted to a publicly accessible website, in whole or in part.

Exercise R: Observing the Orbits and Motions of Mercury and Venus

1. Open the Sky file named ***Inferior.skyx***.

2. After you open the file, click [Go Forward] and observe the motion.

3. Close the file. It is not necessary to save this Sky file.

The second type of planet is called a *superior planet*. Superior planets are planets whose orbits *lie outside* Earth's orbit. Their semimajor axes are *greater than* 1 AU and their periods of revolution are greater than one year. There are now five superior planets, Mars, Jupiter, Saturn, Uranus, Neptune and, of course, the recently demoted Pluto. Of these, Mars, Jupiter, and Saturn are bright enough to be seen with the naked-eye. The Sky file named ***Superior.skyx*** illustrates the orbits and motions of the superior planets with respect to Earth and the Sun as viewed from outside and above the Solar System. **Exercise S** shows the motion of only the planets beyond Earth's orbit. After opening this file, press [Go Forward] to watch the motion of these planets. The viewing perspective can be changed from Earth's sky view to that of a 3-Dimensional view from above to below the plane of the Solar System.

Exercise S: Observing the Orbits and Motions of the Superior Planets

1. Open the Sky file named ***Superior.skyx***.

2. After opening the file, click [Go Forward] and observe the motion previously described.

3. Now, click the [Solar System] button to observe a 3-Dimensional view of the Solar System. Move the slide bar at the right-edge of the window. This allows the user to move from into the plane of the Solar System to a vantage point above it. Look at the orbits! How do they appear? _____.

    Can you draw any conclusions about the "shape" of our Solar System? _____

4. Close the file. It is not necessary to save this Sky file.

All planets move around the Sun in a counterclockwise direction when viewed from above the Solar System. As stated earlier, this motion around the Sun translates to eastward motion in Earth's sky. The sidereal period is a measurement of the time it takes a planet to make one complete revolution around the Sun with respect to the stars. The sidereal periods for the planets are in the same order as their distance from the Sun. The Sky file named ***Sidereal.skyx*** displays all the naked-eye planets moving around the Sun. **Exercise T** illustrates this motion.

Exercise T: Observing the Motions of the Naked-Eye Superior Planets

1. Open the Sky file named ***Sidereal.skyx***.

2. Click [Go Forward] and observe the motion.

© 2011 Cengage Learning. All Rights Reserved. May not be scanned, copied or duplicated, or posted to a publicly accessible website, in whole or in part.

3. Click 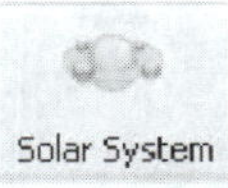

and zoom out to view planets beyond Jupiter.

4. Move the scroll bar at the right edge of the sky window down again into the plane of the Solar System. Look at the orbits! Are they about in the same plane? _____. Would you Say that our Solar System is a flat system? ____________

5. Close the file. It is not necessary to save this Sky file.

A German mathematician finally figured all of this motion out several hundred years ago. In 1609 Johannes Kepler published the book entitled *Astronomia Nova* (*The New Astronomy*). In this book he described planetary motion in the Solar System and formally presented his first two laws of planetary motion.

His first law, known as the *law of ellipses*, described the shape of planetary orbits. After working many years with positional data for the planets, he found that they move in elliptical orbits around the Sun. He also realized that the Sun's position is not exactly in the middle of each planet's orbit. It is off center at a point he called the *focus point*. A schematic diagram of Kepler's first law is shown in **Figure 6-4**.

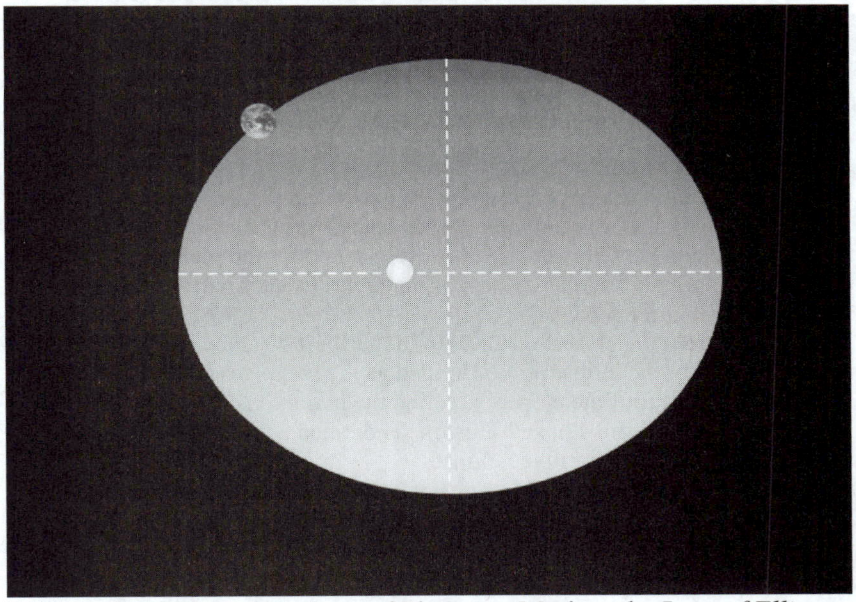

Figure 6-4 Kepler's First Law of Planetary Motion, the L*aw of Ellipses*

Kepler's second law of planetary motion describes the motions of the planets themselves. It is known as the *law of areas*. He found that if you connect a line between the Sun and a planet, and let this line move as the planet moves around the Sun, it sweeps out *equal areas* in space in *equal time intervals*. The line, connecting the Sun and a planet, is called a *radius vector*.

Figure 6-5 illustrates a radius vector connecting the Sun and a planet at points A, B, C, and D. Imagine, as Kepler did, the planet moving in different segments (A to B or C to D, and so on) in its orbit around the Sun. In some time interval, say t, the planet moves from point A to point B in its orbit, sweeping out area P in space. At some later time, same t, the planet moves from point C to point D in its orbit, sweeping out area A. If these two areas are carefully measured, one will find, as Kepler did, that they are indeed equal!

© 2011 Cengage Learning. All Rights Reserved. May not be scanned, copied or duplicated, or posted to a publicly accessible website, in whole or in part.

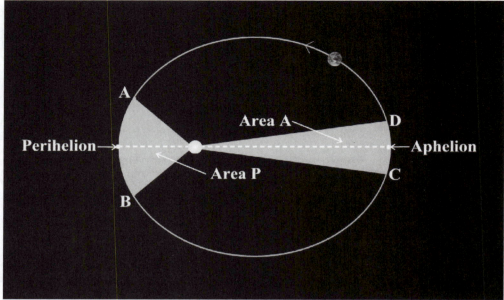

Figure 6-5 Kepler's Second Law of Planetary Motion, the *Law of Areas*

For the second law to be valid, Kepler realized, a planet speeds up and slows down as it orbits the Sun. It moves faster from point A to point B than from point C to point D. A planet travels fastest in its orbit when it is halfway between points A and B and slowest in its orbit when it is halfway between points C and D.

Kepler defined the two points in a planet's orbit where it is *closest* and *farthest* from the Sun as the *perihelion* and *aphelion* points, respectively. So as planets orbit the Sun, their distance varies and they speed up and slow down. The closest distance for the planet lies exactly halfway between points A and B and the farthest distance lies exactly halfway between points C and D and are labeled in **Figure 6-5**.

If a horizontal line is drawn from the closest distance to the farthest distance in **Figure 6-5**, this represents the largest diameter of the ellipse and is defined as the *major axis* of the ellipse. A vertical line drawn through the center of the ellipse, dividing the major axis into two equal parts represents the smallest diameter of the ellipse. This line is defined as the *minor axis* of the ellipse. These two lines are axes of symmetry in the ellipse.

Kepler determined that the average distance that a planet is from the Sun is equal to the length of its *semimajor axis* (half of the largest diameter). He also defined a new unit of distance for the Solar System. This distance unit he called an *astronomical unit* or AU. One AU represents the length of Earth's semimajor axis; that is, Earth's average distance from the Sun. Today this distance is known to be about 93,000,000 miles or 149,600,000 kilometers.

By 1616, Kepler had determined the relationship between average distances of planets from the Sun to their periods of revolution. The period of time he used was the time it took planets to make one circuit of the sky with respect to the stars. This time interval is the planet's *sidereal period*.

The third law of Kepler is known as the *harmonic law*. Simply stated, the square of the sidereal period (*P*) is directly proportional to the cube of the semimajor axis (a) of the planet's orbit.

This law is expressed in a symbolic fashion in Equation 6-1.

$$P^2 \; \alpha \; ka^3 \qquad\qquad\qquad \text{Equation 6-1}$$

k is a constant of proportionality in this expression. If the sidereal period (P) is expressed in years and the average distance (a) is expressed in astronomical units (AUs), then the constant of proportionality, *k*, in Equation 6-1 equals 1. With this judicious choice of units of P and a, Equation 6-1 may be written in the form shown in Equation 6-2.

© 2011 Cengage Learning. All Rights Reserved. May not be scanned, copied or duplicated, or posted to a publicly accessible website, in whole or in part.

$$P^2 = a^3 \qquad\qquad\qquad\qquad \text{Equation 6-2}$$

Kepler's third law may be thought of as the "yardstick" of the Solar System. In other words, if the period of time it takes a planet to complete one revolution around the Sun (in years) can be measured, then it is possible to determine its distance on the average in AUs. Or, if a planet's average distance is known, then it is possible to calculate the planet's period of revolution around the Sun with respect to the stars.

Distances and periods of revolution that are calculated using Kepler's third law are *relative* distances and periods. That is, they are determined in terms of Earth's average distance from the Sun and its period of revolution. This law is very useful in determining distances and periods of revolution of newly discovered objects in the Solar System, such as asteroids and comets! When using *TheSkyX*, one will find the physical and orbital data for the planets very useful.

Configurations of the Planets

Our Solar System is a flat, counterclockwise system. Because the entire Solar System is in motion, it is not hard to imagine that planets can line up with each other as they orbit the Sun. When these alignments occur, they are called *planetary configurations*. Configurations of planets are important when you are trying to understand their motions in the sky or their appearance in a telescope.

Knowing the configuration of a planet actually allows the user to locate where the planet is in its orbit with respect to the Sun and Earth. An easy method of determining the configuration of a planet is to measure an angle. This angle is defined as a planet's *elongation*. It is the angle measured between a line drawn from Earth to the Sun and a line drawn from Earth to the planet. The measurement of elongation is shown in **Figure 6-6**.

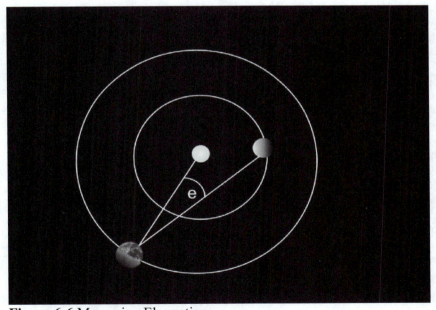

Figure 6-6 Measuring Elongation

In the sky, this angle is measured from an Earth–Sun line to an Earth–planet line along the ecliptic. The elongation of planets varies as they revolve around the Sun. Because inferior planets (closer) move around the Sun faster than Earth, they go through a variety of configurations more quickly than superior (farther) planets do. The configurations of an inferior planet are displayed in **Figure 6-7**.

© 2011 Cengage Learning. All Rights Reserved. May not be scanned, copied or duplicated, or posted to a publicly accessible website, in whole or in part.

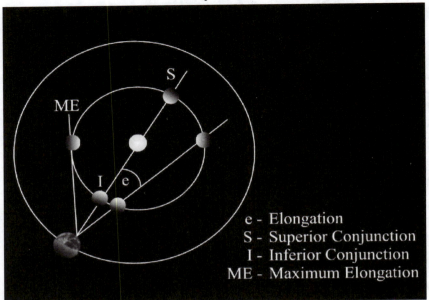

e - Elongation
S - Superior Conjunction
I - Inferior Conjunction
ME - Maximum Elongation

Figure 6-7 Configurations of an Inferior Planet

As inferior planets move around the Sun, their elongations constantly change. In **Figure 6-7** the letter *e* represents the elongation angle. The letters *ME* represent the maximum elongation of an inferior planet. When inferior planets are observed in Earth's sky, they never appear too far, in angle, from the Sun.

Sometimes inferior planets appear east of the Sun in Earth's sky, at other times they appear west of it. When an inferior planet is east of the Sun, it appears to rise after the Sun does and appears to set after the Sun does. It is visible during early evening twilight hours and is often referred to as an "evening star."

When an inferior planet is west of the Sun, it appears to rise before the Sun and appears to set before the Sun. It is visible during the early morning twilight hours and is often referred to as a "morning star."

People are often confused about how to determine directions from the Sun when it comes to measuring elongations in the sky. To put it simply—which way is east or west? To make sense of all this, one must first try to properly orient themselves on the Sun.

To do this, it is helpful to go outside and observe the Sun's position at midday. The Sun's location in the sky, of course, is due South on the local celestial meridian. While standing outside facing south, imagine seeing a planet to the left or right of the Sun. If the planet is left of the Sun in the sky, then it is east of the Sun. If the planet is right of the Sun in the sky, then it is west of the Sun. No matter where the Sun is located on the ecliptic whenever a planet is to the left of it, it is in eastern elongation. Whenever the planet is to the right of the Sun, it is in western elongation.

Figure 6-8 graphically illustrates the direction that a planet may be with respect to the Sun in Earth's sky. When the observer faces south, notice that east is to the left and west is to the right as in this figure.

© 2011 Cengage Learning. All Rights Reserved. May not be scanned, copied or duplicated, or posted to a publicly accessible website, in whole or in part.

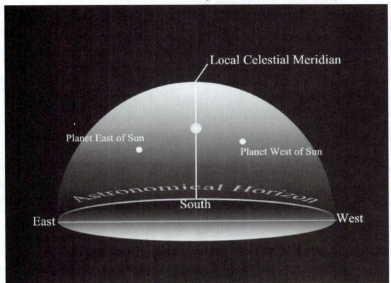

Figure 6-8 Determining the Direction of a Planet with Respect to the Sun

This orientation is true whenever one uses *TheSkyX*. Basically, if a planet is left of the Sun in the sky window, then it is east of the Sun and if it is right of the Sun in the sky window, then it is west of the Sun. In **Figure 6-9** the Sun is on the local celestial meridian and depicts the relationship shown in **Figure 6-8**. Neptune and Mars are east of the Sun while Pluto, Mercury and Venus are west of it.

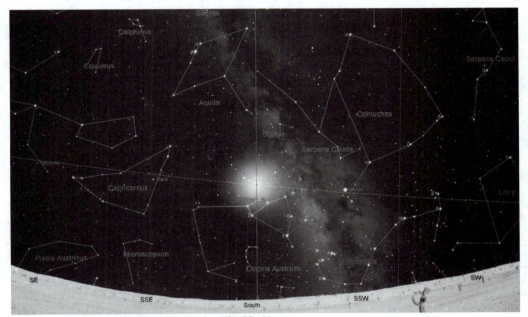

Figure 6-9 Locating the Planets in *TheSkyX*

The elongations of inferior planets have values that range from 0° (in line with the Sun) to some maximum value. The maximum elongation for an inferior planet is defined as the planet's *greatest eastern* or *western elongation*. For the planet Mercury, this angle is about 26° and for Venus it is close to 48°. It must be realized that these angles are for observations made under ideal conditions. The orbits of the planets do not lie exactly in the same plane with each other but are slightly tilted, as you have seen in previous Sky files.

The Sky file named ***VenusGEE.skyx*** illustrates a 3-Dimensional view of the planets Venus and Earth with the Sun. Venus is located at greatest eastern elongation. **Exercise U** demonstrates the position of Earth, Venus, and the Sun when Venus is at its greatest angular

© 2011 Cengage Learning. All Rights Reserved. May not be scanned, copied or duplicated, or posted to a publicly accessible website, in whole or in part.

separation east of the Sun. Notice that Venus appears to the left of the Sun as viewed from Earth's position in the Solar System.

Exercise U: Observing Venus at Greatest Eastern Elongation

1. Open the Sky file named ***VenusGEE.skyx*** and click Solar System, if it is not already displayed. Using the scroll bar at the right edge of the window, observe the configuration from different vantage points above the Solar System.

2. Close the file. It is not necessary to save this Sky file.

The Sky file named ***MercuryGWE.skyx*** is also a 3-Dimensional window illustrating the planets Mercury and Earth with the Sun. Mercury is located at greatest western elongation. **Exercise V** demonstrates the position of Mercury, Earth, and the Sun when Mercury is at its greatest angular separation west of the Sun. Notice that Mercury appears to the right of the Sun as viewed from Earth's position in the Solar System.

Exercise V: Observing Mercury at Greatest Western Elongation

1. Open the Sky file named ***MercuryGWE.skyx*** and click Solar System, if it is not already displayed. Using the scroll bar at the right edge of the window, observe the configuration from different vantage points above the Solar System.

2. Close the file. It is not necessary to save this Sky file.

The greatest eastern and western elongations are the two extreme angles measured for an inferior planet. So any location between these and any alignment with the Sun is usually described as the planet being in either eastern or western elongation.

As an inferior planet orbits the Sun, there are two locations in its orbit where the elongation is 0°. That is, the planet aligns itself with the Sun as viewed from Earth. One location is where the planet is more distant than the Sun and the other is where the planet is between Earth and the Sun. These locations are labeled S and I, respectively, in **Figure 6-7**.

When the planet is located at S, it is aligned with the Sun as viewed from Earth, and its elongation is 0°. This configuration is defined as *superior conjunction* of an inferior planet. The Sky file named ***VenusSupC.skyx*** gives a 3-Dimensional representation of this configuration.

When the planet is located at position I, it is aligned with the Sun again as viewed from Earth, and its elongation is again 0°. This configuration is defined as *inferior conjunction* of an inferior planet. The Sky file named ***VenusInfC.skyx*** is also a 3-Dimensional representation that illustrates this configuration.

At either location, the angular separation from the Sun is 0°. In other words, an inferior planet would have the same right ascension as the Sun and appear to be very close to the Sun in Earth's sky. **Exercise W** illustrates Venus at the configuration of superior conjunction. **Exercise W** illustrates Venus at the configuration of inferior conjunction. In both of these exercises, from Earth's position in the Solar System, Venus appears to be in the direction of the Sun.

© 2011 Cengage Learning. All Rights Reserved. May not be scanned, copied or duplicated, or posted to a publicly accessible website, in whole or in part.

Exercise W: Observing Venus at Superior Conjunction

1. Open the Sky file named ***VenusSupC.skyx*** and click , if it is not already displayed.

2. Using the scroll bar at the right edge and bottom of the window, you can observe this configuration from a variety of vantage points above the Solar System.

3. Click ☒ and notice that Venus and the Sun appear in Earth's sky low in the south southwest. Notice how close they are together in the sky. Click the "Show Daylight" button in the "Display" menu on the Standard toolbar and observe them. Now, click
   ![Solar System] to display Venus' location with respect to the Sun and Earth.

4. Close the file. It is not necessary to save this Sky file.

Exercise X: Observing Venus at Inferior Conjunction

1. Open the Sky file named ***VenusInfC.skyx*** and click Solar System, if it is not already displayed. Using the scroll bar at the right edge of the window, observe the configuration from different vantage points above the Solar System.

2. Click ☒ and notice that Venus and the Sun appear in Earth's sky low in the southwest. Notice too that they are not as close as they were in the previous exercise. It may not be apparent in this sky window that Venus and the Sun ***do have*** about the same right ascension but ***not*** the same declination. Click the equatorial grid and estimate the right ascension of each object. They are very close.

3. If you are still not convinced, then click on the Sun and record its Geocentric Longitude
   _____° _____'.
   Now, click on Venus and record its Geocentric Longitude_____° _____'.

   Compare the difference in their geocentric longitudes = _____° _____'.

4. Click the "Show Menu" to display the menu choices. This menu is shown in **Figure 6-10**. Click "Show Daylight" button and observe them in the day time sky.

© 2011 Cengage Learning. All Rights Reserved. May not be scanned, copied or duplicated, or posted to a publicly accessible website, in whole or in part.

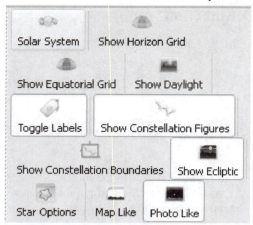

Figure 6-10 Show Daylight Button in *TheSkyX*

5.   Close the file. It is not necessary to save this Sky file.

Once in a great while the declinations of an inferior planet and the Sun are very close to being the same value. In this instance an inferior planet, such as Venus, at inferior conjunction may transit the Sun's disk.

Although among the rarest event, and difficult to predict, when this happens the planet appears as a small black dot projected against the Sun's bright disk. This event seems to be tied to a 243-year cycle. This occurred on June 8, 2004 and lasted about six hours. The previous time since the 2004 transit occurred was in 1882. The next time that Venus will transit the Sun will be on June 5-6, 2012. **Figure 6-11**, rendered in *TheSkyX*, is how this transit should appear on this date.

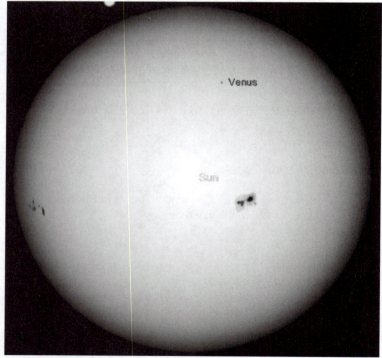

Figure 6-11 The Transit of Venus on June 5 - 6, 2012

The elongations for Mercury and Venus are shown in **Figure 6-12**. Also included in the figure are approximate times when these planets appear to rise and set.

© 2011 Cengage Learning. All Rights Reserved. May not be scanned, copied or duplicated, or posted to a publicly accessible website, in whole or in part.

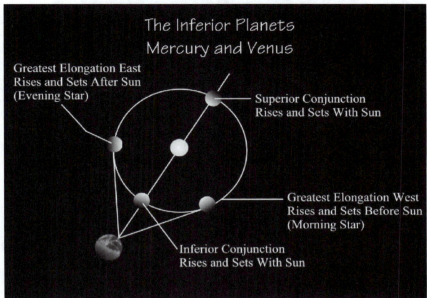

Figure 6-12 Elongations of Mercury and Venus

Even when Mercury is at either of its greatest elongations, it is usually seen in the twilight sky. The reason for this is that it appears to rise and set only one to two hours before and after the Sun does, respectively. Venus, in contrast, is often seen in a totally dark sky. In fact, it's the third brightest object in the sky. When it is at either of its greatest elongations, Venus appears to rise and set approximately three hours before or after the Sun does, respectively.

As superior planets orbit the Sun their orbital positions change, but at a much slower rate than the inferior planets. Therefore their elongations vary as well when they are viewed from Earth. However, because a superior planet's orbit is greater than that of Earth, the elongation can be much greater than that of an inferior planet.

Elongations for superior planets range in values from 0° to 180°. When a superior planet's elongation is 0°, it too has the same right ascension as the Sun. This configuration is defined as *conjunction*. The Sky file named ***JupiterConj.skyx*** illustrates a 3-Dimensional representation of this configuration. It depicts Jupiter's position in the Solar System on May 8, 2000, relative to Earth and the Sun.

Exercise Y illustrates the planets Jupiter and Earth with the Sun in a 3-Dimensional representation of Jupiter in conjunction. You will notice that Jupiter is in the same direction of the Sun as viewed from Earth's position in the Solar System.

Exercise Y: Observing Jupiter in Conjunction

1. Open the Sky file named ***JupiterConj.skyx*** and click Solar System, if it is not already displayed. Using the scroll bar at the right-edge of the window, observe the configuration from different vantage points above the Solar System.

2. Now, click ☒ and notice that Jupiter and the Sun are seen well below the northwestern horizon. Notice too how close together they appear in the sky.

3. Put the cursor on the Sun and note its equatorial coordinates _____° _____'. Now, put the cursor on Jupiter and note its equatorial coordinates _____° _____'. Are they the same as those of the Sun? _____.

© 2011 Cengage Learning. All Rights Reserved. May not be scanned, copied or duplicated, or posted to a publicly accessible website, in whole or in part.

4. Find the angular separation between Jupiter and the Sun by using the Find Result window again. First, click on the Sun and record its Geocentric Longitude _____° _____'. Now, click on Jupiter and record its Geocentric Longitude_____° _____'. Compare the difference in their geocentric longitudes = _____° _____'.

5. Close the file. It is not necessary to save this Sky file.

As a superior planet continues to move counterclockwise around the Sun from conjunction, its elongation continually changes. All the while the planet appears to be left farther of the Sun in the sky, or in eastern elongation. Eventually, it reaches a location in its orbit that is 90° from the Sun. This configuration is defined as a *quadrature*. When a superior planet's position is 90° east of the Sun, its configuration is defined as *eastern quadrature*. The Sky file named *JupiterEQ.skyx* portrays a 3-Dimensional representation of Jupiter at the time of eastern quadrature in January, 2000. **Exercise Z** illustrates the planets Jupiter and Earth with the Sun in the configuration of eastern quadrature. Notice that Jupiter is 90° to the left of the Sun as viewed from Earth's position in the Solar System.

Exercise Z: Observing Jupiter at Eastern Quadrature

1. Open the Sky file named *JupiterEQ.skyx* and click Solar System , if it is not already displayed. Using the scroll bar at the right-edge of the window, observe the configuration from different vantage points above the Solar System.

2. Click ☒ and notice that the Sun is close to the local celestial meridian. Find Jupiter and Center it. Notice that Jupiter has just risen in the east.

3. Put the cursor on the Sun and note its equatorial coordinates _____° _____'. Now, put the cursor on Jupiter and note its equatorial coordinates _____° _____'. Are they about 6^H greater (eastward) than that of the Sun? _____.

4. Find the angular separation between Jupiter and the Sun by using the Find Result window again. First, click on the Sun and record its Geocentric Longitude _____° _____'. Now, click on Jupiter and record its Geocentric Longitude_____° _____'.

    Compare the difference in their geocentric longitudes = _____° _____'.
    Is it about 90°? _____.

5. Close the file. It is not necessary to save this Sky file.

As a superior planet continues its motion in orbit, it eventually reaches a location where it is exactly 180° from the Sun. This configuration is defined as *opposition*. The Sky file named *JupiterOpp.skyx* displays the planets Jupiter and Earth and the Sun at the time of opposition. Earth's location in this window is between Jupiter and the Sun; Jupiter is exactly 180° from the Sun as viewed from Earth.

Whenever a superior planet is at opposition, it appears to rise at the time the Sun appears to set. It is visible all night long. During the interval of time that a superior planet approaches opposition, it also appears to retrograde in Earth's sky. **Exercise AA** illustrates the planets Jupiter and Earth with the Sun in a 3-Dimensional representation at the time of opposition.

© 2011 Cengage Learning. All Rights Reserved. May not be scanned, copied or duplicated, or posted to a publicly accessible website, in whole or in part.

Exercise AA: Observing Jupiter at Opposition

1. Open the Sky file named ***JupiterOpp.skyx*** and click ^{Solar System}, if it is not already displayed. Using the scroll bar at the right edge of the window, observe the configuration from different vantage points above the Solar System.

2. Click ⊠ and notice that the Sun appears to be setting in the west-southwest. Find Jupiter and center it. Did you notice where Jupiter is with respect to the Sun? _____.

3. Put the cursor on the Sun and note its equatorial coordinates _____° _____'. Now, put the cursor on Jupiter and note its equatorial coordinates _____° _____'. Are they about 12^H greater (eastward) than that of the Sun? _____. Remember, RA increases in value along the celestial equator to a ***maximum*** value of 24^H.

4. Find the angular separation between Jupiter and the Sun by using the Find Result window again. First, click on the Sun and record its Geocentric Longitude _____° _____'. Now, click on Jupiter and record its Geocentric Longitude _____° _____'.

   Compare the difference in their geocentric longitudes = _____° _____'.
   Is it about 180°? _____.

5. Close the file. It is not necessary to save this Sky file.

   As the superior planet continues its motion around the Sun, it eventually ends up at a location in its orbit 90° from opposition. It is also 90° from the Sun once again. This time, however, it is now 90° west of the Sun instead of east. When a superior planet is 90° west of the Sun, its configuration is defined as *western quadrature*.
   The Sky file named ***JupiterWQ.skyx*** is a 3-Dimensional representation of Jupiter, Earth, and the Sun at the time of western quadrature. **Exercise BB** illustrates the planets Jupiter and Earth with the Sun at the time when Jupiter is at western quadrature. One will notice that Jupiter is 90° to the right of the Sun as viewed from Earth's position in the Solar System.
   The configurations of a superior planet are shown in **Figure 6-13**.

Exercise BB: Observing Jupiter at Western Quadrature

1. Open the Sky file named ***JupiterWQ.skyx*** and click ^{Solar System}, if it is not already displayed. Using the scroll bar at the right edge of the window, observe the configuration from different vantage points above the Solar System.

2. Click ⊠ and notice that the Sun appears to be rising in the east and that Jupiter appears to be near the local celestial meridian.

3. Put the cursor on the Sun and note its equatorial coordinates _____° _____'. Now, put the cursor on Jupiter and note its equatorial coordinates _____° _____'. Are they about 6^H less (westward) than that of the Sun? _____.

4. Find the angular separation between Jupiter and the Sun by using the Find Result window again. First, click on the Sun and record its Geocentric Longitude _____° _____'. Now, click on Jupiter and record its Geocentric Longitude _____° _____'. Compare the difference in their geocentric longitudes = _____° _____'.

© 2011 Cengage Learning. All Rights Reserved. May not be scanned, copied or duplicated, or posted to a publicly accessible website, in whole or in part.

Is it about 90°? _____.

5.   Close the file. It is not necessary to save this Sky file.

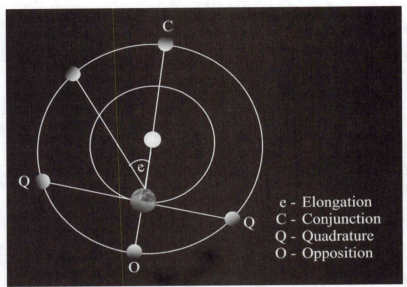

Figure 6-13 Configurations of a Superior Planet

Figure 6-14 shows the elongations of the superior planets as well as their times of visibility.

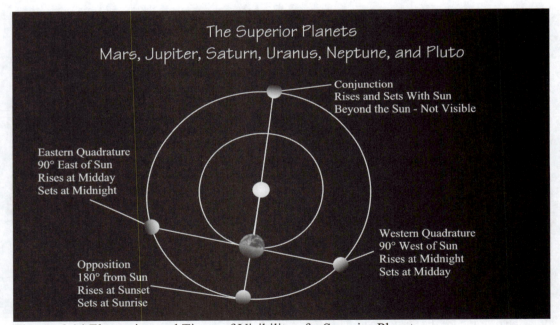

Figure 6-14 Elongation and Times of Visibility of a Superior Planet

Time intervals are also measured between planetary configurations. The interval of time between two successive occurrences of the same configuration is defined as the *synodic period* of a planet. This time interval is measurement of time with respect to the Sun, *not* the stars. It is quite different from the sidereal period of the planet. This period of time includes the motions of Earth and the planet.

© 2011 Cengage Learning. All Rights Reserved. May not be scanned, copied or duplicated, or posted to a publicly accessible website, in whole or in part.

Planetary Phases

As an inferior planet moves around the Sun, not only does its position change with respect to the Sun and Earth but the way sunlight strikes the planet also changes. When observing an inferior planet through telescopes from Earth, it is evident that Mercury and Venus go through phases like our Moon.

This phasing process, observed from Earth, is based on the synodic periods of the inferior planets. When Venus is located at greatest eastern or western elongation, it appears as a half circle or a quarter phase when viewed through a telescope. **Figure 6-15** is ***TheSkyx's*** rendition of Venus at greatest eastern elongation that was done in **Exercise U**. At this point in time it appears as a quarter Venus (phase) and the Sun is shining from the upper right in the figure. The right half of the planet is exposed to the Sun's rays.

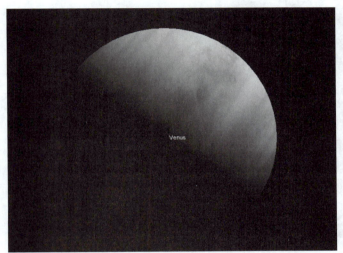

Figure 6-15 Phase of Venus at Greatest Eastern Elongation

As Venus travels from greatest eastern elongation to greatest western elongation in its orbit, it goes through a series of phases. From its quarter phase, at greatest eastern elongation, it goes through many crescent phases that look similar to the image displayed in **Figure 6-16**.

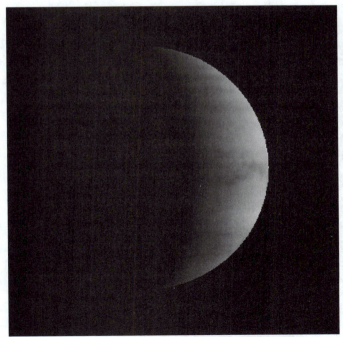

Figure 6-16 Crescent Venus in Eastern Elongation

© 2011 Cengage Learning. All Rights Reserved. May not be scanned, copied or duplicated, or posted to a publicly accessible website, in whole or in part.

During this time period Venus actually passes between Earth and the Sun through inferior conjunction. The closer Venus gets to the Sun, in angle in the sky, the more difficult it becomes to see. When Venus reaches inferior conjunction, it is aligned with the Sun and is not visible from Earth, because its dark hemisphere (night side) is facing Earth.

As Venus continues to move, from inferior conjunction toward the greatest western elongation, it again appears as many crescent phases when viewed from Earth that is similar to the one shown in **Figure 6-17**.

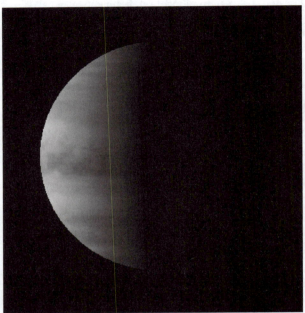

Figure 6-17 Crescent Venus in Western Elongation

If one compares **Figure 6-17** to the image taken by the Hubble Space Telescope in **Figure 6-18**, you will notice that they are very similar in appearance. Venus was in western elongation when this image was taken.

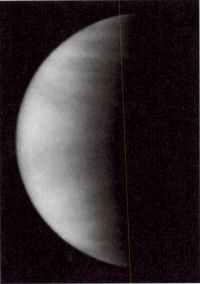

Figure 6-18 Crescent Venus Taken by the Hubble Space Telescope

The Sky file ***VenusGEE_GWE.skyx*** was rendered to display the phases of Venus as it travels from its greatest eastern elongation to its greatest western elongation. It spans a period

© 2011 Cengage Learning. All Rights Reserved. May not be scanned, copied or duplicated, or posted to a publicly accessible website, in whole or in part.

of time from June 6 to October 26, 2015. As Venus moves from greatest eastern elongation to inferior conjunction, its distance decreases as it approaches Earth. After passing inferior conjunction, its distance from Earth increases as it moves toward greatest western elongation.

During this period of time Venus is observed going through a series of phases from Earth. It goes from a quarter phase on the right edge through a series of crescent phases to a new phase when it reaches inferior conjunction. Venus and Earth are now at their closest distance to one another. The angular size of Venus also changes. Venus' angular size is at its greatest when in inferior conjunction.

From the new phase, it continues through a series of crescent phases. The crescent phases are now on the left edge of Venus, similar to the images in **Figure 6-17** and **6-18**. On reaching greatest western elongation, it appears as a quarter phase again but the left side of the planet is exposed to sunlight. This phase is displayed in **Figure 6-19** is *TheSkyX's* rendition of Venus' phase at greatest western elongation. This should have been noticed when was Venus at greatest western elongation in **Exercise U**. The Sun is shining from the upper left in the figure.

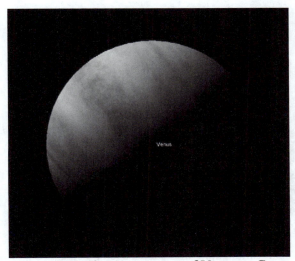

Figure 6-19 The Appearance of Venus at Greatest Western Elongation

Exercise CC demonstrates the phases that Venus exhibits as it moves from greatest eastern elongation to greatest western elongation. In this sky window Venus appears to go through a series of crescent phases through the new phase.

As it moves from greatest eastern elongation to inferior conjunction, the crescent phase is on the right edge of the planet. As it moves from inferior conjunction to greatest western elongation, the crescent phase is on the left edge of the planet. It appears as a new phase when it is at inferior conjunction. The planet appears to get bigger in angular size as it approaches Earth (to inferior conjunction), and smaller as it moves away (from inferior conjunction) from Earth. This is seen in the Sky file named *VenusGEE_GWE.skyx*. The time period of this observation is from approximately June 6, 2015 to October 25, 2015.

Exercise CC: Observing the Phases of Venus from Greatest Eastern to Greatest Western Elongation

1. Open the Sky file named *VenusGEE_GWE.skyx*. Click Go Forward .

2. Click Solar System to show where Venus is with respect to the Sun and Earth.

3. Close the file. It is not necessary to save this Sky file.

© 2011 Cengage Learning. All Rights Reserved. May not be scanned, copied or duplicated, or posted to a publicly accessible website, in whole or in part.

When Venus passes greatest western elongation in its orbit, its distance is actually farther from Earth than the Sun. As Venus travels from greatest western elongation to greatest eastern elongation in its orbit, it continues going through many phases. This time, however, the phases are known as gibbous phases.

A gibbous phase is larger than a quarter phase but smaller than a full phase. A gibbous phase looks like the image displayed in **Figure 6-20**. This image was taken by the Pioneer Venus Orbiter in 1979. One should notice that Venus is slightly out of round on the bottom right-edge.

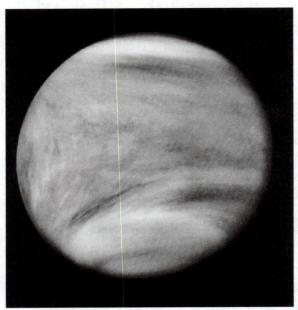

Figure 6-20 Gibbous Venus Taken by Pioneer Venus Orbiter

From its quarter phase, at greatest western elongation, more of the left-edge of Venus becomes visible from Earth, as seen in **Figure 6-20**. Venus was in western elongation when this image was taken. The gibbous phase continues to get larger, or more of Venus becomes visible from Earth. During this time period Venus' distance from Earth continues to increase and the planet appears to get closer to the Sun in the sky (in angle). It becomes more difficult to see from Earth.

Midway through the time period between greatest western elongation and greatest eastern elongation, Venus passes the location that is in direct alignment with the Sun once again (superior conjunction). Venus is also not visible at this time. However, if it could be seen from Earth (perhaps during a total solar eclipse), it would appear like a full Moon phase. At superior conjunction its illuminated hemisphere is toward the Earth. Venus and Earth are now at their farthest distance from one another. The angular size of Venus is at its smallest when in superior conjunction.

As Venus continues to move, from superior conjunction toward the greatest eastern elongation, it again appears as a gibbous phase when viewed from Earth. During this time period, however, the right edge of Venus is illuminated instead of the left. It continues to go through a series of gibbous phases until it reaches greatest eastern elongation. Now it is a quarter phase once more.

The Sky file named **VenusGWE_GEE.skyx** was rendered to display the phases of Venus as it travels from its greatest western elongation to its greatest eastern elongation. As Venus moves from greatest western elongation to superior conjunction, its distance increases as it moves away from Earth. On reaching superior conjunction the distance between Venus and Earth is at its maximum distance and its angular size is the smallest. After passing superior conjunction, Venus' distance from Earth begins to decrease while it moves toward greatest eastern elongation.

During this period of time Venus is observed going through a series of phases from Earth. It goes from a quarter phase on the left edge through a series of gibbous phases and to a full

© 2011 Cengage Learning. All Rights Reserved. May not be scanned, copied or duplicated, or posted to a publicly accessible website, in whole or in part.

phase when it reaches superior conjunction. Venus and Earth are now at their farthest distance from one another. Venus' angular size is at its smallest when at superior conjunction.

From the full phase, it continues through a series of gibbous phases. The gibbous phases are now on the right edge of Venus. On reaching greatest eastern elongation, it appears as a quarter phase on the right edge of the planet.

Exercise DD demonstrates the phases that Venus exhibits as it moves from the configuration of greatest western elongation to greatest eastern elongation. In this sky window Venus appears to move through a series of gibbous phases. As it moves from greatest western elongation to superior conjunction, the gibbous phase is on the left edge of the planet. As it moves from superior conjunction to greatest eastern elongation, the gibbous phase is on the right edge of the planet. It appears as a full phase when it is at superior conjunction. It appears to get smaller in angular size as it moves away from Earth (to superior conjunction) and larger as it approaches Earth (from superior conjunction). The time period for this observation is from approximately October 25, 2015 to January 14, 2017.

Exercise DD: Observing the Phases of Venus from Greatest Western to Greatest Eastern Elongation

1. Open the Sky file named *VenusGWE_GEE.skyx*. Click Go Forward .

2. Click Solar System to show where Venus is with respect to the Sun and Earth.

3. Close the file. It is not necessary to save this Sky file.

The Sky file named *VenusPhases.skyx* demonstrates all the phases of Venus during one revolution around the Sun. **Exercise EE** illustrates what is observed from Earth.

Exercise EE: Observing the Phases of Venus

1. Open the Sky file named *VenusPhases.skyx*. Click Go Forward . Observe Venus as it goes through an entire set of phases through one synodic period ($\sim584^D$). Notice too that its angular size also changes through this time period.

2. Can you estimate Venus' position in its orbit relative to Earth and the Sun at the beginning of this sky window? _____. Where it is in its orbit at the end? _____.

3. Close the file. It is not necessary to save this Sky file.

The Sky file named *VenusOrbit.skyx* displays a 3-Dimensional representation of Venus in the Sky file named *VenusPhase.skyx*. **Exercise FF** demonstrates a 3-Dimensional view of what was seen in the previous exercise.

Exercise FF: 3-Dimensional Representation of VenusPhase.skyx

1. Open the file named *VenusOrbit.skyx* and click Solar System , if it is not already displayed.

   Click Go Forward . Observe Venus as it moves around the Sun through one synodic period.

© 2011 Cengage Learning. All Rights Reserved. May not be scanned, copied or duplicated, or posted to a publicly accessible website, in whole or in part.

2. Close the file. It is not necessary to save this Sky file.

The fact that Venus goes through a complete set of phases is historically significant. In 1610 the Italian astronomer Galileo Galilei was the first person to report his observations of Venus through a small telescope in his book entitled *"Siderius Nuncius."* His hand-made telescope revealed that Venus exhibited phases like that of our Moon. From these observations, he concluded that the Sun was the center of our Solar System, instead of Earth. His observations, he claimed, supported the heliocentric, or sun-centered, "universe" proposed by Nicolas Copernicus and Aristarchus of Samos.

In the case of a superior planet, however, the motion is much slower than that of Mercury or Venus. As a consequence, its change of position in the sky is less conspicuous than that of an inferior planet. The phases of a superior planet are *not as* extreme as an inferior planet. In a small telescope they almost always appear as a full phase. In larger instruments, however, their phases are subtle but more pronounced.

© 2011 Cengage Learning. All Rights Reserved. May not be scanned, copied or duplicated, or posted to a publicly accessible website, in whole or in part.

Chapter 6

TheSkyX Review Exercises

Start *TheSkyX*. You may use your own computer clock for the time in the first exercise.

TheSkyX Exercise 1: Finding Information About the Sun

1. Run *TheSkyX* and open the file *Normal.skyx*.

2. Make sure that the Sun and planet labels are displayed. Go to the "Display" menu on the Standard toolbar at the top of the sky window. After opening the Display menu window, click on "Chart Elements."

3. Click ⊞ next to "Solar System Objects" to expand the list. In order to select planets, it is necessary to click on the "Planets" menu window. It is here that one can either display all or none of the planets in *TheSkyX* desktop window. In order to display Pluto, it too must be clicked on the "Small Solar System Bodies" menu. To change the Local Horizon Fill to being more "transparent," go to the "Horizon & Atmosphere Options" menu and change that as well.

4. If the command toolbars are not displayed in the sky window, then click the "Tools" menu on the Standard toolbar at the top of the screen and then scroll down to "Preferences." In the Preferences menu one may select any, or all, of the command toolbars.

5. Now, find the Sun.

6. After locating the Sun, move through the sky eastward using the directional arrows (blue) on the "Orientation" toolbar. Observe how the Sun appears to move through the stars. What color is this path in *TheSkyX*? _____. Remember, that this color may be changed at any time by the user.

7. What is the name of this apparent path? _____>

8. While moving along the path in Question 6, did you notice any planets? _____.

9. Were any of the planets along this path farther away than ±8°? _____.

10. If the answer in Question 9 was no, why not? _____.

TheSkyX Exercise 2: Finding the Elongation of Jupiter

Run *TheSkyX* and open the file *Normal.skyx* or set *TheSkyX* to the following:

Location: Golden, Colorado
Date: December 31, 2019
Time: 6:00 PM STD

1. In what constellation is the Sun currently located? _____

2. In what constellation is Jupiter currently located? _____

© 2011 Cengage Learning. All Rights Reserved. May not be scanned, copied or duplicated, or posted to a publicly accessible website, in whole or in part.

3. What are the right ascension and declination of Jupiter?

$$R.A._{Jupiter} = \underline{\quad}^H \underline{\quad}^M \qquad Declination_{Jupiter} = \underline{\quad}\,\underline{\quad}°\underline{\quad}' (+, N; -, S)$$

4. Put the cursor on the Sun and note its equatorial coordinates

$$R.A._{Sun} = \underline{\quad}^H \underline{\quad}^M \qquad Declination_{Sun} = \underline{\quad}\,\underline{\quad}°\underline{\quad}' (+, N; -, S)$$

5. What is Jupiter's magnitude? _____.

6. What is Jupiter's elongation (angular separation from the Sun)? _____°.

   Remember, one finds the angular separation between Jupiter and the Sun by using the Find Result window. First, click on the Sun and record its Geocentric Longitude _____° _____'. Next, click on Jupiter and record its Geocentric Longitude_____° _____'.

   Compare the difference in their geocentric longitudes = _____° _____'. This is the elongation!

7. Is Jupiter currently in eastern or western elongation? _____.

TheSkyX Exercise 3: Finding the Elongation of Mars

1. In what constellation is the Sun currently located? _____.

2. In what constellation is Mars currently located? _____.

3. What are the right ascension and declination of Mars?

$$R.A._{Mars} = \underline{\quad}^H \underline{\quad}^M \qquad Declination_{Mars} = \underline{\quad}\,\underline{\quad}°\underline{\quad}' (+, N; -, S)$$

4. Put the cursor on the Sun and note its equatorial coordinates

$$R.A._{Sun} = \underline{\quad}^H \underline{\quad}^M \qquad Declination_{Sun} = \underline{\quad}\,\underline{\quad}°\underline{\quad}' (+, N; -, S)$$

5. What is Mars' magnitude? _____.

6. What is Mars' elongation (angular separation from the Sun)? _____°.

   First, click on the Sun and record its Geocentric Longitude _____° _____'. Next, click on Mars and record its Geocentric Longitude_____° _____'.

   Compare the difference in their geocentric longitudes = _____° _____'. This is the elongation!

7. Is Mars currently in eastern or western elongation? _____

TheSkyX Exercise 4: Finding the Elongation of Saturn

1. In what constellation is the Sun currently located? _____.

2. In what constellation is Saturn currently located? _____.

3. What are the right ascension and declination of Saturn?

$$R.A._{Saturn} = \underline{\quad}^H \underline{\quad}^M \qquad Declination_{Saturn} = \underline{\quad}\,\underline{\quad}°\underline{\quad}' (+, N; -, S)$$

© 2011 Cengage Learning. All Rights Reserved. May not be scanned, copied or duplicated, or posted to a publicly accessible website, in whole or in part.

4. Put the cursor on the Sun and note its equatorial coordinates

    R.A.$_{\text{Sun}}$ = ___^{H} ___^{M} Declination$_{\text{Sun}}$ = __ ___°___′ (+, N; –, S)

5. What is Saturn's magnitude? _____.

6. What is Saturn's elongation (angular separation from the Sun)? _____°.

    First, click on the Sun and record its Geocentric Longitude _____° _____′.
    Next, click on Saturn and record its Geocentric Longitude_____° _____′.

    Compare the difference in their geocentric longitudes = _____° _____′. This is the elongation!

    What is Saturn's elongation (angular separation from the Sun)? _____°.

7. Is Saturn currently in eastern or western elongation? _____

TheSkyX Exercise 5: Finding the Elongation of Jupiter

Now, set *TheSkyX* to the following:
Location: Ball State Observatory
Date: December 31, 2029
Time: 6:00 PM STD

1. In what constellation is the Sun currently located? _____.

2. In what constellation is Jupiter currently located? _____.

3. What are the right ascension and declination of Jupiter?

    R.A. = ___^{H} ___^{M} Declination = __ ___° ___′ (+, N; –, S)

4. What is its magnitude? _____

5. What is Jupiter's elongation (angular separation from the Sun)? _____

6. Is Jupiter currently in eastern or western elongation? _____

TheSkyX Exercise 6: Finding the Elongation of Mars

1. In what constellation is the Sun currently located? _____.

2. In what constellation is Mars currently located? _____.

3. What are the right ascension and declination of Mars?

    R.A. = ___^{H} ___^{M} Declination = __ ___° ___′ (+, N; –, S)

4. What is its magnitude? _____

5. What is Mars' elongation (angular separation from the Sun)? _____°.

6. Is Mars currently in eastern or western elongation? _____

© 2011 Cengage Learning. All Rights Reserved. May not be scanned, copied or duplicated, or posted to a publicly accessible website, in whole or in part.

TheSkyX Exercise 7: Finding the Elongation of Saturn

1. In what constellation is the Sun currently located? _____.

2. In what constellation is Saturn currently located? _____.

3. What are the right ascension and declination of Saturn?

    R.A. = ___^{H} ___^{M} Declination = __ ___° ___' (+, N; –, S)

4. What is its magnitude? _____

5. What is Saturn's elongation? _____

6. Is Saturn currently in eastern or western elongation? _____

TheSkyX Exercise 8: Comparing Measurements with *TheSkyX*

Compare **Exercise2** and **Exercise5**:

1. Were there any differences in the observations from the two locations? _____

2. If so, what were they? _____
    _____

TheSkyX Exercise 9: Comparing Measurements with *TheSkyX*

Compare **Exercise 3** and **Exercise 6**:

1. Were there any differences in the observations from the two locations? _____

2. If so, what were they? _____
    _____

TheSkyX Exercise 10: Comparing Measurements with *TheSkyX*

Compare **Exercise 4** and **Exercise 7**:

1. Were there any differences in the observations from the two locations? _____

2. If so, what were they? _____
    _____

© 2011 Cengage Learning. All Rights Reserved. May not be scanned, copied or duplicated, or posted to a publicly accessible website, in whole or in part.

Chapter 6

TheSkyX Review Questions

Answer the following review questions using *TheSkyX* where necessary.

TheSkyX Review Question 1

1. What are the right ascension and declination of the point in the sky where the ecliptic and the celestial equator intersect in March each year? The Sun crosses this point from the Southern Hemisphere to Northern Hemisphere in the sky.

    R.A. = ___^{H} ___^{M} Declination = __ ___° ___′ (+, N; –, S)

2. What is the name given to this point in space? _____ _____

3. Against what constellation does the Sun appear to be projected in the sky when it is located at this point? _____

4. Now change the view from Earth looking toward the Sun, to that of the Sun looking toward the Earth. Use the blue directional buttons (*Hint*: Look 180° from Sun. Turn on the equatorial grid, if necessary, and move 12^H in right ascension eastward). Against what constellation would the Earth appear to be projected when viewed from the Sun on the date in Question 1? _____

5. What time of day would the constellation in Question 4 appear to be on your local celestial meridian? _____

TheSkyX Review Question 2

1. On what date does the Sun appear to be 23.5° North of the celestial equator? _____

2. What are the right ascension and declination of the point in Question 1?

    R.A. = ___^{H} ___^{M} Declination = __ ___° ___′

3. What is the name of this point in space? _____ _____

4. Against what constellation does the Sun appear to be projected on the date in Question 1?
    _____

5. Now change the view from Earth looking toward the Sun, to that of the Sun looking toward the Earth. Use the blue directional buttons again (*Hint*: Look 12^H in right ascension eastward). Against what constellation would Earth appear to be projected when viewed from the Sun on the date in Question 1? _____

6. What time of day would the constellation in Question 5 appear to be on your local celestial meridian? _____

© 2011 Cengage Learning. All Rights Reserved. May not be scanned, copied or duplicated, or posted to a publicly accessible website, in whole or in part.

TheSkyX Review Question 3

1. What are the right ascension and declination of the point where the ecliptic and celestial equator intersect in September each year? The Sun crosses this point from the Northern Hemisphere to the Southern Hemisphere in the sky?

    R.A. = ___ᴴ ___ᴹ   Declination = __ ___° ___′ (+, N; –, S)

2. What is the name given to this point in space? _____ _____

3. Against what constellation does the Sun appear to be projected when it is located at this point in the sky? _____

4. Now change the view from Earth looking toward the Sun, to that of the Sun looking toward the Earth (use the blue directional buttons again).   Against what constellation does the Earth appear to be projected when viewed from the Sun on the date in Question 1?

    _____

5. What time of day would the constellation in Question 4 appear to be on your local celestial meridian? _____

TheSkyX Review Question 4

1. On what date does the Sun appear to be 23.5° south of the celestial equator?

    _____

2. What are the right ascension and declination of the point in Question 1?

    R.A. = ___ᴴ ___ᴹ   Declination = __ ___° ___′

3. What is the name of this point in space? _____ _____

4. Against what constellation does the Sun appear to be projected on the date in Question 1?

    _____

5. Now change the view from Earth looking toward the Sun, to that of the Sun looking toward the Earth (use the blue directional buttons again). Against what constellation does the Earth appear to be projected when viewed from the Sun on the date in Question 1?

    _____

6. What time of day would the constellation in Question 5 appear to be on the local celestial meridian? _____

© 2011 Cengage Learning. All Rights Reserved. May not be scanned, copied or duplicated, or posted to a publicly accessible website, in whole or in part.

Chapter 7

Keeping Time in *TheSkyX*

One of the beauties of nature that men and women can both truly enjoy is observing the heavens. The diurnal motions of the celestial bodies, changing of constellations with the seasons—the return of old "celestial friends"—are all there for us to observe and enjoy if we only take the time.

Because we inhabit a satellite of the Sun, and it in turn is moving, our planet has several motions. Partly because of these motions, many objects in the sky appear to move in our sky. Ever since the dawn of civilization people have become accustomed to thinking of Earth as being stationary and everything else in the sky as having motion. From our point of view, it does! Nevertheless, looking at our system from a different frame of reference, things look quite different.

One can specify positions of celestial objects at any time in the sky by using the coordinate systems discussed in **Chapter 5**. However, if we wish to follow these objects as they change their positions in the sky we will need one additional parameter, and that is time.

Timekeeping in *TheSkyX*

Over the millennia civilizations have devised ingenious ways to measure time. However, time itself is a difficult concept to define. *You* try it! Define *time* without using terms that involve the concept of time itself—terms such as *between*, *during*, *from*, *to*, or *until*. All these terms imply the passage of time. In fact, physicists and astronomers have literally given up on defining time in any absolute sense. Instead, we use an *operational definition* for time. It is based on an isochronous process, which is something that happens over and over again.

One such process is observing transits (or passages) of some reference object across the local celestial meridian each day. This reference object might possibly be one of several objects. The Sun, of course, is the most obvious one to use. Nevertheless, one could also use the Moon, a star, or simply a point in space. Upon closer inspection, every celestial object appears to move at slightly different rates in the sky, so we must choose an object that doesn't move very much over the duration of a day. Remember that a transit of some reference object across your local celestial meridian, that object's daily motion, is caused by Earth's rotation. **Exercise A** illustrates this motion.

Exercise A: Observing Sun's Transit across the Local Celestial Meridian

1. Run *TheSkyX* and open the file *Normal.skyx*.

2. Open the Sky file named *SunsTransit.skyx*.

3. Click 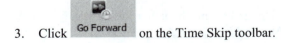 Go Forward on the Time Skip toolbar.

4. Notice that the Sun appears to move from east to west as it crosses the celestial meridian (as well as everything else in the sky). Also, note the time that the Sun transits (crosses) the local celestial meridian. Using the Stop button allows one to read the time in *TheSkyX* window at any instant.

5. Close the file. It is not necessary to save this Sky file.

© 2011 Cengage Learning. All Rights Reserved. May not be scanned, copied or duplicated, or posted to a publicly accessible website, in whole or in part.

Earth's revolution around the Sun makes the Sun appear to move 1° eastward (per day) along the ecliptic relative to the background stars. This motion is easily observed and can be precisely measured. This long-term motion of the Sun is demonstrated in the Sky file named ***SunsAppMotion.skyx***. **Exercise B** illustrates the Sun's eastward motion against the background stars.

Exercise B: Measuring the Sun's Motion in *TheSkyX*

1. Run *TheSkyX* and open the file ***Normal.skyx***.

2. Open the Sky file named ***SunsAppMotion.skyx***.

3. Click [Go Forward] on the Time Skip toolbar. The Time Flow rate is set for a sidereal day. Now, click [Stop].

4. To continue, reset the date and time back to the starting time, that is, back to August 21st at 1:45 PM DST or reopen the file. This is done either by clicking [Go Backward] and stopping the time on the starting date and time. Or by highlighting the date and time numbers on the clock with the mouse and using the wheel to scroll them backward to the starting date and time.

5. One can also measure how far the Sun moves each day in the sky by clicking [Step Forward]. In this exercise, one should notice at the beginning that the Sun moves from just west of the bright star α Leonis to a point in space that is approximately 1° east at the start. One can easily verify this in the next step.

6. Before starting put the cursor anywhere in the desktop and right-click on the mouse. Click Find. Type "Sun" in the "Search For" description line in the Find window. Click "Find Now." In the "Result" tab, the "Object Information Report" is displayed with the following items: the "Object Name," "RA," and "Dec," "Azimuth and Altitude," "Rise, Transit, and Set Times," the "Constellation" in which the Sun is located and its "Geocentric Longitude." [Center] the Sun. Record the Sun's Geocentric Longitude for August 21st: _____° _____'

7. Next, click [Step Forward] again. Repeat Step 6 to see how much the Sun has moved. Record the Sun's Geocentric Longitude for August 22nd. _____° _____'

   This is how far the Sun has moved relative to the background stars (its long-term motion).

8. Do it once again and record the Geocentric Longitude for August 23rd. _____° _____'

9. To observe this motion for a period of several days reset the time and date back to August 21st at 1:45 PM DST. Now, click either [Step Forward] or [Go Forward].

© 2011 Cengage Learning. All Rights Reserved. May not be scanned, copied or duplicated, or posted to a publicly accessible website, in whole or in part.

10.   Record the Geocentric Longitude for September 21st. _____ ° _____ ' _____ "

11.   Compute the difference between September 21st and August 21st. _____ ° _____ '
       Is it about 30°? _____

12.   Close the file. It is not necessary to save this Sky file.

       The Moon, in contrast, revolves around Earth and moves approximately its own angular diameter (½°) per hour. This eastward motion of the Moon amounts to an angle of about 12° per day with respect to the Sun and 13° with respect to the stars. The motion of the Moon can be easily seen in the Sky file named ***MoonsEMotion.skyx***. **Exercise C** illustrates the Moon's eastward motion projected against the distant background stars.

Exercise C: Measuring the Moon's Motion in *TheSkyX*

1.   Run *TheSkyX* and open the file ***Normal.skyx***.

2.   Open the Sky file named ***MoonsEMotion.skyx***.

3.   Click Go Forward on the Time Skip toolbar. The Time Flow rate is again set at one sidereal day. Now, click the Stop button.

4.   To continue, reset the date and time back to the starting time and date, that is, back to January 24th at 8:00 PM STD or reopen the file. This is done the same way as was described in **Exercise B**.

5.   Once, again one can measure how far the Moon moves each day in the sky by clicking the Step Forward button. Do it the same way as it was done for the Sun in **Exercise B**. In this exercise, one should notice at the beginning that the Moon starts at a point in space just west and below α Tauri (Aldebaran). After clicking Step Forward , the motion results in an angle of approximately 13° eastward. This can easily be verified in the next step.

6.   Before starting put the cursor anywhere in the desktop and right-click on the mouse. Click Find. Type "Moon" in the "Search For" description line in the Find window. Click "Find Now." In the "Result" tab, the "Object Information Report" is displayed with the following information: the "Object Name," "RA," and "Dec," "Azimuth and Altitude," "Rise, Transit, and Set Times," the "Constellation" in which the Moon is located and its "Ecliptic Longitude," its "Phase" (in%), and the "Moon's Age" (past Full) .

      Center the Moon. Record the Moon's Ecliptic Longitude for January 24th: _____ ° _____ '

7.   Next, click the Step Forward button again. Repeat Step 6 to see how much the Moon has moved. Record the Moon's Ecliptic Longitude for January 25th: _____ ° _____ '

      This is how far the Moon has moved relative to the background stars (its long-term motion).

8.   Do it once again and record the Moon's Ecliptic Longitude for January 26th: _____ ° _____ '

© 2011 Cengage Learning. All Rights Reserved. May not be scanned, copied or duplicated, or posted to a publicly accessible website, in whole or in part.

9. To observe this motion for a period of several days reset the time back to January 24th at

   8:00 PM STD or reopen the file. Continue clicking 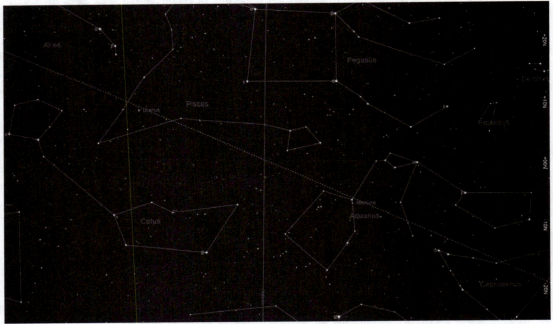 Step Forward or Go Forward.

10. Record the Moon's Ecliptic Longitude for February 20th. _____° _____'

11. Compute the difference between January 24th and February 20th. _____° _____'
    Is it about 360°? _____

12. Close the file. It is not necessary to save this Sky file.

   When using the Moon as a reference object, however, its eastward motion presents us with some minor problems in timekeeping. It moves farther eastward, in angle, through the sky than any other object.

   If we choose a star, in contrast, then the question becomes, which star should we use as our reference object? One might think this isn't a problem, but two people may not agree on the same star to use as the reference object.

   To eliminate confusion in choosing a particular star, astronomers have selected a well-defined point in space. Its position in the sky and its motion are well known. The reference object astronomers use to keep track of time by the stars is the *vernal equinox*. **Figure 7-1** illustrates where the vernal equinox is located in space. It is at the intersection of the ecliptic (dotted line) and celestial equator (0^H, 0° line) where the Sun appears to cross the celestial equator in March. Its current location, as shown in **Figure 7-1**, is directly on the local celestial meridian in the constellation of Pisces. One must turn on the constellation boundaries to confirm this observation.

Figure 7-1 Location of the Vernal Equinox

   We simply don't ignore the fact that the Sun is probably the most obvious object to use for timekeeping. In fact, the time interval between two successive passages of the center of the Sun's disk across your celestial meridian defines an *apparent solar day*. Remember that the Sun only appears to move east to west across the sky! This motion is caused by Earth's rotation and has a period of time equal to 24^H 00^M 00^S.

   When Earth's motion around the Sun is factored in, stars and the vernal equinox appear to move westward ~1° per day with respect to the Sun. Astronomers define the time interval

© 2011 Cengage Learning. All Rights Reserved. May not be scanned, copied or duplicated, or posted to a publicly accessible website, in whole or in part.

between two successive passages of the vernal equinox across an observer's local celestial meridian as a *sidereal day*. This period of time is only $23^H 56^M 4.09^S$. A sidereal day is approximately 4 minutes shorter than a solar day.

The difference in time between a solar day and a sidereal day is actually caused by both Earth's rotation and its orbital motion around the Sun. The sidereal day reflects the *true* rotation period of Earth. This is the actual length of time it takes Earth to rotate once on its axis. The solar day, in contrast, is a time interval that includes both the rotation and revolution of Earth. Each day as Earth rotates once on its axis; it moves approximately 1° in its orbit. This 1° difference amounts to a difference of 4^M in time. Review the angle-to-time conversion discussed in **Chapter 5**.

Local Time

Timekeeping involves measuring an angle between some reference object in the sky and a specific celestial meridian. In fact, it can be measured from any celestial meridian at any location on Earth. The angle itself is measured westward along the celestial equator to the hour circle of the reference object. Recall that an hour circle is a great circle on the celestial sphere passing through the celestial poles and an object on the celestial sphere. This angle is measured in hours and minutes of time, not in degrees and minutes of angle. The angle is appropriately defined as the *hour angle* of the reference object.

Anyone who can tell time already knows how to measure hour angles. That is, when one looks at a watch or a clock on the wall, they are measuring the hour angle of the Sun. This is done from a particular celestial meridian at some location on Earth. Thus, the terms *A.M.* and *P.M.* mean *ante* (before) and *post* (past) meridian, respectively.

For practical reasons, those time systems based on the Sun use the hour angle of the Sun relative to an observer's celestial meridian and *add* 12 hours. This procedure is used so that the calendar date changes at midnight instead of at noon each day. Time kept by a local celestial meridian is defined as *local solar time* or simply *local time*. It is necessary to keep track of local time for events such as the apparent sunrise and sunset or the transits of celestial objects.

Standard (Zone) Time

The system of standard time was established because observers at different longitudes on Earth do not have the same reference line (celestial meridian) in the sky. It is more convenient to standardize time at several locations on Earth rather than keeping track of their own local time. Cities separated by only a few hundred miles could keep the same time on their clocks. In the United States, during the 19th century, railroad companies were instrumental in the implementation of standard time that allowed train schedules to be put into operation within certain regions of the country.

Standard time is time kept by an observer located at a "standard longitude" on the surface of Earth. A standard longitude is a longitude circle that is a multiple of 15° in angle (30°E, 15°E, 0°, 15°W, 30°W). The hour angle of the Sun kept at one of these longitudes is defined as *standard* or "*zone*" *time*. These regions span a zone 15° wide on the surface of Earth and are centered on a longitude circle that is a multiple of 15°.

There are four time zones across the United States. The Eastern Time Zone is centered on the 75th longitude circle west of Greenwich, England, and is very close to Philadelphia, Pennsylvania. The Central Time Zone is centered on the 90th longitude circle west of Greenwich and is very close to Memphis, Tennessee. The Mountain Time Zone is centered on the 105th longitude circle west of Greenwich and is very close to Denver, Colorado. The Pacific Time Zone is centered on the 120th longitude circle west of Greenwich and is very close to Vandenberg Air Force Base, California. Although observers throughout the zone have different local times, they do keep the same zone or standard time on their clocks. **Figure 7-2** depicts the four time zones across the United States. The authors gratefully acknowledge the source of this map as a courtesy of *http://www.theodora.com/maps* and it is used with the permission of that website.

© 2011 Cengage Learning. All Rights Reserved. May not be scanned, copied or duplicated, or posted to a publicly accessible website, in whole or in part.

Figure 7-2 Standard Time Zones Across the United States

One thing should be noted concerning Standard Time. Local events such as the apparent rising or setting of the Sun, Moon, and the planets do not occur at the same instant across the zone. Observers located on the *eastern edge* of the zone will observe events such as the Sun's apparent rising *before* an observer located at the *center* of the zone. Whereas, observers located on the *western edge* of the zone will observe events such as the Sun's apparent rising *after* an observer located at the *center* of the zone. Only their watch times will agree!

Local Time, in contrast, is the time that is kept at one's own location on Earth. It measures the hour angle of the apparent Sun from your local celestial meridian. On a sundial, this is the time that is measured with the *gnomon*, the "marker." The shadow of the gnomon is cast on a plate with hours of time marked on it. It is necessary to keep track of this time so that events such as the apparent rising, transit, and setting of objects visible from your location can be determined. **Figure 7-3** displays a typical sundial. The Sun's shadow is just to the left of the gnomon making the local time approximately 10:10 AM.

Figure 7-3 A Typical Sundial (Image Courtesy of Author)

© 2011 Cengage Learning. All Rights Reserved. May not be scanned, copied or duplicated, or posted to a publicly accessible website, in whole or in part.

Corrections in Time in *TheSkyX*

TheSkyX computes the local time for events that are visible from any observer's location. In order to compute the local time of an astronomical event, one must know the longitude of the observer's location to arcminute precision. The local time is then calculated by either *adding* (if west of the standard longitude) or by *subtracting* (if east of the standard longitude) a time correction to the standard longitude's time of the event in question. This amounts to determining the difference in angle between your longitude circle and the standard longitude circle of your zone. The difference in time is calculated from the difference in angles and applied to the standard time as previously described.

An observer at the Ball State Observatory, longitude = 85° 24′ 38″, is exactly 10° 24′ 38″ west of the 75th longitude circle, which is the central longitude for the Eastern Standard Time zone. This angular difference amounts to a difference of about $41^M 38^S$ in time. This difference in time is then added to the "celestial event" time that occurs at the 75th longitude. In other words, "local noon" occurs at the Ball State Observatory approximately 12:41PM EST every day. However, since Indiana, or most of it, now observes Daylight Savings Time part of the year this is not the case from early spring to late autumn. This is discussed in more detail in "**Determining Solar Transit Times in *TheSkyX***" section later on this chapter.

The reason that it is necessary to keep track of local time is to determine precisely when astronomical objects such as the Sun, Moon, planets, appear to rise, transit your local celestial meridian, and set each day. This consequently determines a time when astronomical twilight begins or ends at a particular location on Earth. Astronomers define *astronomical twilight* as the moment when the center of the Sun's apparent disk is 18° below the astronomical horizon. Astronomers have adopted this definition to determine the instant when the sky is totally dark, unlike civil twilight, which is when the Sun's apparent disk is only 6° below the horizon. Civil twilight is what most states in the United States use to establish a time for turning car headlights on and off.

Universal Time

The last time system involving the Sun that needs addressing is the one that keeps track of time for the world. This time system is kept at the prime meridian (0° longitude) in Greenwich, England. It used to be appropriately called Greenwich Mean Time but today is now known as *Universal Time* (UT). All clocks in the world are synchronized with the "world" clock located at the Royal Naval Observatory in Greenwich, England.

Events such as elongation of planets, lunar phases, lunar eclipses, solar eclipses, or the beginning of the seasons have their times determined in universal time. These times are then later converted to the observer's standard time. Only events involving apparent rising, transits, or setting times of astronomical objects actually need corrections made from standard time to local time.

Events involving phenomena of the Sun, Moon, and planets are found in many resources. One of the most popular resources is *The Astronomical Almanac* that is issued annually by the United States Naval Observatory by direction of the Secretary of the Navy and under the authority of Congress. The U.S. Government Printing Office in Washington, D.C. publishes this almanac. Another excellent resource, for the amateur astronomer, is the *Observer's Handbook*. This resource is also issued annually by the Royal Astronomical Society of Canada. It is printed in Canada by the University of Toronto Press.

Astronomers use universal time extensively rather than using local or zone time. It makes sense to do so, because observations are made from observatories located in remote places on Earth. All observatory clocks worldwide are synchronized to universal time.

© 2011 Cengage Learning. All Rights Reserved. May not be scanned, copied or duplicated, or posted to a publicly accessible website, in whole or in part.

Sidereal Time

The last time system, and perhaps the most important one to astronomers, is *sidereal time*. *Sidereal Time* is defined as the *hour angle* of the vernal equinox. Why is sidereal time so important? Sidereal time is time kept by the stars.

When planning observing sessions, it is very important to know the sidereal time at one's location on Earth. Knowing the sidereal time assists observers in determining what is above the local horizon at any location and consequently what is visible at that time! To accomplish this, the hour angle of the vernal equinox must be determined at the time of the planned observation.

Because sidereal time is defined as the hour angle of the vernal equinox, one must determine where the vernal equinox is located with respect to the local celestial meridian. The hour angle is measured westward, from the local celestial meridian along the celestial equator, to the hour circle passing through the vernal equinox.

Therefore, if the angular distance measured westward along the celestial equator from the local celestial meridian to the hour circle passing through the vernal equinox is known, then, by definition, *is* the sidereal time at a particular location. Measurement of the hour angle of the vernal equinox is illustrated in **Figure 7-4**. The sidereal time in this figure is approximately 03:00, which means the vernal equinox is about three hours west of the local celestial meridian.

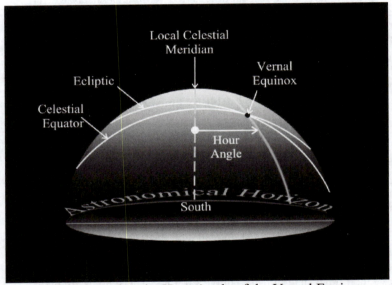

Figure 7-4 Measuring the Hour Angle of the Vernal Equinox

Measuring the hour angle of the vernal equinox from one's location on Earth essentially amounts to determining what the right ascension of objects are on the local celestial meridian. Remember, the coordinate of right ascension is the angle measured eastward from the vernal equinox (R.A. = 0^H) along the celestial equator to the hour circle passing through any celestial object. If the right ascension on the local celestial meridian is known, then the sidereal time is also known!

Determining Longitudinal Differences in Time in *TheSkyX*

Exercises **D**, **E**, **F**, **G** and **H** are designed to demonstrate how *TheSkyX* is used to determine when an object, like a star, transits the local celestial meridian. They have been designed to confirm longitudinal effects of the observer's location on Earth. These effects are observed when an object's transit time across the local celestial meridian is determined. To illustrate that there really is a difference in time, *TheSkyX's* date and time must be set *exactly* in these exercises.

© 2011 Cengage Learning. All Rights Reserved. May not be scanned, copied or duplicated, or posted to a publicly accessible website, in whole or in part.

Exercise D: Observing Longitudinal Differences in Time

1. Run *TheSkyX* and open the file ***Normal.skyx***.

    Set *TheSkyX* as follows:

    Location: Ball State Observatory
    Elevation: 322 meters
    Date: August 25, 2014
    Time: 9:00 PM DST   Time Zone: -5.00 hours
    Daylight Saving option: "U.S. and Canada"

    Orientation: Zoom To: 50° field of view
    Orientation: Navigate – Click the Sky Chart Center Tab

    Enter: R.A.: $18^H 30^M 00^S$ Declination: +40° 30′ 00″
    Epoch: 2014   Click Center on RA/Dec

    Turn off the Equatorial Grid if displayed and, and turn on the Horizon Grid.

    Go to the "Display" menu on the Standard toolbar at the top of the sky desktop window and click on "Labels." Turn off the "Common Star Names" and Click on the "Bayer Letter Designation" box. This ensures that the Bayer letters are displayed in the sky window.

    Find α Lyrae and determine its apparent transit time. Remember? To do this, put the cursor α Lyrae left-click the mouse button.

    Proper name: _____

    Transit time: _____:_____

    R.A. (2014): _____   Declination (2014): _____

    Altitude: _____   Azimuth: _____

2. Change the setup as follows:

    Location: Philadelphia, PA
    Elevation: 11 meters
    Date: August 25, 2014
    Time: 9:00 PM DST   Time Zone: -5.00 hours
    Daylight Saving option: "U.S. and Canada"

    Orientation: Zoom To: 50° field of view
    Orientation: Navigate – Click the Sky Chart Center Tab

    Enter: R.A.: $18^H 30^M 00^S$ Declination: +40° 30′ 00″
    Current Epoch: 2014   Click Center on RA/Dec

    Locate α Lyrae. Determine its apparent transit time.

    Transit time: _____:_____

    R.A. (2014): _____   Declination (2014): _____

    Altitude: _____   Azimuth: _____

© 2011 Cengage Learning. All Rights Reserved. May not be scanned, copied or duplicated, or posted to a publicly accessible website, in whole or in part.

Note the differences between the results of Part 2 and Part 1, and record them in the following spaces:

Difference in transit times: _____ : _____

Were there any differences in the altitude and azimuth? _____ . What were they?

Difference in altitudes: _____   Difference in azimuths: _____

Why do the results for Part 2 and Part 1 differ? _____
_____

3. Change the setup as follows:

Longitude: 75° 00′ 00″
Elevation: 100 meters
Latitude: 40° 11′ 58″ (Same as Ball State Observatory)
Date: August 25, 2014
Time: 9:00 PM DST   Time Zone: -5.00 hours
Daylight Saving option: "U.S. and Canada"

Orientation: Zoom To: 50° field of view
Orientation: Navigate – Click the Sky Chart Center Tab

Enter: R.A.: $18^H 30^M 00^S$ Declination: +40° 30′ 00″
Current Epoch: 2014   Click Center on RA/Dec

Locate α Lyrae. Determine its apparent transit time.

Transit time: _____ : _____

R.A. (2014): _____   Declination (2014): _____

Altitude: _____   Azimuth: _____

Note the differences between the results of Part 3 and Part 1, and record them in the following spaces:

Difference in transit times: _____ : _____

Were there any differences in the altitude and azimuth? _____ . What were they?

Difference in altitudes: _____   Difference in azimuths: _____

Why do the results of Part 3 and Part 1 differ? _____
_____

4. Close the file and do not save, or continue with **Exercise E**.

Exercise E: Observing Longitudinal Differences in Time

1. Reset *TheSkyX* as follows:

Location: Ball State Observatory
Elevation: 322 meters
Date: April 20, 2015

© 2011 Cengage Learning. All Rights Reserved. May not be scanned, copied or duplicated, or posted to a publicly accessible website, in whole or in part.

Time: 8:20 AM DST   Time Zone: -5.00
Daylight Saving option: "U.S. and Canada"

Orientation: Zoom To: 50° field of view
Orientation: Navigate – Click the Sky Chart Center Tab

Enter: R.A.: 20^H 45^M 30^S Declination: +45°30′ 15″
Current Epoch: 2015   Click Center on RA/Dec

Turn the Equatorial Grid off, and turn on the Horizon Grid.

Make sure that the "Common Star Names" are turned off and that the Bayer letters are again displayed in the sky window.

The sky may need to be rotated so that North is oriented toward the top of the desktop window. To do this, click the "Orientation" menu located on the Standard toolbar at the top of the desktop window. Now, click "Navigate" and then the "Rotation" tab. Click on the line with the left mouse button and hold down and move the line so that it is set to 0.00 degrees, click "Set." The Rotation tool is displayed in **Figure 7-5** before and after the change in orientation of the sky.

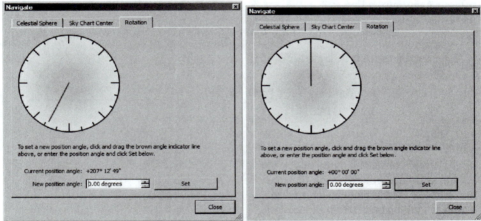

Figure 7-5 Rotation Tool in *TheSkyX* Navigate Menu Before (Left) and After (Right) the Orientation Change

Locate α Cygni. Find it's apparent rising, transit, and setting times.

Proper name: _____

Rising time: _____:_____ Transit time: _____:_____ Setting time: _____:_____

R.A. (2015): _____ Declination (2015): _____

Altitude: _____ Azimuth: _____

2. Change the setup as follows:

Location: Philadelphia, PA
Elevation: 100 meters
Date: April 20, 2015
Time: 8:20 AM DST   Time Zone: -5.00
Daylight Saving option: "U.S. and Canada"

© 2011 Cengage Learning. All Rights Reserved. May not be scanned, copied or duplicated, or posted to a publicly accessible website, in whole or in part.

Orientation: Zoom To: 50° field of view
Orientation: Navigate – Click the Sky Chart Center Tab

Enter: R.A.: 20^{H}45^{M}30^S Declination: +45°30′ 15″
Current Epoch: 2015   Click Center on RA/Dec

The sky may need to be rotated again so that North is oriented toward the top of the desktop window. To do this, follow the directions described in Part 1.

Locate α Cygni. Find it's apparent rising, transit, and setting times.

Rising time: _____:_____ Transit time: _____:_____ Setting time: _____:_____

R.A. (2015): _____ Declination (2015): _____

Altitude: _____ Azimuth: _____

Note the differences between the results of Part 2 and Part 1, and record them in the following spaces:

Differences in rising times: _____ Differences in transit times: _____

Differences in setting times: _____

Were there any differences in the altitude and azimuth? _____. What were they?

Difference in altitudes: _____ Difference in azimuths: _____

Why do the results between Part 2 and Part 1 differ? _____
_____

3. Change the setup as follows:

Longitude: 75° 00′ 00″
Elevation: 100 meters
Latitude: 40° 11′ 59″ (Same as Ball State Observatory)
Date: April 20, 2015
Time: 8:20 AM DST   Time Zone: -5.00
Daylight Saving option: "U.S. and Canada"

Orientation: Zoom To: 50° field of view
Orientation: Navigate – Click the Sky Chart Center Tab

Enter: R.A.: 20^{H}45^{M}30^S Declination: +45°30′ 15″
Current Epoch: 2015   Click Center on RA/Dec

Locate α Cygni. Find it's apparent rising, transit, and setting times.

Rising time: _____:_____ Transit time: _____:_____ Setting time: _____:_____

R.A. (2015): _____ Declination (2015): _____

Altitude: _____ Azimuth: _____

Note the differences between the times in Part 3 and Part 1, and record them in the following spaces:

Differences in rising times: _____ Differences in transit times: _____

© 2011 Cengage Learning. All Rights Reserved. May not be scanned, copied or duplicated, or posted to a publicly accessible website, in whole or in part.

Differences in setting times: _____

Were there any differences in the altitude and azimuth? _____ . What were they?

Difference in altitudes: _____   Difference in azimuths: _____

Why do the results of Part 3 and Part 1 differ? _____
_____

4.   Close the file and do not save, or continue with **Exercise F**.

Exercise F: Observing Longitudinal Differences in Time

1.   Reset *TheSkyX* as follows:

Location: Ball State Observatory
Elevation: 322 meters
Date: August 31, 2020
Time: 9:00 PM DST   Time Zone: -5.00
Daylight Saving option: "U.S. and Canada"

Orientation: Zoom To: 50° field of view
Orientation: Navigate – Click the Sky Chart Center Tab

Enter: R.A.: $18^H 56^M 32^S$ Declination: +40° 11′ 58″
Current Epoch: 2020   Click Center on RA/Dec

Turn the Equatorial Grid off, and turn on the Horizon Grid.

Make sure that the "Common Star Names" are turned off and that the Bayer letters are again displayed in the sky window.

The sky may need to be rotated again so that North is oriented toward the top of the desktop window. To do this, follow the directions described in **EXERCISE E**.

Locate α Lyrae. Find it's apparent rising, transit, and setting times.

Proper name: _____

Rising time: _____:_____ Transit time: _____:_____ Setting time: _____:_____

R.A. (2020): _____ Declination (2020): _____

Altitude: _____ Azimuth: _____

2.   Change the setup as follows:

Location: Memphis, TN
Elevation: 100 meters
Date: August 31, 2020
Time: 9:00 PM DST   Time Zone: -6.00
Daylight Saving option: "U.S. and Canada"

Orientation: Zoom To: 50° field of view
Orientation: Navigate – Click the Sky Chart Center Tab

© 2011 Cengage Learning. All Rights Reserved. May not be scanned, copied or duplicated, or posted to a publicly accessible website, in whole or in part.

Enter: R.A.: $18^H 56^M 32^S$ Declination: $+40° 11' 58''$
Current Epoch: 2020   Click Center on RA/Dec

Locate α Lyrae. Find it's apparent rising, transit, and setting times.

Rising time: _____:_____   Transit time: _____:_____   Setting time: _____:_____

R.A. (2020): _____   Declination (2020): _____

Altitude: _____   Azimuth: _____

Note the differences between the results of Part 2 and Part 1, and record them in the following spaces:

Differences in rising times: _____   Differences in transit times: _____

Differences in setting times: _____

Were there any differences in the altitude and azimuth? _____. What were they?

Difference in altitudes: _____   Difference in azimuths: _____

Why do the results between Part 2 and Part 1 differ? _____
_____

There are a few problems that arise when doing this exercise. One minor issue is the difference in latitudes of the two locations. However, the major issue here is the time zones. If this is issue is not taken into account, the results are misleading and confusing. Even though each location observes Daylight Savings time, Ball State Observatory is in the Eastern Time Zone while Memphis, Tennessee is in the Central Time Zone. Memphis is located at 90° 03' west of Greenwich while Ball State Observatory is located at 85° 24' west of Greenwich. The difference in the two locations' longitudes is 4° 39'. This amounts to a difference of approximately 19^M in time.

Remember, a time zone across the United States is 15° or 1^H wide from one standard longitude center to the next. Therefore across the zone of time in question, that is, from 75° to 90°, Ball State Observatory adds 41^M to transit times determined at the 75th longitude and subtracts 19^M from transits times determined at the 90th longitude. If one then adjusts the difference of one hour (due to time zone difference) in the times recorded, the actual Transit times will be correct. That is, whenever a transit occurs at Ball State Observatory's celestial meridian, the time that it takes place at the 90° longitude celestial meridian is approximately 19^M later. There are a couple of ways to make this adjustment in *TheSkyX* software. One way is to go to the Custom Location tab and changing the Time Zone from -6 to -5 hours for Memphis, Tennessee. This will adjust the rising, transit, and setting times accordingly. Therefore, making this change, the corrected transit time in Memphis becomes 21:55, which is 19^M later than the transit time that occurred at Ball State Observatory. In Part 3 which follows it is suggested to set the Time Zone to -5 hours for the 90th longitude. See if this makes a difference now.

3. Change the setup as follows:

Longitude: 90° 00' 00''
Elevation: 100 meters
Latitude: 40° 11' 58'' (Same as Ball State Observatory)
Date: August 31, 2020
Time: 9:00 PM DST   Time Zone: -5.00
Daylight Saving option: "U.S. and Canada"

Orientation: Zoom To: 50° field of view
Orientation: Navigate – Click the Sky Chart Center Tab

© 2011 Cengage Learning. All Rights Reserved. May not be scanned, copied or duplicated, or posted to a publicly accessible website, in whole or in part.

Enter: R.A.: $18^H 56^M 32^S$ Declination: +40° 11′ 58″
Current Epoch: 2020   Click Center on RA/Dec

Locate α Lyrae. Find it's apparent rising, transit, and setting times.

Rising time: _____:_____   Transit time: _____:_____   Setting time: _____:_____

R.A. (2020): _____   Declination (2020): _____

Altitude: _____   Azimuth: _____

Note the differences between the results of Part 3 and Part 1, and record them in the following spaces:

Differences in rising times: _____   Differences in transit times: _____

Differences in setting times: _____

Were there any differences in the altitude and azimuth? _____. What were they?

Difference in altitudes: _____   Difference in azimuths: _____

Why do the results between Part 3 and Part 1 differ? _____
_____

4. Close the file and do not save, or continue with **Exercise G**.

Exercise G: Observing Longitudinal Differences in Time

1. Reset *TheSkyX* as follows:

Location: Ball State Observatory
Elevation: 322 meters
Date: December 31, 2035
Time: 09:30 PM STD Time Zone: -5.00
Daylight Saving option: "Not Observed"

Orientation: Zoom To: 50° field of view
Orientation: Navigate – Click the Sky Chart Center Tab

Enter: R.A.: $5^H 15^M 00^S$ Declination: +20° 30′ 15″
Current Epoch: 2035   Click Center on RA/Dec

Turn the Equatorial Grid off, and turn on the Horizon Grid.

Make sure that the "Common Star Names" are turned off and that the Bayer letters are again displayed in the sky window.

The sky may need to be rotated again so that North is oriented toward the top of the desktop window. To do this, follow the directions described in **EXERCISE E**.

Locate α Tauri. Find it's apparent rising, transit, and setting times.

Proper name: _____

Rising time: _____:_____   Transit time: _____:_____   Setting time: _____:_____

© 2011 Cengage Learning. All Rights Reserved. May not be scanned, copied or duplicated, or posted to a publicly accessible website, in whole or in part.

R.A. (2035): _____ Declination (2035): _____

Altitude: _____ Azimuth: _____

2. Change the setup as follows:

Location: Golden, CO
Elevation: 1,730 meters
Date: December 31, 2035
Time: 09:30 PM STD Time Zone: -7.00
Daylight Saving option: "Not Observed"

Orientation: Zoom To: 50° field of view
Orientation: Navigate – Click the Sky Chart Center Tab

Enter: R.A.: $5^H 15^M 00^S$ Declination: +20° 30′ 15″
Current Epoch: 2035   Click Center on RA/Dec

Locate α Tauri. Find it's apparent rising, transit, and setting times.

Rising time: _____:_____ Transit time: _____:_____ Setting time: _____:_____

R.A. (2035): _____ Declination (2035): _____

Altitude: _____ Azimuth: _____

Note the differences between the results of Part 2 and Part 1, and record them in the following spaces:

Differences in rising times: _____ Differences in transit times: _____

Differences in setting times: _____

Were there any differences in the altitude and azimuth? _____. What were they?

Difference in altitudes: _____ Difference in azimuths: _____

Why do the results between Part 2 and Part 1 differ? _____
_____

Similar problems arise when doing this exercise too as did in **Exercise F**. Again, one issue is the difference in latitudes of the two locations. However, the major issue once again is the time zones. And, as before, if this is not taken into account, the results are misleading and confusing. Even though each location observes Standard time, Ball State Observatory is in the Eastern Time Zone while Golden, Colorado is in the Mountain Time Zone. Golden is located at 105° 13' west of Greenwich while Ball State Observatory is located at 85° 24' west of Greenwich. The difference in the two locations' longitudes is 19° 49'. This amounts to a difference of approximately $1^H 19^M$ in time. This is how much later that α Tauri crosses the local celestial meridian in Golden, Colorado relative to the time it occurs at Ball State Observatory. In Part 3 which follows it is suggested to set the Time Zone to -5 hours for the 105th longitude. See if this makes a difference now.

3. Change the setup as follows:

Longitude: 105° 00′ 00″
Elevation: 1,676 meters
Latitude: 40° 11′ 58″ (Same as Ball State Observatory)
Date: December 31, 2035

© 2011 Cengage Learning. All Rights Reserved. May not be scanned, copied or duplicated, or posted to a publicly accessible website, in whole or in part.

Time: 09:30 PM STD   Time Zone: -5.00
Daylight Saving option: "Not Observed"

Orientation: Zoom To: 50° field of view
Orientation: Navigate – Click the Sky Chart Center Tab

Enter: R.A.: $5^H 15^M 00^S$ Declination: +20° 30′ 15″
Current Epoch: 2035   Click Center on RA/Dec

Locate α Tauri. Find it's apparent rising, transit, and setting times.

Rising time: _____:_____   Transit time: _____:_____   Setting time: _____:_____

R.A. (2035): _____   Declination (2035): _____

Altitude: _____   Azimuth: _____

Note the differences between the results of Part 3 and Part 1, and record them in the following spaces:

Differences in rising times: _____   Differences in transit times: _____

Differences in setting times: _____

The difference in transit times should be close to the value mentioned at the end of Part 2.

Were there any differences in the altitude and azimuth? _____. What were they?

Difference in altitudes: _____   Difference in azimuths: _____

Why do the results between Part 3 and Part 1 differ? _____
_____

4. Close the file and do not save, or continue with **Exercise H**.

Exercise H: Observing Longitudinal Differences in Time

1. Reset *TheSkyX* as follows:

    Location: Denver, CO
    Elevation: 1609 meters
    Date: November 26, 2047
    Time: 9:30 PM STD   Time Zone: -7.00
    Daylight Saving option: "Not Observed"

    Orientation: Zoom To: 50° field of view
    Orientation: Navigate – Click the Sky Chart Center Tab

    Enter: R.A.: $05^H 58^M 00^S$ Declination: +7° 25′ 00″
    Current Epoch: 2047   Click Center on RA/Dec
    Turn the Equatorial Grid off, and turn on the Horizon Grid.

    Make sure that the "Common Star Names" are turned off and that the Bayer letters are again displayed in the sky window.

    Locate α Orionis. Find its apparent transit time.

© 2011 Cengage Learning. All Rights Reserved. May not be scanned, copied or duplicated, or posted to a publicly accessible website, in whole or in part.

Proper name: _____

Rising time: _____:_____ Transit time: _____:_____ Setting time: _____:_____

R.A. (2047): _____ Declination (2047): _____

Altitude: _____ Azimuth: _____

2. Change the setup as follows:

Location: Salt Lake City, UT
Elevation: 1288 meters
Date: November 26, 2047
Time: 9:30 PM STD   Time Zone: -7.00
Daylight Saving option: "Not Observed"

Orientation: Zoom To: 50° field of view
Orientation: Navigate – Click the Sky Chart Center Tab

Enter: R.A.: $05^H 58^M 00^S$ Declination: +7° 25′ 00″
Current Epoch: 2047   Click Center on RA/Dec

Locate α Orionis. Find its apparent transit time.

Rising time: _____:_____ Transit time: _____:_____ Setting time: _____:_____

R.A. (2047): _____ Declination (2047): _____

Altitude: _____ Azimuth: _____

Note the differences between the results of Part 2 and Part 1 and record them in the following spaces:

Differences in rising times: _____   Differences in transit times: _____
Differences in setting times: _____

Were there any differences in the altitude and azimuth? _____. **What were they?**

Difference in altitudes: _____   Difference in azimuths: _____

Why do the results between Part 2 and Part 1 differ? _____
_____

3. Change the setup as follows:

Longitude: 105° 00′ 00″
Elevation: 1300 meters
Latitude: 40° 11′ 58″ (Same as Ball State Observatory)
Date: November 26, 2047
Time: 9:30 PM STD   Time Zone: -7.00
Daylight Saving option: "Not Observed"

Orientation: Zoom To: 50° field of view
Orientation: Navigate – Click the Sky Chart Center Tab

© 2011 Cengage Learning. All Rights Reserved. May not be scanned, copied or duplicated, or posted to a publicly accessible website, in whole or in part.

Enter: R.A.: $05^H 58^M 00^S$ Declination: +7° 25′ 00″
Current Epoch: 2047   Click Center on RA/Dec

Locate α Orionis. Find its apparent transit time.

Rising time: _____ : _____   Transit time: _____ : _____   Setting time: _____ : _____

R.A. (2047): _____   Declination (2047): _____

Altitude: _____   Azimuth: _____

Note the differences between the results of Part 3 and Part 1 **and record them in the following spaces:**

Differences in rising times: _____   Differences in transit times: _____
Differences in setting times: _____

Were there any differences in the altitude and azimuth? _____. **What were they?**

Difference in altitudes: _____   Difference in azimuths: _____

Why do the results between Part 3 and Part 1 differ? _____
_____

4. Close the file and do not save it.

     If you answered that the results differed between Part 1 and Part 2, and Part 1 and Part 3, in the preceding exercises because of the different longitudes of the observers, then you are correct!

Determining Sidereal Times in *TheSkyX*

     Exercises **I**, **J**, **K**, **L**, and **M** are designed to assist in determining the sidereal time. One can determine the sidereal time for any location, for any time of day, and for day of the year. Settings for time zone setting, elevation, and Daylight Savings Time (except **Exercise M**) remain the same as in the previous examples. In actuality, these settings would change to reflect a particular location on Earth. The settings for these exercises will not be changed, in order to demonstrate the longitudinal effects.

Exercise I: Determining the Sidereal Time at Ball State Observatory

1. Run *TheSkyX* and open the file *Normal.skyx* or set *TheSkyX* as follows:

Location: Ball State Observatory
Elevation: 322 meters
Date: February 2, 2015
Time: 9:00 PM STD   Time Zone: -5.00 hours
Daylight Saving option: "Not Observed"

Orientation: Zoom To: 50° field of view
Orientation: Navigate – Click the Sky Chart Center Tab

Enter: R.A.: $05^H 15^M 00^S$ Declination: +49° 30′ 00″
Current Epoch: 2015   Click Center on RA/Dec

© 2011 Cengage Learning. All Rights Reserved. May not be scanned, copied or duplicated, or posted to a publicly accessible website, in whole or in part.

Turn both the Equatorial and Horizon Grids off and display the "Chart Status" window. Go to the Standard toolbar at the top of the sky window and click "Display" then click on "Chart Status." If the cursor's RA/Dec location is not displayed, follow the directions described in **Part 1A**. The Chart Status window may be modified to display a variety of parameters in *TheSkyX* window on the desktop such as: the Location, Date, Time, and the Cursor RA/Dec location in the sky desktop window.

What is the name of the red line that passes through the field of view?

_____   _____   _____

Find the coordinate readout located in the "Chart Status" box in the window. It is similar to the one displayed in **Figure 7-6**.

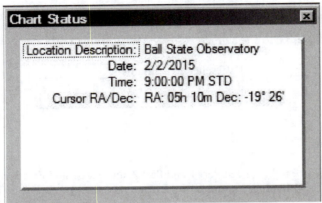

Figure 7-6 Right Ascension Readout in Chart Status Box in *TheSkyX*

It displays Location, the Date and Time, and the right ascension and declination of the cursor's position in the window. If one puts the cursor on the vertical red line (local celestial meridian for an observer at Ball State Observatory), then this displays the coordinates of any object located on the local celestial meridian at any instant in time. The sidereal time *is*, by definition, the right ascension of any object on the local celestial meridian. Put the cursor on the red line! What is the sidereal time at the Ball State Observatory?

Sidereal Time = ____ : ____ : ____

There is another way to estimate the sidereal time at this location. Are there any bright stars located close to the local celestial meridian? If so does it have a proper name?

Proper name: _____

What are its equatorial coordinates?

R.A. (2015): ____ H ____ M ____ S   Declination (2015): __ ____ ° ____ ' ____ "

Can they help anyone to determine the sidereal time? _____ . If so, how? _____
_____ .

© 2011 Cengage Learning. All Rights Reserved. May not be scanned, copied or duplicated, or posted to a publicly accessible website, in whole or in part.

1A. If the cursor's right ascension and declination are not displayed in the Chart Status window, then one must complete the follow in order to display them.

   a.   Click "Tools" on the Standard toolbar at the top of the sky window.

   b.   Click on "Preferences."

   c.   Click on the "Chart Status" icon. **Figure 7-7** displays the Chart Status menu options on the left side in the Preferences window.

   d.   Click on "Cursor RA/Dec" and then on 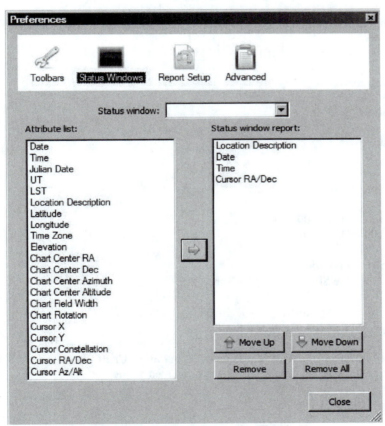. This adds this parameter to list of items in the Status Chart window. By clicking **Move Up** or **Move Down** in this menu , one can prioritize the list of Chart Status options chosen. This can be used to add more options as needed from this menu at a later time. The "Location Description" is at the top of the "Status window report" in **Figure 7-7** followed by the "Date" and "Time" and finally the "Cursor RA/Dec" location.

Figure 7-7 Chart Status Options Menu in *TheSkyX*

   e.   Click close or ⊠.

Now, *redo* Part 1 and continue to Part 2.

© 2011 Cengage Learning. All Rights Reserved. May not be scanned, copied or duplicated, or posted to a publicly accessible website, in whole or in part.

2.   Reset *TheSkyX* to the following location:

Location: Philadelphia, PA
Elevation: 100 meters
Date: February 2, 2015
Time: 9:00 PM STD   Time Zone: -5.00 hours
Daylight Saving option: "Not Observed"

Orientation: Zoom To: 50° field of view
Orientation: Navigate – Click the Sky Chart Center Tab

Enter: R.A.: $05^H\,15^M\,0^S$ Declination: +49° 30′ 0″
Current Epoch: 2015   Click Center on RA/Dec

Put your cursor on the red line again. Use the coordinate readout in the Chart Status window again to determine the sidereal time for an observer located in Philadelphia, PA.

Sidereal Time = _____:_____:_____

Now compute the difference in the sidereal times between the two locations.

Difference in sidereal time (Part 2 minus Part 1) = _____:_____

Why are the two times different? _____

3.   Reset *TheSkyX* to the following location:

Location: 75th Longitude
Longitude: 75° 00′ 00″
Elevation: 100 meters
Latitude: 40° 11′ 58″
Date: February 2, 2015
Time: 9:00 PM STD   Time Zone: -5.00 hours
Daylight Saving option: "Not Observed"

Orientation: Zoom To: 50° field of view
Orientation: Navigate – Click the Sky Chart Center Tab

Enter: R.A.: $05^H\,15^M\,0^S$ Declination: +49° 30′ 0″
Current Epoch: 2015   Click Center on RA/Dec

Put your cursor on the red line again. Use the coordinate readout in the Chart Status window again to determine the sidereal time for an observer located on the 75th longitude circle.

Sidereal Time = _____:_____:_____

Now compute the difference in time between the locations.

Difference in Sidereal Time (Part 3 minus Part 1) = _____:_____

Difference in Sidereal Time (Part 3 minus Part 2) = _____:_____

Why are the two time differences different? _____

4.   Close the file and do not save, or continue with **Exercise J**.

© 2011 Cengage Learning. All Rights Reserved. May not be scanned, copied or duplicated, or posted to a publicly accessible website, in whole or in part.

Exercise J: Determining Sidereal Time for Your Location on Earth

1. Run *TheSkyX* and open the file ***Normal.skyx*** or reset *TheSkyX* as follows:

    Location: Ball State Observatory
    Elevation: 322 meters
    Date: March 15, 2025
    Time: 9:45 PM DST   Time Zone: -5.00 hours
    Daylight Saving option: "U.S. and Canada"

    Orientation: Zoom To: 50° field of view
    Orientation: Navigate – Click the Sky Chart Center Tab

    Enter: R.A.: $07^H 45^M 38.7^S$ Declination: +28° 00′ 57.8″
    Current Epoch: 2025   Click Center on RA/Dec

    What is the name of the red line that passes through the field of view?

    _____   _____   _____

    If you wish to change the orientation of this line, repeat what was done in **Exercise E, Part 2**. This will make the red line appear to be in a north-south direction in the sky if it is not so already.

    Put your cursor on the red line again. Use the coordinate readout in the Chart Status window again to determine the sidereal time for an observer located at the Ball State Observatory.
    Sidereal Time = _____:_____:_____

    Are there any bright stars located close to the local celestial meridian near the zenith? If so, does it have a proper name? _____

    Proper Name: _____

    What are its equatorial coordinates?

    R.A. (2025): ____^{H} ___^{M} ____^{S} Declination (2025): __ ____° ____′ ____″

2. Reset *TheSkyX* to the following location:

    Location: Memphis, TN
    Elevation: 100 meters
    Date: March 15, 2025
    Time: 9:45 PM DST   Time Zone: -6.00 hours
    Daylight Saving option: "U.S. and Canada"

    Orientation: Zoom To: 50° field of view
    Orientation: Navigate – Click the Sky Chart Center Tab

    Enter: R.A.: $07^H 45^M 38.7^S$ Declination: +28° 00′ 57.8″
    Current Epoch: 2025   Click Center on RA/Dec

    Put your cursor on the red line again. Use the coordinate readout in the Chart Status window again to determine the sidereal time for an observer located in Memphis, Tennessee.
    Sidereal Time = _____:_____:_____

© 2011 Cengage Learning. All Rights Reserved. May not be scanned, copied or duplicated, or posted to a publicly accessible website, in whole or in part.

Now compute the difference in time between the two locations.

Difference in Sidereal Time (Part 2 minus Part 1) = _____ : _____

Why are the two times different? _____

3. Reset *TheSkyX* to the following location:

Location: 90th Longitude

Longitude: 90° 00' 00″
Elevation: 100 meters
Latitude: 40° 11' 58″

Date: March 15, 2025
Time: 9:45 PM DST   Time Zone: -6.00 hours
Daylight Saving option: "U.S. and Canada"

Orientation: Zoom To: 50° field of view
Orientation: Navigate – Click the Sky Chart Center Tab

Enter: R.A.: 07^H 45^M 38.7^S Declination: +28° 00′ 57.8″
Current Epoch: 2025   Click Center on RA/Dec

Put your cursor on the red line again. Use the coordinate readout in the Chart Status window again to determine the sidereal time for an observer located on the 90th longitude circle.

Sidereal Time = _____ : _____ : _____

Now compute the difference in time between the two locations.

Difference in Sidereal Time (Part 3 minus Part 1) = _____ : _____

Difference in Sidereal Time (Part 3 minus Part 2) = _____ : _____

Why are the two time differences different? _____

4. Close the file and do not save, or continue with **Exercise K**.

Exercise K: Determining Sidereal Time for Your Location on Earth

1. Run *TheSkyX* and open the file ***Normal.skyx*** or reset *TheSkyX* as follows:

Location: Ball State Observatory
Elevation: 322 meters
Date: July 4, 2030
Time: 9:15 PM DST   Time Zone: -5.00 hours
Daylight Saving option: "U.S. and Canada"

Orientation: Zoom To: 50° field of view
Orientation: Navigate – Click the Sky Chart Center Tab

Enter: R.A.: 14^H 35^M 45.0^S Declination: +19° 30′ 5.0″
Current Epoch: 2030   Click Center on RA/Dec

© 2011 Cengage Learning. All Rights Reserved. May not be scanned, copied or duplicated, or posted to a publicly accessible website, in whole or in part.

What is the name of the red line that passes through this field of view?

_____ _____ _____

Put your cursor on the red line again. Use the coordinate readout in the **Chart Status** window again to determine the sidereal time for an observer located at the **Ball State Observatory**.

Sidereal Time = _____:_____:_____

Are there any bright stars located close to the local celestial meridian **near the zenith**? If so, does it have a proper name? _____

Proper Name: _____

What are its equatorial coordinates?

R.A. (2030): ____^{H} ___^{M} ____^{S} Declination (2030): __ ____° ____′ ____″

2. Reset *TheSkyX* to the following location:

Location: Denver, CO
Elevation: 100 meters
Date: July 4, 2030
Time: 9:15 PM DST   Time Zone: -7.00 hours
Daylight Saving option: "U.S. and Canada"

Orientation: Zoom To: 50° field of view
Orientation: Navigate – Click the Sky Chart Center Tab

Enter: R.A.: 14^H 35^M 45.0^S Declination: +19° 30′ 5.0″
Current Epoch: 2030   Click Center on RA/Dec

Put your cursor on the red line again. Use the coordinate readout in the **Chart Status** window again to determine the sidereal time for an observer located in **Denver, Colorado**.

Sidereal Time = _____:_____:_____

Now compute the difference in time between the two locations.

Difference in Sidereal Time (Part 2 minus Part 1) = ____:____

Why are the two times different? _____

3. Reset *TheSkyX* to the following location:

Set Location: 105th Longitude

Longitude: 105° 00′ 00″
Elevation: 100 meters
Latitude: 40° 11′ 58″

Date: July 4, 2030
Time: 9:15 PM DST   Time Zone: -7.00 hours
Daylight Saving option: "U.S. and Canada"

Orientation: Zoom To: 50° field of view
Orientation: Navigate – Click the Sky Chart Center Tab

© 2011 Cengage Learning. All Rights Reserved. May not be scanned, copied or duplicated, or posted to a publicly accessible website, in whole or in part.

Enter: R.A.: $14^H 35^M 45.0^S$ Declination: $+19° 30' 5.0''$
Current Epoch: 2030   Click Center on RA/Dec

Put your cursor on the red line again. Use the coordinate readout in the Chart Status window again to determine the sidereal time for an observer located on the 105th longitude circle.

Sidereal Time = _____:_____:_____

Now compute the difference in time between the two locations.

Difference in Sidereal Time (Part 3 minus Part 1) = ____:____

Difference in Sidereal Time (Part 3 minus Part 2) = ____:____

Why are the two time differences different? _____

4. Close the file and do not save, or continue with **Exercise L**.

Exercise L: Determining Sidereal Time for Your Location on Earth

1. Run *TheSkyX* and open the file ***Normal.skyx*** or reset *TheSkyX* as follows:

Location: Ball State Observatory
Elevation: 322 meters
Date: April 20, 2019
Time: 6:15 AM DST   Time Zone: -5.00 hours
Daylight Saving option: "U.S. and Canada"

Orientation: Zoom To: 50° field of view
Orientation: Navigate – Click the Sky Chart Center Tab

Enter: R.A.: $18^H 56^M 32.1^S$ Declination: $+40° 11' 8.7''$
Current Epoch: 2019   Click Center on RA/Dec

If you wish to change the orientation of this line, repeat what was done in **Exercise E, Part 2**. This will make the red line appear to be in a north-south direction in the sky if it is not so already.

What is the name of the red line that passes through this field of view?

_____ _____ _____

Put your cursor on the red line again. Use the coordinate readout in the Chart Status window again to determine the sidereal time for an observer located at the Ball State Observatory.

Sidereal Time = _____:_____:_____

Are there any bright stars located close to the local celestial meridian near the zenith?

If so, does it have a proper name? _____

Proper Name: _____

What are its equatorial coordinates?

R.A. (2019): ____^{H} ____^{M} ____^{S} Declination (2019): __ ____° ____' ____''

© 2011 Cengage Learning. All Rights Reserved. May not be scanned, copied or duplicated, or posted to a publicly accessible website, in whole or in part.

2.   Reset *TheSkyX* to the following location:

Location: New York City, NY
Elevation: 100 meters
Date: April 20, 2019
Time: 6:15 AM DST   Time Zone: -5.00 hours
Daylight Saving option: "U.S. and Canada"

Orientation: Zoom To: 50° field of view
Orientation: Navigate – Click the Sky Chart Center Tab

Enter: R.A.: $18^H 56^M 32.1^S$ Declination: +40° 11′ 8.7″
Current Epoch: 2019   Click Center on RA/Dec

Put your cursor on the red line again. Use the coordinate readout in the Chart Status
window again to determine the sidereal time for an observer located in New York City, NY.

Sidereal Time = _____:_____:_____

Now compute the difference in time between the two locations.

Difference in Sidereal Time (Part 2 minus Part 1) = ____: ____

Why are the two times different? _____

3.   Now reset *TheSkyX* to the following location.

Set Location: 75th Longitude

Longitude: 75° 00′ 00″
Elevation: 100 meters
Latitude: 40° 11′ 58″

Date: April 20, 2019
Time: 6:15 AM DST   Time Zone: -5.00 hours
Daylight Saving option: "U.S. and Canada"

Orientation: Zoom To: 50° field of view
Orientation: Navigate – Click the Sky Chart Center Tab

Enter: R.A.: $18^H 56^M 32.1^S$ Declination: +40° 11′ 8.7″
Current Epoch: 2019   Click Center on RA/Dec

Put your cursor on the red line again. Use the coordinate readout in the Chart Status
window again to determine the sidereal time for an observer located on the 75th longitude
circle.

Sidereal Time = _____:_____:_____

Now compute the difference in time between the two locations.

Difference in Sidereal Time (Part 3 minus Part 1) = ____:____

Difference in Sidereal Time (Part 3 minus Part 2) = ____:____

Why are the two time differences different? _____

4.   Close the file and do not save , or continue with **Exercise M**.

© 2011 Cengage Learning. All Rights Reserved. May not be scanned, copied or duplicated, or posted to a publicly accessible website, in whole or in part.

Exercise M: Determining Sidereal Time for Your Location on Earth

1. Run *TheSkyX* and open the file ***Normal.skyx*** or reset *TheSkyX* as follows:

   Location: Vandenberg Air Force Base, CA
   Longitude: 120° 34′ 37″ W   Latitude: 34° 43′ 47″ N
   Elevation: 112 meters
   Date: December 8, 2018
   Time: 11:30 AM STD   Time Zone: -8.00 hours
   Daylight Saving option: "U.S. and Canada"

   Orientation: Zoom To: 50° field of view
   Orientation: Navigate – Click the Sky Chart Center Tab

   Enter: R.A.: 17^H 30^M 30.5^S Declination: +35° 00′ 00″
   Current Epoch: 2018   Click Center on RA/Dec

   What is the name of the red line that passes through this field of view?

   _____   _____   _____

   If you wish to change the orientation of this line, repeat what was done in **Exercise E, Part 2**. This will make the red line appear to be in a north-south direction in the sky if it is not so already.

   Put your cursor on the red line again. Use the coordinate readout in the Chart Status window again to determine the sidereal time for an observer located at Vandenberg Air Force Base, CA.
   Sidereal Time = _____:_____:_____

   Are there any bright stars located close to the local celestial meridian near the zenith? If so, does it have a proper name? _____

   Proper Name: _____

   What are its equatorial coordinates?

   R.A. (2018): ____^{H} ___^{M} ____^{S} Declination (2018): __ ____° ____′ ____″

2. Reset *TheSkyX* to the following location:

   Location: Los Angles, CA
   Longitude: 118° 16′ 48″ W   Latitude: 34° 01′ 06″ N
   Elevation: 100 meters
   Date: December 8, 2018
   Time: 11:30 AM STD   Time Zone: -8.00 hours
   Daylight Saving option: "U.S. and Canada"

   Orientation: Zoom To: 50° field of view
   Orientation: Navigate – Click the Sky Chart Center Tab

   Enter: R.A.: 17^H 30^M 30.5^S Declination: +35° 00′ 00″
   Current Epoch: 2018   Click Center on RA/Dec

© 2011 Cengage Learning. All Rights Reserved. May not be scanned, copied or duplicated, or posted to a publicly accessible website, in whole or in part.

Put your cursor on the red line again. Use the coordinate readout in the Chart Status window again to determine the sidereal time for an observer located in Los Angeles, CA.

Sidereal Time = _____:_____:_____

Now compute the difference in time between the two locations.

Difference in Sidereal Time (Part 2 minus Part 1) = ____:____

Why are the two times different? _____

3.   Reset *TheSkyX* to the following location:

Set Location: 120th Longitude

Longitude: 120° 00' 00"
Elevation: 100 meters
Latitude: 40° 11' 58"

Date: December 8, 2018
Time: 11:30 AM STD   Time Zone: -8.00 hours
Daylight Saving option: "U.S. and Canada"

Orientation: Zoom To: 50° field of view
Orientation: Navigate – Click the Sky Chart Center Tab

Enter: R.A.: $17^H 30^M 30.5^S$ Declination: +35° 00' 00"
Current Epoch: 2018   Click Center on RA/Dec

Put your cursor on the red line again. Use the coordinate readout in the Chart Status window again to determine the sidereal time for an observer located on the 120th longitude circle.

Sidereal Time = _____:_____:_____

Now compute the difference in time between the two locations.

Difference in Sidereal Time (Part 3 minus Part 1) = ____:____

Difference in Sidereal Time (Part 3 minus Part 2) = ____:____

Is there any significance of the two values obtained in the previous two questions.

Why are the two time differences different? _____

4. Close the file and do not save it.

If your answers to why the results between Part 1 and Part 2, and Part 1 and Part 3, in the previous exercises were because of longitude differences of the observers, then you are right again!

It should become apparent by now that if one knows the right ascension of objects on the local celestial meridian, then one *will* know the sidereal time! As it turns out, it does not matter where the observer is located on Earth, the day of the year, or even the year itself…one can always determine sidereal time for a particular location. Using *TheSkyX* to ascertain sidereal times for a particular location is a very simple and useful task. The previous exercises were designed to illustrate this and may have seemed repetitious. But, repetition is sometimes necessary to become proficient.

© 2011 Cengage Learning. All Rights Reserved. May not be scanned, copied or duplicated, or posted to a publicly accessible website, in whole or in part.

If on the other hand, one has a star chart, then look for a bright star or some constellation that is on or near the local celestial meridian. This is another quick way to estimate the sidereal time too but *TheSkyX* is the best way!

Knowing sidereal time is crucial in determining what constellation, star, planet, or any other celestial object might be visible for a given date at a specific location on Earth. *TheSkyX* makes determining sidereal time for any date and time simple and easy. Planning observing sessions also becomes easier. If an object's right ascension is within $\pm 6^H$ of the sidereal time, then one knows that the object is above the horizon at that location. Knowing the sidereal time before observing makes the experience more productive and pleasant.

Determining Solar Transit Times in *TheSkyX*

Time differences seen in the transits of astronomical objects are not limited to the objects seen just in the night sky. *TheSkyX* may be used to observe transits of the most obvious astronomical objects in the sky as well. Not only do we observe time differences in the transits of stars at different locations on Earth, but we also observe that the Sun itself does not transit the local celestial meridian at exactly noon each day. In fact, it does but only about four times during the year.

There are time differences in solar transits because of a variety of reasons. First of all Earth's orbit is not circular, but elliptical. As Earth revolves around the Sun, its distance from the Sun constantly changes. The Sun exerts a net force on Earth that is also constantly changing. Earth responds to this net force exerted on it by the Sun by accelerating. That is, Earth speeds up and slows down as it orbits the Sun.

Earth moves faster when it is closer to the Sun and slower when it is farther away. It *moves fastest* in its orbit when it is *closest* to the Sun (*perihelion*), which occurs on or about January 3rd each year. It moves *slowest* in its orbit when it is *farthest away* from the Sun (*aphelion*), which occurs on or about July 3rd each year.

The motion and tilt of Earth affect our observations of the Sun and how we determine solar time. As Earth moves about the Sun the Sun appears to move eastward along the ecliptic, it reflects the motion of Earth around it. That is, the Sun's motion in the sky appears to speed up during January and slow down during July each year.

The 1° eastward motion discussed in **Chapter 5** is an average value. During January each year, the Sun appears to move a bit more than 1° per day, whereas during July each year the Sun appears to move a bit less than 1° per day along the ecliptic. The tilt of Earth's axis makes the Sun appear to move north and south of the celestial equator. When all this is factored into determining the solar time, it turns out that Earth is a poor timekeeper!

Astronomers and timekeepers have avoided this problem by inventing a fictitious Sun called the ***mean Sun***. This *mean Sun* moves along the celestial equator at a uniform rate rather than along the ecliptic. The rate of movement, at which this fictitious Sun moves along the celestial equator, is an average value of the apparent Sun's motion during the course of a year along the ecliptic. The mean Sun eliminates any irregularities observed in the apparent Sun's motion caused by the motions of Earth. The time interval between two successive passages of the mean Sun across a celestial meridian is defined as a *mean solar day*.

One manifestation of this irregular motion of Earth ***is*** the time at which the Sun appears to transit the local celestial meridian. **Exercises N, O, P, Q** and **R** are designed to help assist in determining differences in transit times of the Sun across an observer's local celestial meridian. *When doing these exercises, it is **imperative** that you **set** the **date, time,** and **Daylight savings adjustment** option as **indicated in** each exercise.*

© 2011 Cengage Learning. All Rights Reserved. May not be scanned, copied or duplicated, or posted to a publicly accessible website, in whole or in part.

Exercise N: Determining the Solar Transit Time at Ball State Observatory

1. Run *TheSkyX* and open the file ***Normal.skyx*** or reset *TheSkyX* as follows:

   Location: Ball State Observatory
   Elevation: 322 meters
   Date: January 15, 2021
   Time: 12:00 (PM) STD – Noon   Time Zone: -5.00 hours
   Daylight Saving option: "U.S. and Canada"

   Find the Sun and [⊙ Center] it in the sky window.

   What does the solid blue line through the Sun represent? _____

   What time does the Sun appear to transit? _____:_____.

   Why isn't it 12:00 Noon? _____

2. Reset *TheSkyX* to the following location:

   Location: Philadelphia, PA
   Date: January 15, 2021
   Time: 12:00 (PM) STD – Noon   Time Zone: -5.00 hours
   Daylight Saving option: "U.S. and Canada"

   Find the Sun and [⊙ Center] it in the sky window.

   What time does the Sun appear to transit? _____:_____.

   Why isn't it 12:00 Noon again? _____

   What is the difference in the transit time (12:00 − observed transit time)? _____:_____

   Is the difference in time close to the equation of time value for this date in **Table 7-1 (p. 222)**? _____

3. Close the file and do not save, or continue with **Exercise O**.

Exercise O: Determining Solar Transit Time for Your Location on Earth

1. Run *TheSkyX* and open the file ***Normal.skyx*** or reset *TheSkyX* as follows:

   Location: Ball State Observatory
   Date: March 15, 2021
   Time: 12:00 (PM) DST – Noon   Time Zone: -5.00 hours
   Daylight Saving option: "U.S. and Canada"

   Find the Sun and [⊙ Center] it in the sky window.

   What does the solid blue line through the Sun represent? _____

   What time does the Sun appear to transit? _____:_____.

© 2011 Cengage Learning. All Rights Reserved. May not be scanned, copied or duplicated, or posted to a publicly accessible website, in whole or in part.

A comment must be made here!

Before determining the difference between the observed transit time and 12:00, you must allow for an additional 1^H due to Daylight Savings time that is in effect at this time of year. So, one must subtract 1^H from the observed transit time. Correct for this by subtracting 1^H from *TheSkyX's* transit time _____:_____.

Why isn't it 12:00 Noon? _____

2. Reset *TheSkyX* to the following location:

    Location: Philadelphia, PA
    Date: March 15, 2021
    Time: 12:00 (PM) STD – Noon   Time Zone: -5.00 hours
    Daylight Saving option: "U.S. and Canada"

    Find the Sun and [Center] it in the sky window.

    What time does the Sun appear to transit? _____:_____. Again, due to Daylight Savings Time being in effect, one must subtract 1^H from the observed transit time. Correct for this by subtracting 1^H from *TheSkyX's* transit time _____:_____.

    Why isn't it 12:00 Noon again? _____

    What is the difference in the transit time (12:00 – observed transit time)? _____:_____

    Is the difference in time close to the equation of time value for this date in **Table 7-1** (p. 222)? _____

3. Close the file and do not save, or continue with **Exercise P**.

Exercise P: Determining Solar Transit Time for Your Location on Earth

1. Run *TheSkyX* and open the file ***Normal.skyx*** or reset *TheSkyX* as follows:

    Location: Ball State Observatory
    Date: June 1, 2019
    Time: 12:00 (PM) DST – Noon   Time Zone: -5.00 hours
    Daylight Saving option: "U.S. and Canada"

    Find the Sun and [Center] it in the sky window.

    What does the solid blue line through the Sun represent? _____

    What time does the Sun appear to transit? _____:_____. Again, due to Daylight Savings Time being in effect, one must subtract 1^H from the observed transit time. Correct for this by subtracting 1^H from *TheSkyX's* transit time _____:_____.

    Why isn't it 12:00 Noon? _____

2. Reset *TheSkyX* to the following location:

    Location: Memphis, TN
    Date: June 1, 2019

© 2011 Cengage Learning. All Rights Reserved. May not be scanned, copied or duplicated, or posted to a publicly accessible website, in whole or in part.

Time: 12:00 (PM) DST – Noon   Time Zone: -6.00 hours
Daylight Saving option: "U.S. and Canada"

Find the Sun and [Center] it in the sky window.

What time does the Sun appear to transit? _____:_____. Again, due to Daylight Savings Time being in effect, one must subtract 1^H from the observed transit time.
Correct for this by subtracting 1^H from *TheSkyX's* transit time _____:_____.

Why isn't it 12:00 Noon again? _____

What is the difference in the transit time (12:00 – observed transit time)? _____

Is the difference in time close to the equation of time value for this date in **Table 7-1** (p. 222)?
_____.

3.   Close the file and do not save, or continue with **Exercise Q**.

Exercise Q: Determining Solar Transit Time for Your Location on Earth

1.   Run *TheSkyX* and open the file ***Normal.skyx*** or reset *TheSkyX* as follows:

Location: Ball State Observatory
Date: August 15, 2029
Time: 12:00 (PM) DST – Noon   Time Zone: -5.00 hours
Daylight Saving option: "U.S. and Canada"

Find the Sun and [Center] it in the sky window.

What does the solid blue line through the Sun represent? _____

What time does the Sun appear to transit? _____:_____.

Note: Did you subtract one hour from the observed transit time again?

Why isn't it 12:00 Noon? _____

2.   Reset *TheSkyX* to the following location:

Location: Denver, CO
Date: August 15, 2029
Time: 12:00 (PM) DST – Noon   Time Zone: -7.00 hours
Daylight Saving option: "U.S. and Canada"

Find the Sun and [Center] it in the sky window.

What time does the Sun appear to transit? _____:_____.

Why isn't it 12:00 Noon again? _____

What is the difference in the transit time (12:00 – observed transit time)? _____

© 2011 Cengage Learning. All Rights Reserved. May not be scanned, copied or duplicated, or posted to a publicly accessible website, in whole or in part.

Note: Did you subtract one hour from the observed transit time again?

Is the difference in time close to the equation of time value for this date in **Table 7-1** (p. 222)? _____

3. Close the file and do not save it.

Exercise R: Determining Solar Transit Time for Your Location on Earth

1. Run *TheSkyX* and open the file ***Normal.skyx*** or reset *TheSkyX* as follows:

    Location: Ball State Observatory
    Date: December 1, 2023
    Time: 12:00 (PM) STD – Noon   Time Zone: -5.00 hours
    Daylight Saving option: "U.S. and Canada"

    Find the Sun and [Center] it in the sky window.

    What does the solid blue line through the Sun represent? _____

    What time does the Sun appear to transit? _____:_____.

    Why isn't it 12:00 Noon? _____

2. Reset *TheSkyX* to the following location:

    Location: Memphis, TN
    Date: December 1, 2023
    Time: 12:00 (PM) STD – Noon   Time Zone: -6.00 hours
    Daylight Saving option: "U.S. and Canada"

    Find the Sun and [Center] it in the sky window.

    What time does the Sun appear to transit? _____:_____.

    Why isn't it 12:00 Noon again? _____

    What is the difference in the transit time (12:00 – observed transit time)? _____

    Is the difference in time close to the equation of time value for this date in **Table 7-1** (p.222)? _____

3. Close the file and do not save it.

These differences in transit times of the Sun can be determined for any of the standard longitudes for any day of the year. As previously mentioned, Earth is a lousy timekeeper. It speeds up and slows down as it orbits the Sun; its tilt makes the Sun appear to move north and south of the celestial equator. All these factors combined make these differences in time, noted in **Table 7-1** on page 222 more apparent.

It should be apparent from doing the previous exercises that the State of Indiana is actually closer to the Central Standard Time (CST) zone, longitude (90° W) rather than the Eastern Standard Time (EST) zone, longitude (75° W). The State of Indiana used to keep its clocks set on Eastern Standard Time (EST) all year round.

In fact, it was only one of two states (Arizona), in the entire continental United States that remained on the same time (at least most of the State did) all year long. That is, it chose to

© 2011 Cengage Learning. All Rights Reserved. May not be scanned, copied or duplicated, or posted to a publicly accessible website, in whole or in part.

remain on Eastern Standard Time throughout the entire year! Unlike Arizona which keeps it clocks statewide on the same time (MST), Indianapolis, the State Capital, was on Eastern Standard Time all year round while the rest of the State was *never* on the same time on any given date during the year.

For example, the counties in the southern part of the State, along the Ohio River, were on *Eastern Daylight Savings Time* during the summer months and on *Eastern Standard Time* during the winter months. The counties in the northwestern part of the State, near Chicago, were just the opposite. They observed *Central Daylight Savings Time* (same as EST in Indianapolis) during the summer months, and during the winter months these counties were one hour behind the rest of the State of Indiana, on Central Standard Time instead of Eastern Standard Time.

In April 2006, after many years of debate and approval of the Department of Transportation (which regulates highway right of ways and time zones in the United States) the Indiana State legislature elected to have the State of Indiana observe Daylight Savings Time for the first time in many years during the summer months and standard time in the winter months. I might add that this was promoted by many businesses rather than the State population.

However, there is a slight quirk in this recent adoption of Daylight Savings Time for the State of Indiana in 2006. **Rather than** choosing to have the **entire** State of Indiana observe Daylight Savings Time during the summer months and Standard Time in the winter months the State allowed counties that bordered surrounding states to choose the zone time most suited to their needs. Given that choice, many counties in the northwestern and west central part of the State did not choose to observe Daylight Savings Time in the summer months. They do, however, observe Standard Time in the winter time so again the State of Indiana does not entirely remain on the same time throughout the entire year.

Now that the observance of Daylight Savings Time has been implemented back into the State of Indiana it appears that some of these counties may have regretted the choice of remaining on Standard Time in the summer while the rest of the State of Indiana changes to Daylight Savings Time. Leaping ahead, falling behind, or is it falling ahead, leaping behind—*confusing*? No wonder people have so much difficulty with time! One can only imagine what further confusion lies ahead for the citizens of the State of Indiana!

Another manifestation of the anomalies of Earth's motion is the position of the Sun in the sky at the time it appears to transit celestial meridians. During the year, the mean Sun sometimes runs ahead of the apparent Sun and at other times during the year the mean Sun runs behind the apparent Sun.

During the course of a year, this difference varies from −14.1 minutes to +16.3 minutes. The *difference* between *apparent solar time* and *mean solar time* is defined as the *equation of time*. These differences have been calculated for the year 2009 and are displayed in **Table 7-1**. High-precision formulae taken from the *2009 Astronomical Almanac* were used to determine these values. Calculations done for every day of the year appear at either of the two following websites:

http://web006.pavilion.net/users/aghelyar/sundat.htm

or at

http://freepages.pavilion.net/users/aghelyar/sundat.htm

These sites also have more information concerning how these values are determined.

© 2011 Cengage Learning. All Rights Reserved. May not be scanned, copied or duplicated, or posted to a publicly accessible website, in whole or in part.

Table 7-1

The Equation of Time

(Mean Solar Time − Apparent Solar Time)

Date	Minutes
January 1	−3.2
15	−9.3
February 1	−13.6
15	−14.1
March 1	−12.2
15	−8.6
April 1	−3.5
15	+0.2
May 1	+3.0
15	+3.7
June 1	+2.2
15	−0.4
July 1	−3.7
15	−5.8
August 1	−6.4
15	−4.9
September 1	−0.8
15	+3.8
October 1	+9.4
15	+13.6
November 1	+16.3
15	+15.8
December 1	+11.7
15	+5.6

If one graphs the differences in time plotted against the declination of the Sun for several dates throughout the year, the resulting graph is known as an *Analemma*. **Figure 7-8** depicts *TheSkyX's* rendition of the Sun's Analemma throughout the year 2010 as observed from the Ball State Observatory. In fact, with persistence one can photograph this phenomenon by taking multiple exposures on a single frame of photographic film.

© 2011 Cengage Learning. All Rights Reserved. May not be scanned, copied or duplicated, or posted to a publicly accessible website, in whole or in part.

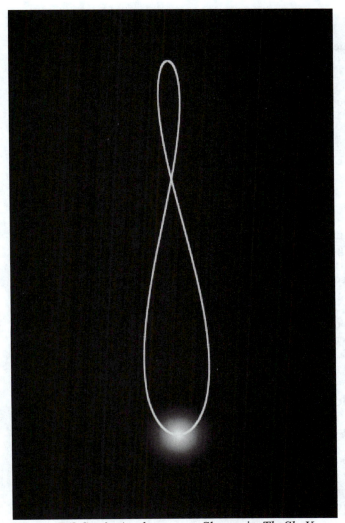

Figure 7-8 Sun's Analemma as Shown in *TheSkyX*

To do so, a permanently mounted camera is needed so that a picture of the Sun's position in the sky may be taken at regular intervals throughout the year. The photograph must be taken at the same time of day. An excellent and detailed explanation of the analemma appears at the following website: www.analemma.com.

In *TheSkyX* the Sky file named ***Analemma.skyx*** illustrates an animated version of the analemma displayed in **Figure 7-8**. **Exercise S** displays the sky window with the Sun's location at the same time of day throughout an entire year for an observer in Golden, Colorado. The dates for this sky window are from approximately December 21, 2007 to December 20, 2008. The approximate time of day is 12:00 noon Mountain Time.

Exercise S: Observing the Analemma from Golden, Colorado

1. Run *TheSkyX* and open the file ***Normal.skyx***.

2. Go to "Display" on the standard toolbar at the top of the sky window and click "Tours."

3. Now, open the file named ***Analemma.skyx***.

4. Click "Start" in the open Tour window.

© 2011 Cengage Learning. All Rights Reserved. May not be scanned, copied or duplicated, or posted to a publicly accessible website, in whole or in part.

5. To see more of the Analemma, click **Move Up** on the Orientation toolbar once.

6. Turn on the equatorial grid by clicking **Show Equatorial Grid** and watch the Sun's equatorial coordinates change throughout the year

7. Now, turn on the horizon grid by clicking **Show Horizon Grid** and watch the Sun's meridian transits throughout the year.

8. Close the file and do not save it, or leave it open click "Restore" to continue with **Exercise T**.

It is interesting to change the location and observe the analemma. The analemma is not something that happens at a particular latitude and longitude. It can be seen, for the most part, from every location on Earth. When one turns on the horizon grid while observing the analemma, it becomes apparent that the altitude of the Sun changes for any observer throughout the year. When one turns on the equatorial grid, they notice that the right ascension and declination of the Sun also change throughout the year.

Exercises T, U, V, W and **X** views the Analemma from different locations on Earth. **Exercise T** takes the observer to the north geographic pole while **Exercise U** takes the observer to the south geographic pole. **Exercise V** takes the observer to the geographic equator while **Exercise W** takes the observer to the Tropic of Cancer. Finally, **Exercise X** takes one to the Tropic of Capricorn.

Exercise T: Observing the Analemma from Earth's North Pole

1. Run *TheSkyX* and open the file ***Normal.skyx***.

2. Open the Sky file named ***AnalemmaNP.skyx***.

3. Click **Go Forward** to watch the animation.

4. **Figure 7-9** displays the Sun's Analemma as viewed from the North Geographic Pole. In this view the elapsed time interval is from approximately June 21, 2015 through June 22, 2016.

Figure 7-9 The Sun's Analemma as Viewed from the North Geographic Pole

© 2011 Cengage Learning. All Rights Reserved. May not be scanned, copied or duplicated, or posted to a publicly accessible website, in whole or in part.

5. Click **Show Equatorial Grid** and **Show Horizon Grid** to watch the Sun's coordinates and meridian transits change throughout the year.

6. Close the file and do not save it, or leave it open and continue with **Exercise U**.

What one sees from the North Pole is that portion of the analemma that is above the observer's horizon between mid-March and mid-September. The Sun's declination has positive values between 0° and +23½°. For about five and a half months, the North Pole is in total darkness throughout the year. For about two weeks or so, it is in twilight and the rest of the time the Sun is above the horizon for the entire day (24^H). The date at which this animation begins is June 21st, 2015.

Exercise U: Observing the Analemma from Earth's South Pole

1. Run *TheSkyX* and open the file *Normal.skyx*.

2. Open the Sky file named *AnalemmaSP.skyx*.

3. Click **Go Forward** to watch the animation.

4. **Figure 7-10** displays the Sun's Analemma as viewed from the South Geographic Pole. In this view the elapsed time interval is from approximately December 22, 2015 through December 20, 2016.

Figure 7-10 The Sun's Analemma as Viewed from the South Geographic Pole

5. Click **Show Equatorial Grid** and **Show Horizon Grid** to watch the Sun's coordinates and meridian transits change throughout the year.

6. Close the file and do not save it, or leave it open and continue with **Exercise V**.

What one sees at the South Pole is the portion of the analemma that is above an observer's horizon between mid-September and mid-March. The Sun's declination has negative values between 0° to −23½°. For about five and a half months, the South Pole is in total darkness

© 2011 Cengage Learning. All Rights Reserved. May not be scanned, copied or duplicated, or posted to a publicly accessible website, in whole or in part.

throughout the year. For about two weeks or so, it is in twilight and the rest of the time the Sun is above the horizon for the entire day (24^H). The date at which this animation begins is December 22nd, 2015.

Exercise V: Observing the Analemma from Earth's Equator

1. Run *TheSkyX* and open the file ***Normal.skyx***.

2. Open the Sky file named ***AnalemmaGEQ.skyx***.

3. Click 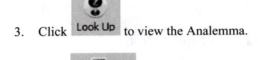 to view the Analemma.

4. Click ![Go Forward] to watch the animation.

5. **Figure 7-11** displays the Sun's position in the sky as viewed from the Geographic Equator. The grid that is displayed in this figure is the horizontal grid. The Sun appears at your zenith on this date. Do you know what day it is? _____. What time of day? _____. In this view the observer is looking at the zenith.

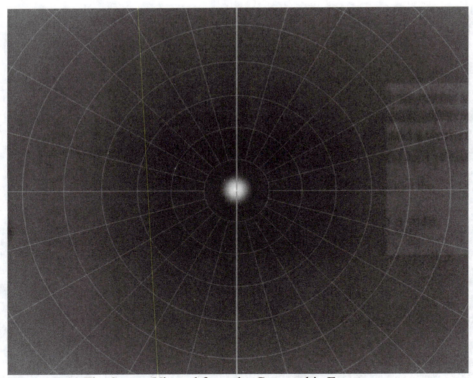

Figure 7-11 The Sun as Viewed from the Geographic Equator

6. Turn off the horizon grid and click ![Show Equatorial Grid] to watch the Sun's coordinates change throughout the year.

    What is the Sun's RA when it is at your zenith? _____:_____. The elapsed time interval in

© 2011 Cengage Learning. All Rights Reserved. May not be scanned, copied or duplicated, or posted to a publicly accessible website, in whole or in part.

the view is from approximately March 21, 2015 through March 22, 2016.

7. Close the file and do not save it, or leave it open and continue with **Exercise W**.

For an observer at Earth's equator the Sun is at the zenith twice a year, once at the vernal equinox and once at the autumnal equinox. At other times of the year, it is either north or south of the zenith point. The Sun appears to be 23° 26′ north of the zenith point on the first day of summer and 23° 26′ south of the zenith point on the first day of winter. Turn on the equatorial grid again and use either the

button or the ⬛ Go Forward button to verify this. Remember that the Analemma is something that can be viewed over an entire year from anywhere on Earth.

If one cannot find a particular location in the "List of Locations," then simply add it to the "Location Database." This is done by clicking the "Input" button on the Standard toolbar at the top of the sky window then "Location." Click the "Custom" tab and enter the Name of the Location, its Latitude and Longitude and Time Zone, and its Elevation if known. In order to change the new location to the current location click the "List of Locations" tab then click "Set As My Location." Since the geographic equator is in *TheSkyX's* database, it won't have to be added it to the database.

Exercise W allows you to view the analemma from Havana, Cuba, or Syene, Egypt. These places are located near the Tropic of Cancer, 23 ½° north of the geographic equator.

Exercise W: Observing the Analemma from Near the Tropic of Cancer

1. Run *TheSkyX* and open the file *Normal.skyx*.

2. Open the Sky file named *AnalemmaTCAN.skyx*.

3. Click ⬛ Go Forward to watch the animation.

4. **Figure 7-12** displays the Sun's Analemma as viewed from Syene, Egypt whose latitude is 23° 26′ N. The grid that is displayed in this figure is the equatorial grid.

   What is the Sun's approximate Declination on this date? _____°. In this view the observer is facing North.

5. After clicking ⬛ Go Forward, notice how the Sun's coordinates change throughout an entire year.

6. Click 🔵 Show Horizon Grid to observe that the Sun is at the zenith one time during this time span of a year.

   On what date does the Sun appear at the zenith at this location? _____. In this view the elapsed time interval is from approximately June 21, 2015 through June 22, 2016.

© 2011 Cengage Learning. All Rights Reserved. May not be scanned, copied or duplicated, or posted to a publicly accessible website, in whole or in part.

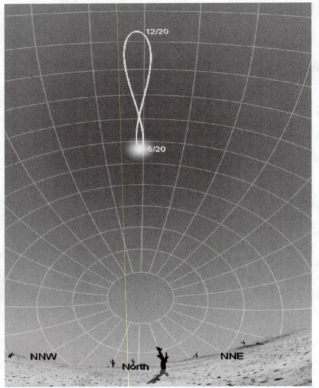

Figure 7-12 The Sun's Analemma as Viewed for Syene, Egypt

7.   Close the file and do not save it, or leave it open and continue with **Exercise X**.

      If an observer travels to the Tropic of Cancer (Syene, Egypt, for instance), the Sun's position in the sky is directly overhead at the zenith on the summer solstice at midday. The Sun is at its northernmost declination on the analemma on this date. For any observer on the Tropic of Cancer, the Sun is at the zenith only once during the year, at the summer solstice. Havana, Cuba is the only location in the database that is close to the latitude of Syene, Egypt or the Tropic of Cancer.

Exercise X: Observing the Analemma from near the Tropic of Capricorn

1.   Run *TheSkyX* and open the file *Normal.skyx*.

2.   Open the Sky file named *AnalemmaTCAP.skyx*.

3.   Click 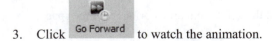 to watch the animation.

4.   **Figure 7-13** displays the Sun's Analemma as viewed from Antofagasta, Chile whose latitude is 23° 26' S. The grid that is displayed in this figure is the equatorial grid once again.

      What is the Sun's approximate Declination on this date? _____°. In this view, the observer is facing North.

© 2011 Cengage Learning. All Rights Reserved. May not be scanned, copied or duplicated, or posted to a publicly accessible website, in whole or in part.

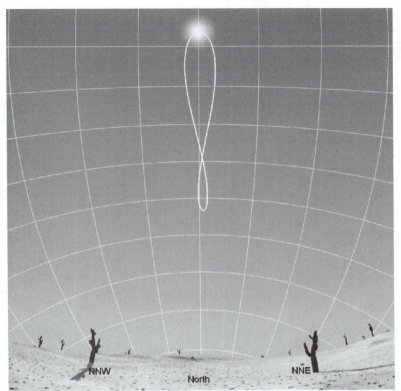

Figure 7-13 The Sun's Analemma as Viewed for Antofagasta, Chile

5.   After clicking Go Forward , notice how the Sun's coordinates change throughout an entire year.

6.   Click Show Horizon Grid , to observe that the Sun at the zenith one time during this time span of a year.

On what date does the Sun appear at the zenith at this location? _____.
In this view the elapsed time interval is from approximately December 21, 2015 through December 22, 2016.

7.   Close the file and do not save it.

If an observer travels to the *Tropic of Capricorn*, the Sun's position in the sky is directly overhead at the zenith, on the winter solstice at midday. The Sun is at its southernmost declination on the analemma on this date. For any observer on the Tropic of Capricorn, the Sun is at the zenith only once during the year, at the winter solstice.

The main thing to remember while exploring this window or any other sky window is to be certain not to save this particular sky window when you decide to close it. If you do want to save this or any other sky window, remember to use the *Save As* option in the Standard toolbar at the top of the sky window. The file then can be safely saved under a different filename.

© 2011 Cengage Learning. All Rights Reserved. May not be scanned, copied or duplicated, or posted to a publicly accessible website, in whole or in part.

Chapter 7

TheSkyX Review Exercises

Use *TheSkyX* to answer the following questions:

TheSkyX Exercise 1: Measuring Solar Time

Run *TheSkyX* and open the file *Normal.skyx*.

Date: January 15, 2013
Time: 12:10 PM STD
Daylight Saving option: "U.S. and Canada"
Location: 75° 00′ 00″ W   40° 00′ 00″ N
Time Zone: -5.00 hours
Elevation: 100 meters

Turn off the daytime sky if it is on.

Find the Sun and Center it. Click Look South.

Notice that the Sun is aligned with the star α Aquilae which is about 30° N of the Sun.

Set the Time Flow rate to 23^H 56^M 04^S (1 Sidereal Day = 23.934 Hours).

1. Use Step Forward to advance the sky forward in time by 1 sidereal day.

2. How much does the Sun appear to move in the sky on this date? _____° _____′ _____″

3. What direction does the Sun appear to move with respect to the background stars? _____

4. Which object appears to cross the local celestial meridian at the 75th longitude circle on January 16th, the Sun or Altair? _____

5. How much earlier does this object cross the local celestial meridian each day? _____

6. What time does the Sun transit the local celestial meridian at this location? _____:_____

7. Is the difference in transit time in Question 5 (12:00 – observed transit time) consistent with the value in **Table 7-1** (p. 222)? _____

© 2011 Cengage Learning. All Rights Reserved. May not be scanned, copied or duplicated, or posted to a publicly accessible website, in whole or in part.

TheSkyX Exercise 2: Measuring Solar Time

Run *TheSkyX* and open the file *Normal.skyx*, or make the following changes:

Date: January 15, 2013
Time: 12:10 PM STD
Daylight Saving option: "U.S. and Canada"
Location: 90° 00′ 00″ W   40° 00′ 00″ N
Time Zone: -6.00 hours
Elevation: 100 meters

Find the Sun and [Center] it.   Click [Look South].

1. [Step Forward] to advance the sky forward in time by 1 sidereal day.

2. Is the Sun aligned with the star α Aquilae from this location? _____

3. How much does the Sun appear to move in the sky on this date from this location?

    _____ ° _____ ′ _____ ″

3. In what direction does the Sun appear to move with respect to the background stars from this location? _____

4. Which object appears to cross the local celestial meridian at this location on January 16th, the Sun or Altair? _____

5. How much earlier does this object cross the local celestial meridian each day? _____

6. What time does the Sun transit the local celestial meridian at this location? _____:_____

7. Is the difference in transit time in Question 6 (12:00 – observed transit time) consistent with the value in **Table 7-1** (p.175)? _____

© 2011 Cengage Learning. All Rights Reserved. May not be scanned, copied or duplicated, or posted to a publicly accessible website, in whole or in part.

TheSkyX Exercise 3: Measuring Solar Time

Run *TheSkyX* and open the file *Normal.skyx*, or make the following changes:

Date: July 15, 2040
Time: 12:45 PM DST
Daylight Saving option: "U.S. and Canada"
Location: 75° 00′ 00″ W   40° 00′ 00″ N
Time Zone: -5.00 hours
Elevation: 100 meters

Find the Sun and [Center] it.   Click [Look South].

1.   [Step Forward] to advance the sky forward in time by 1 sidereal day.

2.   Is the Sun aligned with the star α Aquilae from this location? _____

3.   How much does the Sun appear to move in the sky on this date from this location?

_____ ° _____ ′ _____ ″

3.   In what direction does the Sun appear to move with respect to the background stars from this location? _____

4.   Which object appears to cross the local celestial meridian at this location on January 16th, the Sun or Altair? _____

5.   How much earlier does this object cross the local celestial meridian each day? _____

6.   What time does the Sun transit the local celestial meridian at this location? _____:_____
     Did you remember the time change to Daylight Savings time?

7.   Is the difference in transit time in Question 6 (12:00 – observed transit time) still consistent with the value in **Table 7-1** (p.175)? _____

© 2011 Cengage Learning. All Rights Reserved. May not be scanned, copied or duplicated, or posted to a publicly accessible website, in whole or in part.

TheSkyX Exercise 4: Measuring Solar Time

Run *TheSkyX* and open the file ***Normal.skyx***, or make the following changes:

Date: July 15, 2040
Time: 12:45 PM DST
Daylight Saving option: "U.S. and Canada"
Location: 90° 00′ 00″ W   40° 00′ 00″ N
Time Zone: -6.00 hours
Elevation: 100 meters

Find the Sun and [Center] it.   Click [Look South].

1. [Step Forward] to advance the sky forward in time by 1 sidereal day.

2. Is the Sun aligned with the star α Aquilae from this location? _____

3. How much does the Sun appear to move in the sky on this date from this location?

    _____ ° _____ ′ _____ ″

3. In what direction does the Sun appear to move with respect to the background stars from this location? _____

4. Which object appears to cross the local celestial meridian at this location on January 16th, the Sun or Altair? _____

5. How much earlier does this object cross the local celestial meridian each day? _____

6. What time does the Sun transit the local celestial meridian at this location? _____:_____
    Did you remember the time change to Daylight Savings time?

7. Is the difference in transit time in Question 6 (12:00 – observed transit time) still consistent with the value in **Table 7-1** (p.175)? _____

© 2011 Cengage Learning. All Rights Reserved. May not be scanned, copied or duplicated, or posted to a publicly accessible website, in whole or in part.

Chapter 7

TheSkyX Review Questions

1. The great circle passing through the south point on the horizon, the zenith, and the north point on the horizon is called the _____ _____ _____

2. What object in the sky do astronomer use to define apparent solar time? _____

3. Which point in the sky do astronomers use to define sidereal time? _____

4. An hour circle on the celestial sphere is exactly like a(n) _____ on the surface of the Earth.

5. If two locations on the surface of the Earth have the same local times, they can be said to have the same _____.

6. If the Sun and a star cross your local celestial meridian at exactly noon today, which object will appear to cross your local celestial meridian *first* at exactly noon tomorrow? _____

7. How much earlier will the transit take place tomorrow in Question 6? _____

8. Do the constellations (stars) appear to shift east or west each day as the seasons go by? _____

9. How much do the stars in Question 8 appear to shift each day? _____

10. Why do the stars in Question 8 appear to shift? _____

11. Because the Earth is curved, would an observer on the eastern edge of the Eastern Standard Time zone see an object appear to rise before or after an observer at the 75th longitude circle? _____

12. Similarly, does an observer on the western edge of the Eastern Standard Time zone see an object appear to rise before or after an observer at the 75th longitude circle? _____

   Although objects appear to rise or set at approximately the same local solar time for each observer, their watch times will differ for each observer at these different locations.

13. The hour angle of the apparent Sun is defined as _____ time.

14. The hour angle of the vernal equinox ais defined as _____ time.

© 2011 Cengage Learning. All Rights Reserved. May not be scanned, copied or duplicated, or posted to a publicly accessible website, in whole or in part.

Chapter 8

Seasons in *TheSkyX*

The Seasons

Hopefully by now the complex topic of time seems clearer now. There are many things that one must know in order to make observations productive. The beauty of using *TheSkyX* is that it does most of the hard work for you.

Now it's time to ask practical questions, like when does spring or autumn begin in Cleveland or when does the summer or winter solstice happen in Sidney this year? Usually, people associate seasons with the location of the Sun in the sky. Of course, this is not necessarily true…the seasons are determined not by where the Sun is located in the sky but by where Earth is in its orbit. The Sun's position in the sky is one of several manifestations that occur due to Earth's physical location in space. The location of Earth in its orbit and its tilt *do*, in fact, determine where the Sun *is* on the ecliptic.

Many people associate the seasons with climatic changes on Earth. Our most reliable resource for finding the day of the year is the calendar, not the weather conditions. This is apparent for someone who lives in the southern and mid-western parts of the United States in spring time. Many cities in these parts enjoy springtime flowers and flowering trees, while at the same time those who live in the most northerly regions are subjected to cold and snowy conditions.

In **Chapter 5**, we discussed seasons as points in space. They were given celestial coordinates or positions of right ascension and declination on the celestial sphere. Now we will discuss them as points in time.

If one observed Earth from some distance away in space, they would easily see its motion around the Sun. As it travels around the Sun, its axis of rotation clearly remains oriented in the same fashion in space. From the perspective of space, Earth as it travels in its orbit around the Sun is not oriented upright. It appears "tilted" by an angle of 23° 26′.

The main cause of the seasons on Earth is this 23° 26′ angular difference between Earth's axis of rotation and its orbital plane. This tilt is known as the *obliquity of the ecliptic*. Astronomers measure this angle as the angle between Earth's equator and the plane of its orbit around the Sun. This angle is also the angle between the ecliptic north pole (point on the celestial sphere where Sun's axis of rotation points) and the north celestial pole (point on the celestial sphere where Earth's axis of rotation points). Earth's orientation in space and its orbital motion around the Sun are illustrated in **Figure 8-1**.

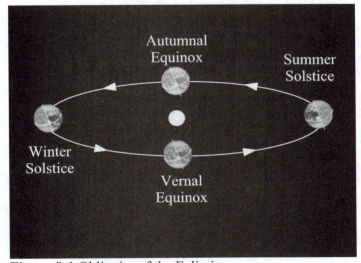

Figure 8-1 Obliquity of the Ecliptic

© 2011 Cengage Learning. All Rights Reserved. May not be scanned, copied or duplicated, or posted to a publicly accessible website, in whole or in part.

As Earth revolves around the Sun, its axis of rotation remains rigid or fixed in space. That is, Earth's axis of rotation always points in the same direction in space, toward the north celestial pole. As a consequence of this, it points very close to the bright star Polaris, our North Star.

Earth's orientation in space does change, however, over very long periods of time. The Earth spins like a top about its axis. And, like a top, it wobbles and changes its orientation.

The wobbling of Earth's axis is called *nutation*. If one looks at an Earth globe, right away they will notice that the continental masses on the surface of Earth are not evenly distributed around the globe. As a result of this, Earth is unbalanced. Like an unbalanced tire on a car, it wobbles and shakes the car—in this case, Earth. The wobbling of Earth's axis takes about 18.6 years to complete one cycle. The amount of wobble is only 9″ in angle projected onto the celestial sphere. As a result of Earth's nutation, the celestial sphere appears to wobble. This motion, of course, is so subtle that it is not observed except in large telescopes.

Another motion of Earth is the change in orientation of its axis of rotation. This motion is called *precession*. The fact that Earth rotates on its axis causes a bulge to occur in its oceans in the equatorial region. The Moon and Sun gravitationally tug on this bulge. Earth responds and would fall over if it were not spinning. Precession is simply a manifestation of Earth's stability in its orientation in space as it orbits the Sun. The time period for Earth to complete one change in orientation is 26,000 years.

These motions are two of Earth's lesser-known *true* motions in addition to rotation and revolution. Earth's precession has a very profound effect on the equatorial coordinates of right ascension and declination of objects in the sky and on calendars. *TheSkyX* lets the observer move ahead or backward in time. In doing so, it is easy to see the effects of precession. By moving ahead in time several thousand years, it becomes obvious that Polaris **will not always be** our North Star.

Try to imagine, while standing on Earth, the motion produced in the sky while Earth moves around the Sun. Revolution is the motion that is responsible for the apparent motion of the Sun through the constellations of the zodiac. This apparent motion is eastward in the sky at a rate of about 1° per day.

The Sun appears to make one circuit of the sky along the ecliptic each year. The tilt of Earth causes the Sun to appear to move both north and south of the celestial equator during this period of time. It has taken humanity thousands of years to fully understand these motions. But, only in the last few hundred years have people fully begun to understand the nature of these motions and how they are related to the motions in the sky and the seasons on Earth.

Vernal Equinox

Recall as we traced the ecliptic through the stars it eventually intersects the celestial equator at two points in space. These points are known as the equinoxes. One intersection or equinox is where the Sun moves from the Southern Hemisphere of the sky into the Northern Hemisphere. This event takes place on or about March 21 each year and is known as the *vernal equinox*.

If the Sun happens to be located at the vernal equinox, then this point in space actually becomes a point in time. That is, it *is* the first day of spring in the Northern Hemisphere. The equatorial coordinates of the Sun on the celestial sphere at the time of the vernal equinox are right ascension 00^H 00^M 00^S and declination of $00° 00′ 00″$. In other words, the Sun's location on the celestial sphere is precisely on the celestial equator at the 0^H of right ascension and 0° of declination.

Figure 8-2 clearly shows the location of the vernal equinox in *TheSkyX*. The view is from the Ball State Observatory with the vernal equinox positioned on the local celestial meridian. The altitude of this point is approximately 50° and its azimuth is 180°. At the bottom and right edge of the window the right ascension and declination coordinates are displayed, respectively.

© 2011 Cengage Learning. All Rights Reserved. May not be scanned, copied or duplicated, or posted to a publicly accessible website, in whole or in part.

If one toggles on the constellation boundaries, this point appears to be located in the western part of the constellation of Pisces. In about 700 or 800 years, this point will move into the constellation of Aquarius and will remain in this constellation for roughly 2000 years. This apparent motion is a result of Earth's precession.

Figure 8-2 Location of the Vernal Equinox as Shown in *TheSkyX*

Remember that the vernal equinox is the *zero point* for the coordinate of right ascension. All the equatorial coordinates of stars and deep sky objects are measured with respect to this point. If it moves, then the coordinates of objects in the sky change. *TheSkyX* calculates these new coordinates automatically. Only when a comparison is made between the coordinates in different epochs does it become apparent that Earth is changing its orientation in space. The sky window in **Figure 8-2** may be viewed in the Sky file named ***VernalEquinox.skyx***. It is interesting and fun to move ahead or back in time and to imagine how the sky appeared to our ancestors, or how it will appear in the future. It will also let us compare the coordinates of objects.

Exercises A, **B**, **C**, **D**, **E**, **F**, **G**, and **H**, which follow are designed to assist in determining when spring occurs in the Northern Hemisphere and how the equatorial coordinates of objects are affected by Earth's precession for a particular year.

Exercise A: Determining When the Vernal Equinox Will Occur in 2029 C.E.

1. Run *TheSkyX* and set as follows:

   Set Date: January 1, 2029
   Set Time: 12:00 PM STD
   Time Zone: -5 hours
   Daylight Saving option: "U.S. and Canada"

2. Find the Sun and  .

3. Click "Input" at the top of the Standard toolbar window, then the "Date and Time" menu button.

© 2011 Cengage Learning. All Rights Reserved. May not be scanned, copied or duplicated, or posted to a publicly accessible website, in whole or in part.

4. Now, click on the "Set Specific Time" button as shown in **Figure 8-3**.

Set Specific Time	▼

Figure 8-3 The Set Specific Time Button

5. After clicking the Set Specific Time button a menu opens like the one shown in **Figure 8-4**.

Now
Sunrise
Noon
Sunset
Midnight
Morning (Begin Twilight)
Evening (End Twilight)
New Moon
First Quarter
Last Quarter
Full Moon
Moonrise
Moonset
Vernal Equinox
Summer Solstice
Autumnal Equinox
Winter Solstice
Enter Julian Date...

Figure 8-4 The Set Specific Time Menu

6. Click "Vernal Equinox" in the Set Specific Time menu.

7. When will the vernal equinox occur in 2029 C.E.? _____.
(It's March 20th at 3:02 AM STD)

8. Close the file and do not save it, or continue with **Exercise B**.

Exercise B: Determining When the Vernal Equinox Occurred in 52 C.E.

1. Run *TheSkyX* and set as follows:

   Set Date: June 1, 52
   Set Time: 12:00 Noon STD
   Time Zone: -5 hours
   Daylight Saving option: "Not Observed"

2. Find the Sun and  .

3. Now, click on the "Set Specific Time" button as was shown in **Figure 8-3**.

4. When did the vernal equinox occur in 52 C.E.? _____.
(It's March 22nd at 4:04 AM EST

© 2011 Cengage Learning. All Rights Reserved. May not be scanned, copied or duplicated, or posted to a publicly accessible website, in whole or in part.

5. Close the file and do no*t* save it, or continue with **Exercise C**.

Exercise C: Determining When the Vernal Equinox Occurred in 2735 B.C.E.

1. Run *TheSkyX* and set as follows:

   Set Date: October 4, −2734
   Set Time: 12:00 Noon STD
   Time Zone: -5 hours
   Daylight Saving option: "Not Observed"

2. Find the Sun and .

3. When did the vernal equinox occur in 2735 B.C.E.? _____

4. Close the file and do not save it, or continue with **Exercise D**.

Exercise D: Determining When the Vernal Equinox Will Occur in 9999 C.E.

1. Run *TheSkyX* and set as follows:

   Set Date: September 15, 9999
   Set Time: 3:00 PM DST
   Time Zone: -5 hours
   Daylight Saving option: "U.S. and Canada"

2. Find the Sun and 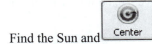 .

3. When will the vernal equinox occur in 9999 C.E.? _____

4. Close the file and do not save it, or continue with **Exercise E**.

Exercise E: Determining the Equatorial Coordinates in 2029 C.E.

1. Run *TheSkyX* and set as follows:

   Set Date: January 10, 2029
   Set Time: 8:00 PM STD
   Time Zone: -5 hours
   Daylight Saving option: "U.S. and Canada"

2. Find Sirius and  .

3. What are the right ascension and declination of Sirius?

   R.A. (2029) = ____^{H} ____^{M} ____^{S} Declination (2029) = __ ____° ____' ____"

© 2011 Cengage Learning. All Rights Reserved. May not be scanned, copied or duplicated, or posted to a publicly accessible website, in whole or in part.

4. Now, find Polaris and [Center] it in your desktop. Click [Go Forward] to observe the motion. Be sure to have set the time flow to at least 1 minute. Turn on the equatorial grid. Describe what you see _____. Where is Polaris in relationship to the equatorial grid? _____

5. Close the file and do not save it, or continue with **Exercise F**.

Exercise F: Determining the Equatorial Coordinates in 52 C.E.

1. Run *TheSkyX* and set as follows:

   Set Date: January 10, 52
   Set Time: 8:00 PM STD
   Time Zone: -5 hours
   Daylight Saving option: "Not Observed"

2. Turn off the equatorial grid if it is still turned on. Find Sirius and [Center] it.

3. What are the right ascension and declination of Sirius?

   R.A. (52) = _____ H _____ M _____ S   Declination (52) = __ _____ ° _____ ' _____ "
4. How do the coordinates compare with **Exercises E**? _____

5. Are they the same or different? _____

6. Now find Polaris and center it. Click [Go Forward] to observe the motion. Turn on the equatorial grid.

7. Describe what you see. _____

8. Where is Polaris in relationship to the equatorial grid? _____

9. Close the file and do not save it, or continue with **Exercise G**.

Exercise G: Determining the Equatorial Coordinates in 9999 C.E.

1. Run *TheSkyX* and set as follows:

   Set Date: January 10, 9999
   Set Time: 8:00 PM STD
   Time Zone: -5 hours
   Daylight Saving option: "Not Observed"

2. Turn off the equatorial grid if it is still turned on. Find Sirius and [Center] it.

3. What are the right ascension and declination of Sirius?

   R.A. (9999) = _____ H _____ M _____ S   Declination (9999) = __ _____ ° _____ ' _____ "

© 2011 Cengage Learning. All Rights Reserved. May not be scanned, copied or duplicated, or posted to a publicly accessible website, in whole or in part.

4. How do the coordinates compare with those in **Exercises E** and **F**? _____
   _____

5. Are they the same or different? _____

6. What do you notice about the location of Sirius with respect to your horizon? _____
   _____

7. Now find Polaris and center it. Click Go Forward to observe the motion. **Turn on the
   equatorial grid.** Describe what you see. _____
   _____. Where is Polaris in
   relationship to the equatorial grid? _____

8. Close the file and do not save it, or continue with **Exercise H**.

Exercise H: Determining the Equatorial Coordinates in 2735 B.C.E

1. Run *TheSkyX* and set as follows:

   Set Date: January 10, −2734
   Set Time: 8:00 PM STD
   Time Zone: -5 hours
   Daylight Saving option: "Not Observed"

2. Find Sirius and Center it.

3. What are the right ascension and declination of Sirius?

   R.A. (-2735) = _____^{H} _____^{M} _____^{S} Declination (-2735) = __ _____$^\circ$ _____$'$ _____$''$

4. How do the coordinates compare with those in **Exercises E, F,** and **G**? _____
   _____

5. Are they the same or different? _____

6. What do you notice about the location of Sirius with respect to your horizon? _____
   _____.

7. Find Polaris and center it. Click Go Forward to observe the motion. **Turn on the
   equatorial grid.** Describe what you see. _____
   _____

8. Is the motion of the sky about Polaris? _____

9. What star is located near the north celestial pole? _____

   Note: The Egyptians used this star as their North Star.

10. Close the file and do not save it.

© 2011 Cengage Learning. All Rights Reserved. May not be scanned, copied or duplicated, or posted to a publicly accessible website, in whole or in part.

The position of the Sun on the celestial sphere is directly over Earth's equator at the time of either equinox. The Sun's position in the sky on the vernal equinox in March is shown in **Figure 8-5**.

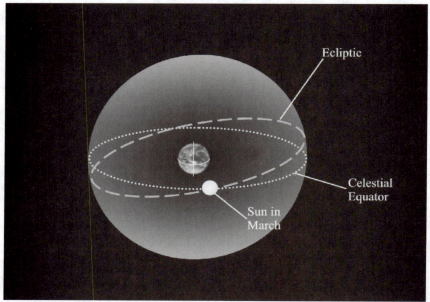

Figure 8-5 Position of Sun on the Celestial Sphere on the Vernal Equinox

If Earth is viewed from space at the time of the vernal equinox, the Sun neither shines directly on the northern or southern hemisphere. In fact, it shines directly on the Earth's equator and equally on both hemispheres, as shown in **Figure 8-6**. The length of days and nights are nearly equal at the time of the vernal equinox.

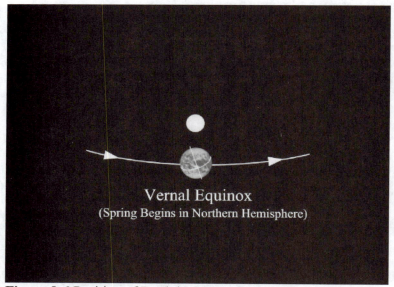

Figure 8-6 Position of Earth in Space on the Vernal Equinox

Two factors contribute to seasons on Earth throughout the year. The first is the length of time the Sun spends above an observer's horizon. The second is the Sun's angular distance above the horizon, or its altitude. Both are a direct result of Earth's tilt.

© 2011 Cengage Learning. All Rights Reserved. May not be scanned, copied or duplicated, or posted to a publicly accessible website, in whole or in part.

From the time that the Sun appears to rise in the east each day, its altitude constantly changes in the sky. At midday, it reaches its highest altitude above the horizon. At this point, it is due south on the local celestial meridian. After midday, the Sun's altitude gets lower until it appears to set in the west. If one watches the apparent sunrise and sunset at the time of the vernal equinox, the Sun appears to rise *due east* and appears to set *due west* on the horizon.

Figure 8-7 illustrates the Sun's position on the eastern horizon on March 20, 2055 for an observer located at the Ball State Observatory at 7:45 AM DST. The Sun reaches the vernal equinox point at precisely 11:30 AM DST on this date. Notice that the equatorial grid is displayed. The sky window plainly shows a dotted-line, representing the ecliptic, intersecting the celestial equator near the East Point on the horizon. If one travels to Earth's equator and observes the apparent sunrise there, the Sun appears to rise due east at that location too.

Figure 8-7 Apparent Sunrise on March 20, 2055 from the Ball State Observatory

Figure 8-8 illustrates the apparent sunrise for an observer located at the geographic equator.

Figure 8-8 Apparent Sunrise on March 20, 2055 at the Geographic Equator

© 2011 Cengage Learning. All Rights Reserved. May not be scanned, copied or duplicated, or posted to a publicly accessible website, in whole or in part.

The altitude of the Sun at midday on the vernal or autumnal equinox is about 50° for an observer located at the 40°north latitude circle. The altitude of the Sun at midday can be found for *any* observer in the Northern Hemisphere by applying the following relationship:

$$\text{Altitude}_{\text{Sun}} = 90° - \text{Latitude}_{\text{observer}}$$

The position of the Sun on March 20, 2055 at midday at is illustrated in **Figure 8-9**. Notice that the horizon grid is displayed. The Sun's position on the local celestial meridian is near the 50° altitude circle. For an observer located at the Ball State Observatory at 1:49 PM DST, the actual altitude of the Sun is 49° 50′ 14″ as is found in the Find Result window.

Figure 8-9 Altitude of Sun at Midday March 20, 2055 at the Ball State Observatory

The position of the Sun for an observer at a different location on Earth on March 20, 2055 at midday is illustrated in **Figure 8-10**. Notice that the horizon grid is once again displayed. It is readily seen that the Sun's position on the local celestial meridian is near the 90° altitude circle. Where is the observer located?

© 2011 Cengage Learning. All Rights Reserved. May not be scanned, copied or duplicated, or posted to a publicly accessible website, in whole or in part.

Figure 8-10 Altitude of Sun at Midday, at a Different Location than in **Figure 8-9**

The answer, of course, is the geographic equator!

Summer Solstice

Each day that follows the vernal equinox, the Sun appears to move farther into the Northern Hemisphere of the sky. A few days after the vernal equinox, the Sun's motion is both eastward and northward along the ecliptic. In the Northern Hemisphere, days become longer and warmer, and those in the Southern Hemisphere become shorter and cooler. It soon becomes apparent that the rising point of the Sun appears farther north on the eastern horizon. To view this change of position in the Sun's rising point on the eastern horizon, complete **Exercise I**, which follows.

Exercise I: Observing the Sun's Apparent Rising Point on Eastern Horizon

1. Run *TheSkyX*.

2. Open the Sky file named ***Season1.skyx***.

3. Click **Step Forward**. The time step is a 2-day interval.

4. After several clicks, observe the position of the dotted-line on the horizon. Describe what happened. _____ _____.

5. Close the file.

© 2011 Cengage Learning. All Rights Reserved. May not be scanned, copied or duplicated, or posted to a publicly accessible website, in whole or in part.

The time sequence, in **Exercise I**, begins at the apparent sunrise on March 20, 2055 and should continue through June 21, 2055. As one continues to advance the time step it also becomes apparent that the Sun's altitude gets higher in the sky almost immediately. This is because the Sun appears to rise earlier each day from March until June. The ecliptic appears to move northward along the eastern horizon—watch the dotted-line.

Viewed from space during this time, the northern half of Earth's axis of rotation begins to align itself on the Sun. Earth has moved $90°$ in its orbit from the vernal equinox. The sunlight begins to shine more directly on the Northern Hemisphere, making the days longer and warmer.

The track of Earth's position during the time from the vernal equinox to the summer solstice can be seen in **Figure 8-11**. As Earth revolves around the Sun, it does *not* lean toward the Sun. Its orientation in space has remained this way all along.

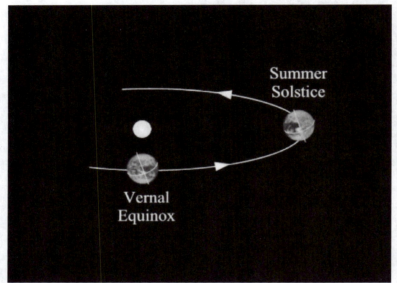

Figure 8-11 Track of Earth's Position from the Vernal Equinox to the Summer Solstice

This position of Earth in space at the time of the summer solstice is illustrated in **Figure 8-12**. As Earth moves in its orbit from the vernal equinox to the location of the summer solstice, an interesting phenomenon appears on the eastern horizon. The apparent rising point of the Sun changes throughout this time interval—it begins to move farther northward. It reaches its most northerly position on the horizon on or about June 21st each year. At the most northerly point on the eastern horizon, the Sun appears to stop its northward motion. This point on the horizon where the Sun appears to stop is called the solstice point. The word *solstice* comes from a Greek word meaning, "Sun standing still" on the horizon. This is seen in the Sky file labeled ***Season1.skyx***.

© 2011 Cengage Learning. All Rights Reserved. May not be scanned, copied or duplicated, or posted to a publicly accessible website, in whole or in part.

Figure 8-12 Position of Earth in Space on the Summer Solstice

The apparent sunrise is illustrated in **Figure 8-13** for an observer located at the **Ball State Observatory** on June 21, 2055 at 6:12 AM DST. The Sun reached the summer solstice point at precisely 4:40 AM DST on this date. At that instant, the Sun's equatorial coordinates on the celestial sphere are right ascension $6^H 00^M 00^S$ and declination $+23° 26'$. Notice that the dotted-line, representing the ecliptic, is approximately 31° north of the East Point at this location (Note: with the Horizon Grid turned-on, one simply subtracts 59° from 90° -- due East).

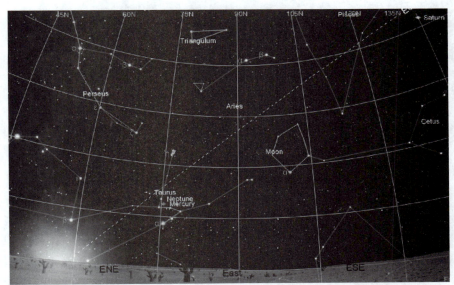

Figure 8-13 Apparent Sunrise on June 21, 2055 at the Ball State Observatory

If an observer is located at the geographic equator, however, the apparent rising point of the Sun is exactly 23° 26′ north of the East Point. The apparent sunrise for an observer located at the geographic equator is displayed in **Figure 8-14**.

© 2011 Cengage Learning. All Rights Reserved. May not be scanned, copied or duplicated, or posted to a publicly accessible website, in whole or in part.

Figure 8-14 Apparent Sunrise at Geographic Equator on June 21, 2055

Figure 8-15 illustrates the position of the Sun on the celestial sphere on the summer solstice in June. It is obvious that sunlight shines most directly on the Northern Hemisphere in this figure.

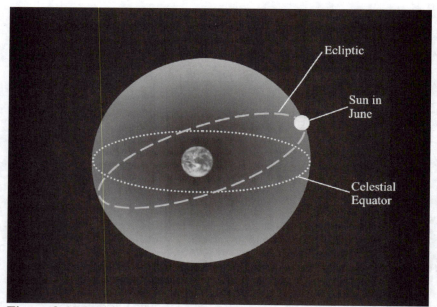

Figure 8-15 Position of Sun on the Celestial Sphere on the Summer Solstice

The Sun has the highest altitude at midday on the summer solstice for any observer located in the Northern Hemisphere. **Figure 8-16** shows the Sun at midday for an observer located at the Ball State Observatory on June 21, 2055. Notice that the horizon grid is displayed. The altitude of the Sun is approximately 73°on June 21st on this date. If one looks in the Find Result window for this date, one finds the altitude of the Sun at midday.

© 2011 Cengage Learning. All Rights Reserved. May not be scanned, copied or duplicated, or posted to a publicly accessible website, in whole or in part.

Is it approximately 73° 13'at this location? _____. The location of the summer solstice point may be viewed in the Sky file named ***SummerSolstice.skyx***.

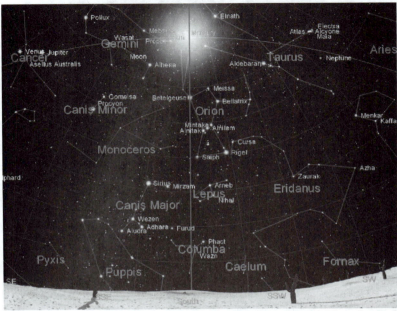

Figure 8-16 Altitude of Sun at Midday on June 21, 2055 at the Ball State Observatory

In fact, for any observer who is located on the 23° 26′ north latitude circle (Tropic of Cancer), at midday, the Sun is located at their zenith. **Exercise J** illustrates the position of the Sun at midday on the Tropic of Cancer on June 21, 2055.

Exercise J: Observing the Sun at Midday on the Tropic of Cancer

1. Run *TheSkyX*.

2. Open the Sky file named ***SummerSolstice.skyx***.

3. Set time: 1:43 PM DST

4. Set latitude: 23° 26′ N

5. Turn on the horizon grid.

6. Find the Sun and ⟨Center⟩.

7. Does your computer screen look like the one displayed in **Figure 8-17**?

© 2011 Cengage Learning. All Rights Reserved. May not be scanned, copied or duplicated, or posted to a publicly accessible website, in whole or in part.

Figure 8-17 Sun's Location in the Sky at Midday on June 21, 2055 on the Tropic of Cancer

8. What is the altitude of the Sun? _____°

9. Close the file and do not save it.

The **Exercises K**, **L**, **M**, and **N**, that follow are designed to assist in determining when summer occurs in the Northern Hemisphere for a particular year.

Exercise K: Determining When the Summer Solstice Will Occur in 2029 C.E.

1. Run *TheSkyX* and open the file *Normal.skyx* and set it follows:

   Set Date: February 2, 2029
   Set Time: 8:00 PM STD
   Time Zone: - 5 hours
   Daylight Saving option: "U.S. and Canada"

2. Find the Sun and  .

3. Click "Input" at the top of the Standard toolbar window, then the "Date and Time" menu button.

4. Now, click on the "Set Specific Time" button that was shown in **Figure 8-3**.

5. Click "Summer Solstice" in the Set Specific Time menu.

6. When does the summer solstice occur in 2029 C.E.? _____
   (June 21ˢᵗ at 9:49 PM DST

7. Close the file and do not save it, or continue with **Exercise L**.

© 2011 Cengage Learning. All Rights Reserved. May not be scanned, copied or duplicated, or posted to a publicly accessible website, in whole or in part.

Exercise L: Determining When the Summer Solstice Occurred in 52 C.E.

1. Run *TheSkyX* and open the file *Normal.skyx*, or set it follows:

   Set Date: March 1, 52
   Set Time: 8:00 AM STD
   Time Zone: - 5 hours
   Daylight Saving option: "Not Observed"

2. Find the Sun and .

3. Now, click on the "Set Specific Time" button that was shown in **Figure 8-3**.

4. When did the summer solstice occur in 52 C.E.?
   (June 24th at 2:43 AM STD

5. Close the file and do not save it, or continue with **Exercise M**.

Exercise M: Determining When the Summer Solstice Occurred in 2735 B.C.E.

1. Run *TheSkyX* and open the file *Normal.skyx*, or set it follows:

   Set Date: July 4, −2734
   Set Time: 12:00 Noon STD
   Time Zone: - 5 hours
   Daylight Saving option: "Not Observed"

2. Find the Sun and .

3. When did the summer solstice occur in 2735 B.C.E.? _____

4. Close the file and do not save it, or continue with **Exercise N**.

Exercise N: Determining When the Summer Solstice Will Occur in 9999 C.E.

1. Run *TheSkyX* and open the file *Normal.skyx*, or set it follows:

   Set Date: September 15, 9999.
   Set Time: 3:00 PM DST
   Time Zone: - 5 hours
   Daylight Saving option: "U.S. and Canada"

2. Find the Sun and  .

3. When does the summer solstice occur in 9999 C.E.? _____

4. Close the file and do not save it.

© 2011 Cengage Learning. All Rights Reserved. May not be scanned, copied or duplicated, or posted to a publicly accessible website, in whole or in part.

Autumnal Equinox

Each day that follows the summer solstice, the Sun continues its eastward motion along the ecliptic. A few days after the summer solstice the Sun's apparent motion is both eastward and southward along the ecliptic. It becomes evident once again that the point where the ecliptic intersects the eastern horizon appears to be continually moving…but this time the motion is southward.

The Sun eventually crosses the celestial equator and moves from the Northern Hemisphere of the sky into the Southern Hemisphere. This occurs on or about September 21st each year and is known as the *autumnal equinox*. If the Sun happens to be located at the autumnal equinox, then this point in space becomes a point in time. That is, it is the first day of autumn in the Northern Hemisphere. The equatorial coordinates of the Sun on the celestial sphere at the time of the autumnal equinox are right ascension $12^H 00^M 00^S$ and declination $00° 00' 00''$. In other words, the Sun's location on the celestial sphere is precisely on the celestial equator once again.

Figure 8-18 displays the location of the autumnal equinox in *TheSkyX*. The view is from the Ball State Observatory.

Figure 8-18 Location of the Autumnal Equinox as Shown in *TheSkyX*

The autumnal equinox is positioned on the local celestial meridian once again. The altitude of this point is about 50°, and its azimuth is 180°. At the bottom and right edge of the window the right ascension and declination coordinates are displayed, respectively.

If one toggles on the constellation boundaries, this point appears to be located near the western edge of the constellation of Virgo. In about 700 to 800 years from now, this point will move into the constellation of Leo and will take nearly 2000 years to move through it. This apparent motion is the result of Earth's precession. The sky window in **Figure 8-18** may be viewed in the Sky file named ***AutumnalEquinox.skyx***.

As Earth moves in its orbit from the summer solstice to the autumnal equinox, the days in the Northern Hemisphere become shorter and cooler and those in the Southern Hemisphere become longer and warmer. The apparent rising point of the Sun continues to change position along the eastern horizon. It appears to be moving southward now. To view the change of the Sun's position on the eastern horizon, complete **Exercise O**.

© 2011 Cengage Learning. All Rights Reserved. May not be scanned, copied or duplicated, or posted to a publicly accessible website, in whole or in part.

Exercise O: Observing the Sun's Apparent Rising Point on Eastern Horizon

1. Run *TheSkyX*.

2. Open the Sky file named ***Season2.skyx***.

3. Click 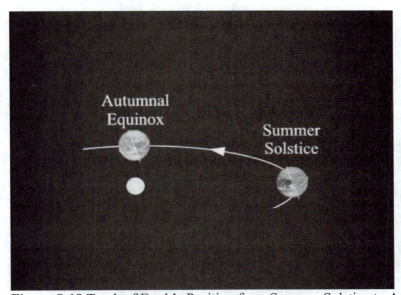 Step Forward. The time step is a 2-day interval.

4. After several clicks, observe the position of the dotted-line on the horizon. Describe what happened. _____.

    If you are daring, you may click Go Forward .

5. Close the file and do not save it.

   The time sequence, in **Exercise O**, begins with the apparent sunrise on June 21, 2055 and should continue through September 22, 2055 for an observer located at the Ball State Observatory. Notice that the Sun immediately disappears below the eastern horizon. This is because the Sun appears to rise later each day from June until September. The ecliptic appears to move southward along the eastern horizon—watch the dotted-line again.

   Earth has moved an additional 90° from the vernal equinox. The sunlight begins to shine less on the Northern Hemisphere and more so on the Southern Hemisphere. The track of Earth's position during the time from the summer solstice to the autumnal equinox can be seen in **Figure 8-19**.

Autumnal Equinox

Summer Solstice

Figure 8-19 Track of Earth's Position from Summer Solstice to Autumnal Equinox

   As Earth continues its journey around the Sun, it does *not* lean toward or away from the Sun. Its orientation in space remains the same as before. The Sun shines directly onto Earth's equatorial region again and equally on both hemispheres, as shown in **Figure 8-20**.

   If Earth is viewed from space at the time of the autumnal equinox, the Sun neither shines on the northern or southern hemisphere directly once again. Earth now is exactly 180° in its orbit from the time of the vernal equinox. The lengths of days and nights are nearly equal at the time of the autumnal equinox. The autumnal equinox occurs at precisely 8:50 PM DST at the Ball State Observatory on September 22, 2055. The Sun's equatorial

© 2011 Cengage Learning. All Rights Reserved. May not be scanned, copied or duplicated, or posted to a publicly accessible website, in whole or in part.

coordinates on the celestial sphere at the time of the autumnal equinox are right ascension $12^H 00^M 00^S$ and declination $00° 00' 00''$. That is, the Sun is precisely on the celestial equator exactly 180° from the vernal equinox.

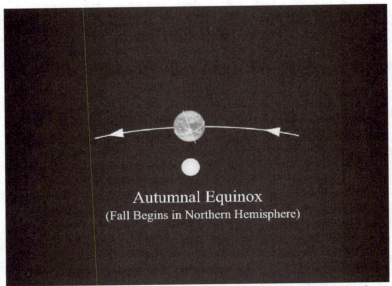

Figure 8-20 Position of Earth in Space on the Autumnal Equinox

The position of the apparent rising point of the Sun once again changes throughout this time sequence. It reaches the East Point on the horizon on or about September 21st each year, and the Sun appears to rise due east as it did on the vernal equinox as was shown in **Figure 8-7**. At the beginning of this sequence, June 21st, the Sun appears to rise at 6:12 AM DST while at the end of the sequence, September 22nd; it appears to rise at 7:10 AM DST.

The apparent sunrise for an observer located at the Ball State Observatory on September 22, 2055 is illustrated in **Figure 8-21**. Notice that the equatorial grid is displayed once again. The Sun reaches the autumnal equinox point at precisely 8:50 PM DST on this date. The sky window again clearly shows the dotted-line, representing the ecliptic, intersecting the celestial equator near the East Point on the horizon.

Figure 8-21 Apparent Sunrise, September 22, 2055 from the Ball State Observatory

© 2011 Cengage Learning. All Rights Reserved. May not be scanned, copied or duplicated, or posted to a publicly accessible website, in whole or in part.

If one travels to the geographic equator and observe the apparent sunrise there, the Sun once again appears to rise due east at that location. **Figure 8-22** illustrates the apparent sunrise for an observer at the geographic equator. The equatorial grid is displayed in this sky window.

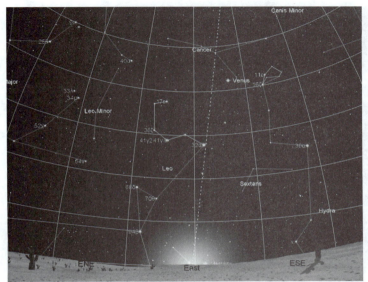

Figure 8-22 Apparent Sunrise, September 22, 2055 at the Geographic Equator

At the Ball State Observatory, the Sun's altitude constantly changes throughout the morning hours, and it eventually crosses the local celestial meridian. At midday, the Sun's altitude while crossing the local celestial meridian is the same as it was on the vernal equinox. The Sun's position in the sky looks exactly as it did in **Figure 8-9**.

Figure 8-23 shows the Sun's position in the sky at midday for an observer at the Ball State Observatory at 1:34 PM DST. Notice that the horizon grid is displayed again. The Sun's position on the local celestial meridian is near the 50° altitude circle once again. For an observer located at the Ball State Observatory the actual altitude of the Sun is 49°55′ 01″ as is found in the Find Result window.

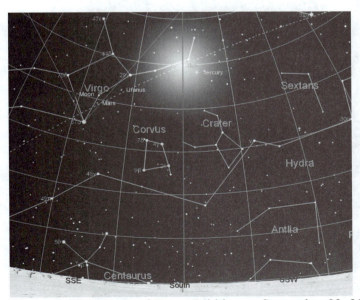

Figure 8-23 Altitude of Sun at Midday on September 22, 2055 at the Ball State Observatory

© 2011 Cengage Learning. All Rights Reserved. May not be scanned, copied or duplicated, or posted to a publicly accessible website, in whole or in part.

The position of the Sun for an observer at a different location on Earth on September 22, 2055, at midday is illustrated in **Figure 8-24**. Notice that the horizon grid is once again displayed. Where is the observer's location on Earth? _____.

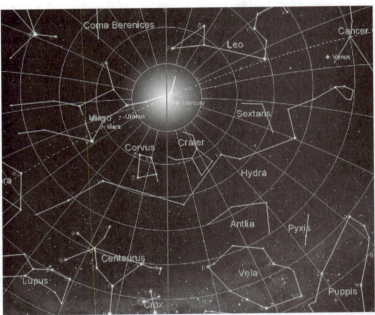

Figure 8-24 Altitude of the Sun at Midday on September 22, 2055 from an Undisclosed Location on Earth

The location of the observer in **Figure 8-24** is, of course, the geographic equator.

The position of the Sun on the celestial sphere is directly over Earth's equator at the time of the autumnal equinox. The Sun's position in the sky on the autumnal equinox in September is shown in **Figure 8-25**. It is obvious that the Sun shines directly on both hemispheres in September, as it did in March.

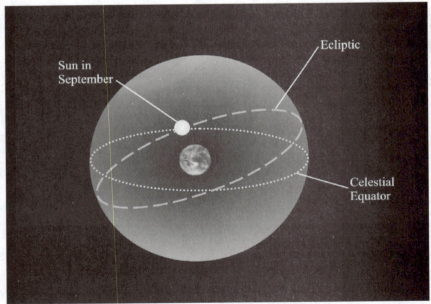

Figure 8-25 Position of the Sun on the Celestial Sphere on the Autumnal Equinox

© 2011 Cengage Learning. All Rights Reserved. May not be scanned, copied or duplicated, or posted to a publicly accessible website, in whole or in part.

Exercises **P**, **Q**, **R**, and **S**, which follow, are designed to assist in determining when autumn occurs in the Northern Hemisphere for a particular year.

Exercise P: Determining When the Autumnal Equinox Will Occur in 2292 C.E.

1. Run *TheSkyX* and set as follows:

   Set Date: March 15, 2292
   Set Time: 10:00 AM DST
   Time Zone: -5 hours
   Daylight Saving option: "U.S. and Canada"

2. Find the Sun and .

3. Now, click on the "Set Specific Time" button that was shown in **Figure 8-3**.

4. When does the autumnal equinox occur in 2292 C.E.? _____
   (The answer is September 22nd at 4:55 AM DST)

5. Close the file and do not save it, or continue with **Exercise Q**.

Exercise Q: Determining When the Autumnal Equinox Occurred in 52 C.E.

1. Run *TheSkyX* and set as follows:

   Set Date: March 1, 52 C.E.
   Set Time: 8:00 AM STD
   Time Zone: -5 hours
   Daylight Saving option: "Not Observed"

2. Find the Sun and .

3. Now, click on the "Set Specific Time" button that was shown in **Figure 8-3**.

4. When did the autumnal equinox occur in 52 C.E.? _____
   (The answer is September 24th at 2:31 PM STD

5. Close the file and do not save it, or continue with **Exercise R**.

Exercise R: Determining When the Autumnal Equinox Occurred in 2735 B.C.E.

1. Run *TheSkyX* and set as follows:

   Set Date: July 4, −2734
   Set Time: 12:00 Noon STD
   Time Zone: -5 hours
   Daylight Saving option: "Not Observed"

© 2011 Cengage Learning. All Rights Reserved. May not be scanned, copied or duplicated, or posted to a publicly accessible website, in whole or in part.

2. Find the Sun and .

3. When did the autumnal equinox occur in 2735 B.C.E.? _____

4. Close the file and do not save it, or continue with **Exercise S**.

Exercise S: Determining When the Autumnal Equinox Will Occur in 9999 C.E.

1. Run *TheSkyX* and set as follows:

   Set Date: September 15, 9999
   Set Time: 3:00 PM DST
   Time Zone: -5 hours
   Daylight Saving option: "U.S. and Canada"

2. Find the Sun and .

3. When does the autumnal equinox occur in 9999 C.E.? _____

4. Close the file and do not save it.

Winter Solstice

   After the Sun crosses the celestial equator on the autumnal equinox it begins to move into the Southern Hemisphere of the sky. Each day that follows the autumnal equinox, the Sun's motion appears to carry it farther and farther into the Southern Hemisphere of the sky. Several days after the autumnal equinox the Sun's motion continues to be eastward and southward.

   If Earth is viewed from space during this time, the southern half of its axis of rotation begins to align itself on the Sun. Earth has now moved 90° in its orbit from the autumnal equinox and 270° from the vernal equinox. Sunlight begins to shine more directly on the Southern Hemisphere and less so on the Northern Hemisphere. The days in the Northern Hemisphere become shorter and cooler, and those in the Southern Hemisphere become longer and warmer.

   The track of Earth's position during the time from the autumnal equinox to the winter solstice can be seen in **Figure 8-26**. While Earth moves around the Sun during this time period, it does *not* lean away from the Sun.

© 2011 Cengage Learning. All Rights Reserved. May not be scanned, copied or duplicated, or posted to a publicly accessible website, in whole or in part.

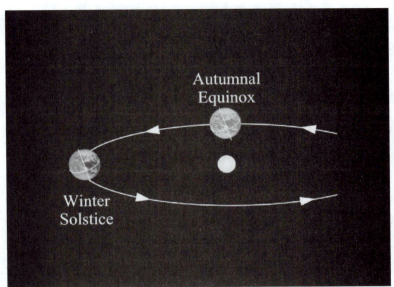

Figure 8-26 Track of Earth's Position from Autumnal Equinox to Winter Solstice

Its orientation in space has remained the same way as before. The position of Earth in space at the time of the winter solstice is shown in **Figure 8-27**.

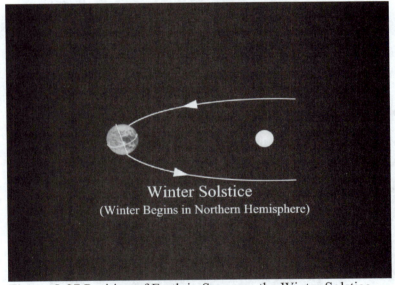

Figure 8-27 Position of Earth in Space on the Winter Solstice

As before, it becomes apparent once again, the rising point of the Sun appears to be moving farther South on the eastern horizon. To view this change of position in the Sun's rising point on the eastern horizon, complete **Exercise T**.

Exercise T: Observing the Sun's Apparent Rising Point on Eastern Horizon

1. Run *TheSkyX*.

2. Open the Sky file named ***Season3.skyx***.

3. Click Step Forward . The time step is a 2-day interval.

© 2011 Cengage Learning. All Rights Reserved. May not be scanned, copied or duplicated, or posted to a publicly accessible website, in whole or in part.

4. After several clicks, observe the position of the dotted-line on the horizon. Describe what happened. _____.

   If you are daring, you may click Go Forward.

5. Close the file and do not save it.

The time sequence, in **Exercise T**, begins at the apparent sunrise on September 22, 2055 and continues to December 21, 2055 for an observer located at the Ball State Observatory. Almost immediately the Sun disappears below the eastern horizon. This is because the Sun appears to rise later each day from September until December. The ecliptic appears to continue its southward motion along the eastern horizon—watch the dotted-line along the eastern horizon.

As Earth moves in its orbit from the autumnal equinox throughout this time interval, the intersection of the ecliptic and the horizon continues to change its position along the eastern horizon. The apparent rising point of the Sun also appears to change its position on the eastern horizon throughout this period of time. It is now moving southward along the horizon. It reaches its most southerly position on the horizon on or about December 21st each year.

At the most southerly point on the eastern horizon, the Sun appears to stop its southward motion. This is the other solstice point. That is, the Sun "stands still" on the horizon once again. The Sun has reached the winter solstice point. At the beginning of the sequence in *Season3.skyx*, September 22nd, the Sun appears to rise at 7:29 AM DST, and at the end of the sequence, December 21st, the Sun appears to rise at 8:00 AM STD. During this time interval, Earth does *not* lean away from the Sun. Its orientation in space has always been this way as before!

The apparent sunrise is shown in **Figure 8-28** for an observer located at the Ball State Observatory on December 21, 2055 at 8:01 AM STD. The dotted-line, representing the Sun's apparent path, is located approximately 31° south of due east at this location. The winter solstice occurs at precisely at 4:57 PM STD on this date. Notice that the horizon grid is displayed. The Sun's equatorial coordinates on the celestial sphere at the time of the winter solstice are right ascension $18^H 00^M 00^S$ and declination $-23° 26'$. This means, of course, that the Sun is an additional 90° from the autumnal equinox and 270° from the vernal equinox.

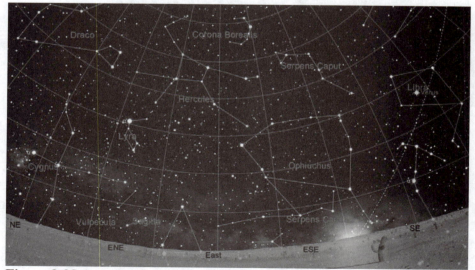

Figure 8-28 Apparent Sunrise on December 21, 2055 at the Ball State Observatory

© 2011 Cengage Learning. All Rights Reserved. May not be scanned, copied or duplicated, or posted to a publicly accessible website, in whole or in part.

If an observer is located at the geographic equator, the apparent rising point of the Sun is exactly 23° 26′ south of the East Point. The apparent sunrise for an observer located at the geographic equator is displayed in **Figure 8-29**. Once again the horizon grid displayed.

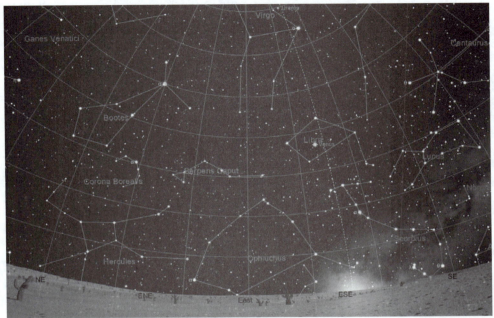

Figure 8-29 Apparent Sunrise on December 21, 2055 at the Geographic Equator

On the winter solstice, the Sun has the lowest altitude at midday for any observer located in the Northern Hemisphere. **Figure 8-30** shows the position of the Sun at midday for an observer located at the Ball State Observatory on December 21, 2055. Notice that the horizon grid is displayed. The Sun has an altitude of about 26° on December 21st at the Ball State Observatory. If one looks in the Find Result window for this date, one will find the altitude of the Sun at midday. Is it 26°22′04′ at this location? _____. The location of the winter solstice point may be viewed in the Sky file named *WinterSolstice.skyx*.

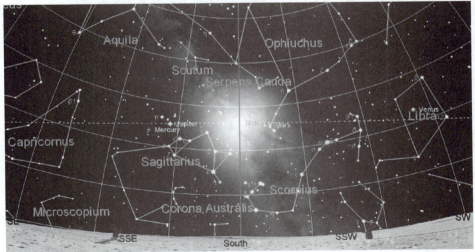

Figure 8-30 Altitude of the Sun at Midday on December 21, 2055 at Ball State Observatory

© 2011 Cengage Learning. All Rights Reserved. May not be scanned, copied or duplicated, or posted to a publicly accessible website, in whole or in part.

For any observer located on the geographic equator, the Sun's position at midday is exactly 23°26′ south of their zenith. In fact, the Sun is located at the zenith of observers who live on the 23°26′ south latitude circle (Tropic of Capricorn) at midday. To view the Sun from the Tropic of Capricorn on December 21, 2055, complete **Exercise U**.

Exercise U: Observing the Sun at Midday on the Tropic of Capricorn

1. Run *TheSkyX*.

2. Open the Sky file named ***WinterSolstice.skyx***.

3. Set time: 12:40 PM STD

4. Set Latitude: 23° 26′ S.

5. Click Show Horizon Grid .

6. Find the Sun and Center it.

7. Does your computer screen look like the one displayed in **Figure 8-31**? _____.

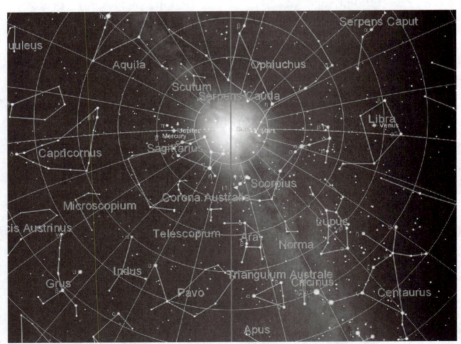

Figure 8-31 Sun's Location in the Sky at Midday on December 21, 2055 on the Tropic of Capricorn

8. What is the altitude of the Sun? _____°

9. Close the file and do not save it.

© 2011 Cengage Learning. All Rights Reserved. May not be scanned, copied or duplicated, or posted to a publicly accessible website, in whole or in part.

Figure 8-32 illustrates the position of the Sun on the celestial sphere on the winter solstice. It is obvious that sunlight shines most directly on the Southern Hemisphere in December.

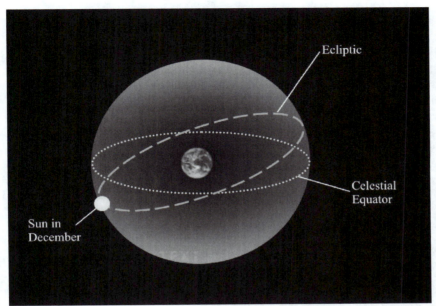

Figure 8-32 Position of Sun on the Celestial Sphere on the Winter Solstice

Exercises V, W, X, and **Y** are designed to assist in determining when winter occurs for a particular year in the Northern Hemisphere.

Exercise V: Determining When the Winter Solstice Will Occur in 2292 C.E.

1. Run *TheSkyX* and set as follows:

    Set Date: November 15, 2292
    Set Time: 10:00 AM STD
    Time Zone: -5 hours
    Daylight Saving option: "U.S. and Canada"

2. Find the Sun and ⟨Center⟩.

3. Now, click on the "Set Specific Time" button that was shown in **Figure 8-3**.

4. When does the winter solstice occur in 2292 C.E.? _____
    (The answer is December 21ˢᵗ at 5:21 AM STD)

5. Close the file and do not save it, or continue with **Exercise W**.

Exercise W: Determining When the Winter Solstice Occurred in 52 C.E.

1. Run *TheSkyX* and set as follows:

    Set Date: October 12, 52 C.E.

© 2011 Cengage Learning. All Rights Reserved. May not be scanned, copied or duplicated, or posted to a publicly accessible website, in whole or in part.

Set Time: 8:00 AM STD
Time Zone: -5 hours
Daylight Saving option: "Not Observed"

2. Find the Sun and .

3. Now, click on the "Set Specific Time" button that was shown in **Figure 8-3**.

4. When did the winter solstice occur in 52 C.E.? _____
   (The answer is December 22nd at 7:42 AM STD)

5. Close the file and do not save it, or continue with **Exercise X**.

Exercise X: Determining When the Winter Solstice Occurred in 2735 B.C.E.

1. Run *TheSkyX* and set as follows:

   Set Date: April 1, −2735
   Set Time: 12:00 Noon STD
   Time Zone: -5 hours
   Daylight Saving option: "Not Observed"

2. Find the Sun and .

3. When does the winter solstice occur in 2735 B.C.E.? _____

4. Close the file and do not save it, or continue with **Exercise Y**.

Exercise Y: Determining When the Winter Solstice Will Occur in 9999 C.E.

1. Run *TheSkyX* and set as follows:

   Set Date: September 29, 9999
   Set Time: 3:00 PM DST
   Time Zone: -5 hours
   Daylight Saving option: "U.S. and Canada"

2. Find the Sun and .

3. When does the winter solstice occur in 9999 C.E.? _____

4. Close the file and do not save it, or continue with **Exercise Z**.

Each day that follows the winter solstice, the Sun appears to continue its motion eastward in the sky. A few days after the winter solstice the Sun's motion is both eastward and northward along the ecliptic. The days become longer and warmer in the Northern Hemisphere and those in the Southern Hemisphere become shorter and cooler.

© 2011 Cengage Learning. All Rights Reserved. May not be scanned, copied or duplicated, or posted to a publicly accessible website, in whole or in part.

It becomes apparent once more that the rising point of the Sun appears to be moving farther north on the eastern horizon. To view the rising point of the Sun change its position on the eastern horizon during this time interval, complete **Exercise Z**.

Exercise Z: Observing the Sun's Apparent Rising Point on Eastern Horizon

1. Run *TheSkyX*.

2. Open the Sky file named ***Season4.skyx***.

3. After several clicks, observe the position of the dotted-line on the horizon. Describe what happened. _____.

   If you are daring, you may click the [Go Forward] button.

4. Close the file and do not save it.

The time sequence in **Exercise Z** begins at the apparent sunrise on December 21, 2055 and continues through March 19, 2056 for an observer at the Ball State Observatory. The horizon grid is displayed and the apparent sunrise is approximately 31° south of the East

Point at the beginning of this sequence. After clicking [Step Backward] once or twice, the observer notices right away that the Sun's altitude once again gets higher in the sky. This is because the Sun appears to rise earlier each day from December until March. Once again the ecliptic appears to continue its northward motion along the eastern horizon. That is, the apparent rising point of the Sun continues to change throughout this time sequence as well.

The Sun appears to be heading back toward the celestial equator. It will take three months for it to reach the celestial equator and the vernal equinox once again. The Sun finally reaches the East Point on the horizon on March 19, 2056. At the beginning of the sequence on December 21, 2055 the Sun appears to rise at 8:01 AM STD whereas at the end of the sequence on March 19, 2056 the Sun appears to rise at 7:45 AM DST for an observer at the Ball State Observatory. The Sun reaches the vernal equinox point at precisely 5:12 PM DST on this date. The exact location of the Sun on the celestial sphere is once again at right ascension $00^H 00^M 00^S$ and declination $00° 00' 00''$. The apparent rising point of the Sun is shown in **Figure 8-33** for an observer located at the Ball State Observatory on March 19, 2056.

© 2011 Cengage Learning. All Rights Reserved. May not be scanned, copied or duplicated, or posted to a publicly accessible website, in whole or in part.

Figure 8-33 Apparent sunrise on March 19, 2056 from the Ball State Observatory

If Earth is viewed from space during this interval of time, its southern half of its axis of rotation begins to move away from its alignment on the Sun. The Earth has moved 90° in its orbit from the winter solstice and 360° from the vernal equinox point where it started in 2055. The Sun has finally come full circle to where we began our journey and our discussion of the seasons.

The track of Earth's position during the time from the winter solstice to the vernal equinox can be seen in **Figure 8-34**. As Earth continues its motion around the Sun, once again it does *not* lean away from the Sun; its orientation in space has remained the same as it has throughout the year.

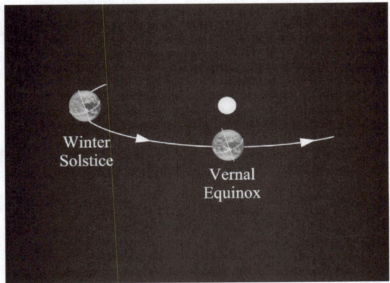

Figure 8-34 Track of Earth's Position from Winter Solstice to Vernal Equinox

The Sun once again shines directly onto the Earth's equatorial region and equally on both hemispheres, as shown in **Figure 8-35**. The lengths of days and nights are nearly

© 2011 Cengage Learning. All Rights Reserved. May not be scanned, copied or duplicated, or posted to a publicly accessible website, in whole or in part.

equal, as they were when we started a year ago. It is springtime once more in the Northern Hemisphere! And so it goes, year after year, after year.

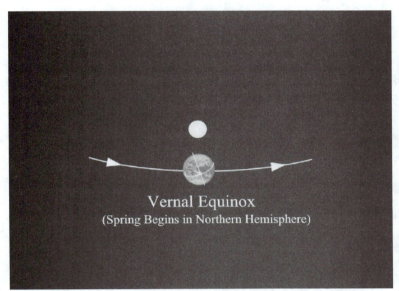

Figure 8-35 Position of Earth in Space on the Vernal Equinox

Observers who will watch the sky over the decades between 2001 and 2055 will notice the same constellations appear, disappear, and reappear at precisely the same time of year, each year. Even though the dates (days) of the seasons change throughout this period of time, it soon becomes apparent that the seasons occur when the Sun unequivocally reaches the same point on the celestial sphere each year. The appearance of these seasonal constellations was the basis for ancient time systems and calendars.

One such ancient calendar is Stonehenge; perhaps the most famous of the ancient calendars (See **Figure 8-36**). If you stand in the center of these stones, you can accurately ascertain the date by observing the position of the Sun's apparent rising point on the eastern horizon. Stonehenge, like many other assemblies of megaliths in England, was built by prehistoric Celts thousands of years ago. Most such calendars are as accurate today as they were in ancient times.

Figure 8-36 Stonehenge near Salisbury, England (Photo Courtesy of Elle Smith)

The North American continent is no stranger to ancient calendars. Native Americans apparently built calendars similar to Stonehenge on this continent too. Their scale pales in

© 2011 Cengage Learning. All Rights Reserved. May not be scanned, copied or duplicated, or posted to a publicly accessible website, in whole or in part.

comparison to the megalith calendars in Europe, but they were quite accurate nonetheless. **Figure 8-37** shows the Bighorn Medicine Wheel on Medicine Mountain, located west of Sheridan, Wyoming, built about 7500 years ago.

Figure 8-37 Bighorn Medicine Wheel near Sheridan, Wyoming,
(Photo Courtesy of U.S. Forest Service).

Modern calendars have incorporated the motion of Earth's precession into their production. The seasons we recognize today *will* begin on about the same dates in the future as they do in this epoch. And the holidays we celebrate today will probably be celebrated on the same dates in the future. That is, most people will celebrate Christmas on December 25th and *Easter* will be (as it is today) the *first* Sunday *after* the *first* full Moon *after* the *vernal equinox* each year.

The nighttime sky, however, will tell a different story because of Earth's precession. The constellations you see today in your summer sky will be some of the ones people will see in the winter sky thousands of years from now, while those in your winter sky today will be some of the summer constellations seen in 13,000 years. Halfway through the precessional cycle of Earth, the bright star Vega will become our North Star, as it was in 11,000 B.C.E. *TheSkyX* lets you view the projected results of Earth's precession.

As one becomes more and more proficient using *TheSkyX*, it is possible to take trips into the future or back into the ancient past. It is left as an exercise for the reader to make these trips and to explore the capabilities of *TheSkyX*. Who knows what ancient mysteries *you* might discover or uncover in pursuing time travel explorations using *TheSkyX*? *TheSkyX* lets you set the date of the sky window from 4713 B.C.E. to 9,999 C.E. This involves a lot of events in human history.

© 2011 Cengage Learning. All Rights Reserved. May not be scanned, copied or duplicated, or posted to a publicly accessible website, in whole or in part.

Chapter 8

TheSkyX Review Exercises

Answer these questions using *TheSkyX* software

TheSkyX Review Exercise 1: Determining the Date of the Vernal Equinox

1. On what date does the vernal equinox occur in the year 2025C.E.? _____

2. What time does the event occur in Question 1 in Washington, D.C.? _____:_____

3. What time does the event occur in Question 1 in Waterloo, IA? _____: _____

4. What time does the event occur in Question 1 in Seattle, Washington? _____:_____

5. What time does the event occur in Question 1 in Cleveland, Ohio? _____:_____

6. In what constellation does the Sun appear to be located in Question 1?
    _____

7. In what constellation does the Sun appear to be located in the year 2825 C.E.?
    _____

8. Does the Sun appear to be located in the same constellation in Questions 6 and 7? _____.
    If not, why not? _____.

TheSkyX Review Exercise 2: Determining the Date of the Summer Solstice

1. On what date does the summer solstice occur in the year 6500 C.E.? _____

2. What time does the event in Question 1 occur in Greenwich, England? _____:_____

3. What time does the event in Question 1 occur in New York City, NY? _____:_____

4. What time does the event in Question 1 occur at the Ball State Observatory? _____:_____

5. What time does the event in Question 1 occur in Los Angeles, CA? _____:_____

6. What time does the event in Question 1 occur in Sydney, Australia? _____:_____

7. In what constellation does the Sun appear to be located on the date in Question 1?
    _____

8. In what constellation doe the Sun appear to be located in 4500 C.E.?
    _____

9. Does the Sun appear to be located in the same constellation in Questions 7 and 8? _____.
    If not, why not? _____.

© 2011 Cengage Learning. All Rights Reserved. May not be scanned, copied or duplicated, or posted to a publicly accessible website, in whole or in part.

TheSkyX **Review Exercise 3: Determining the Date of the Autumnal Equinox**

1. On what date does the autumnal equinox occur in the year 2215 C.E.? _____

2. What time does the event in Question 1 occur in Greenwich, England? _____:_____

3. What time does the event in Question 1 occur in Columbus, GA? _____:_____

4. What time does the event in Question 1 occur in Miami, FL? _____:_____

5. What time does the event in Question 1 occur in Cleveland, OH? _____:_____

6. In what constellation does the Sun appear to be located on the date in Question 1?
    _____

7. In what constellation does the Sun appear to be located in the year 5515 C.E.?
    _____

8. Does the Sun appear to be located in the same constellation in Questions 7 and 8? _____.
    If not, why not? _____.

TheSkyX **Review Exercise 4: Determining the Date of the Winter Solstice**

1. On what date does the winter solstice occur in the year 8592 C.E.? _____

2. At what time does the event in Question 1 occur in Atlanta, GA? _____:_____

3. At what time does the event in Question 1 occur in Athens, Greece? _____:_____

4. At what time does the event in Question 1 occur in Sydney, Australia? _____:_____

5. In what constellation does the Sun appear to be located in Question 1?

6. In what constellation does the Sun appear to be located in the year 1492 C.E.?

7. Does the Sun appear to be located in the same constellations in Questions 5 and 6? _____.
    If not, why not? _____.

© 2011 Cengage Learning. All Rights Reserved. May not be scanned, copied or duplicated, or posted to a publicly accessible website, in whole or in part.

Chapter 8

TheSkyX Review Questions

Answer these following review questions concerning the material in the chapter.

1. What is the name of the point in space where the Sun appears to cross the celestial equator from South to North? _____ _____

2. On what date does the Sun appear to cross the point on the celestial equator in Question 1? _____

3. What are the Sun's right ascension and declination at the vernal equinox?

    R.A. = _____^{H} _____^{M} Declination = __ _____$^°$ _____$'$

4. On what date does the Sun appear to be 23½° North of the celestial equator? _____

5. What is the name of the point on the celestial sphere in Question 4? _____ _____

6. What are the right ascension and declination of the Sun on approximately June 21st?

    R.A. = _____^{H} _____^{M} Declination = __ _____$^°$ _____$'$

7. On what date does the Sun appear to cross the celestial equator from North to South? _____

8. What is the name of the point on the celestial sphere in Question 7? _____ _____

9. What are the right ascension and declination of the Sun at the time of the autumnal equinox?

    R.A. = _____^{H} _____^{M} Declination = __ _____$^°$ _____$'$

10. On what date does the Sun appear to be 23½° South of the celestial equator? _____

11. What is the name of the point on the celestial sphere in Question 10? _____ _____

12. What are the right ascension and declination of the Sun on approximately December 21st?

    R.A. = _____^{H} _____^{M} Declination = __ _____$^°$ _____$'$

© 2011 Cengage Learning. All Rights Reserved. May not be scanned, copied or duplicated, or posted to a publicly accessible website, in whole or in part.

Phases and Eclipses in *TheSkyX*

Motion of the Moon in *TheSkyX*

One of the most obvious objects that people notice in the night sky is the Moon. For millennia humans have observed it and constructed calendars by it. It really is something that is straightforward to observe. It is an easy task to observe the Moon—only time must be taken to do it.

The ancient Greeks considered the Moon a planet because, it moved with respect to the constellations. However, today we realize that it is not a planet but merely a satellite of Earth. Like, the Sun, it moves on a well-defined path in the sky near the ecliptic. Its path, however, is tilted 5° 8′ 43″ with respect to the Sun's apparent path. As the Moon revolves around Earth, its declination varies from the declination of the ecliptic in the sky during the course of a month. Sometimes the Moon is above the ecliptic, and sometimes it is below.

The Moon's motion among the stars is easy to observe, because it moves its own diameter every hour. Its eastward motion amounts to approximately ½° per hour, which is approximately 13° each day relative to the stars. In fact, its angular motion is by far greater than any other astronomical object observed in the sky. In **Exercise A**, the path and motion of the Moon is illustrated with respect to the stars and the ecliptic.

Exercise A: Motion of the Moon

1. Run *TheSkyX*.

2. Open the Sky file named ***MoonMotion.skyx***.

3. In this Sky file the Moon is located 3° north of the bright star Aldebaran.

4. Click [Go Forward] to observe the Moon's motion relative to the background stars. The motion is in 1-hour time intervals.

5. Click [Move Left] on the Orientation toolbar to follow the Moon through a complete cycle in the sky.

6. Notice too that the Moon not only changes its position in the sky but also goes through a phasing process.

7. Change the Time Flow rate from 1 hour to 1 sidereal day (23.934^H).

8. Click [Step Forward] to observe how much the Moon moves each day and how its phase changes.

9. Close the file and do not save it.

The Moon takes about a month to complete one circuit of the sky. This is the Moon's sidereal period. The actual time is 27.322 days. During this time, as it moves counterclockwise around Earth its position with respect to the Sun constantly changes. As a result of this motion, different amounts of the lunar surface are illuminated and become visible from Earth's surface. In other words, as one observes the Moon from Earth, it appears to go through a series of phases.

© 2011 Cengage Learning. All Rights Reserved. May not be scanned, copied or duplicated, or posted to a publicly accessible website, in whole or in part.

Phases of the Moon in *TheSkyX*

The Moon's phasing period has a time interval that is longer than its sidereal period. The phase period is a time interval *with respect to* the Sun, not to the stars. This period of time is also about a month. Actually, it takes the Moon 29.531 days to go through a complete phasing cycle. This period of time is known as the Moon's synodic (phase) period. Timekeepers have been aware of this cycle for hundreds or perhaps thousands of years.

Figure 9-1 is a schematic diagram of the Moon's orbit around Earth as viewed from some distance away in space. The motion of the Moon about Earth in this figure is counterclockwise as viewed from above the Earth-Moon System.

The arrows at the right-hand side in **Figure 9-1** represent sunlight passing through the Earth-Moon System. The Sun's distance is so great in this figure that its rays passing through the system essentially are parallel to one another. Earth and Moon are equally illuminated on the half toward the Sun. The lighter areas on the right-hand side of Earth and Moon represent the dayside while the dark shaded areas opposite the Sun's rays represent the nightside on Earth and Moon.

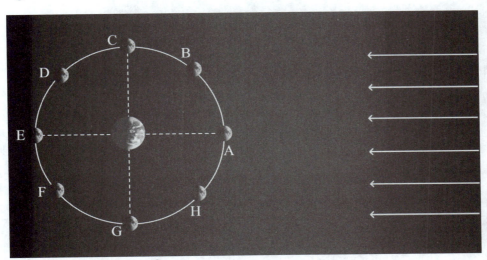

Figure 9-1 Earth-Moon System

New Moon

As viewed from Earth, the Moon is aligned with the Sun at Point A. If the Moon's angular separation (elongation) is measured from the Sun at this point, it is 0°. In other words, the right ascensions of the Sun and Moon are almost the same when the Moon is located at Point A. On careful inspection of **Figure 9-1**, one should notice that at Point A the Moon's nightside faces Earth. This means that *none* of the Moon's illuminated surface can be seen from Earth's surface. The phase of the Moon at this point is called a *New Moon*. The Moon appears to rise and set with the Sun and is not visible to us on Earth, because it is above the horizon with the Sun in the daytime sky.

Figure 9-2 displays *TheSkyX's* view of a New Moon as viewed by an observer from the Ball State Observatory on March 24, 2020, at 4:29 AM DST in the constellation Pisces. The Moon is located directly below the Sun in **Figure 9-2**. In actuality the Moon cannot be seen when its phase is a New Moon. **Figure 9-3**, however, illustrates how the Moon would look when viewed through a telescope on this date. Since the Moon is between Earth and the Sun its night side is facing Earth and is not visible.

© 2011 Cengage Learning. All Rights Reserved. May not be scanned, copied or duplicated, or posted to a publicly accessible website, in whole or in part.

Figure 9-2 New Moon in Pisces on March 24, 2020

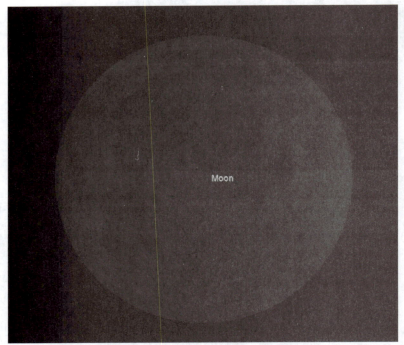

Figure 9-3 Appearance of the New Moon Phase in Earth's Sky in *TheSkyX*

Figure 9-4 shows the Object Information Report window for the Moon on March 24th in the **Find Result** window. *TheSkyX*, rather than naming lunar phases, indicates a percentage of the surface of the Moon that is visible from Earth. The New Moon's phase can be found in the **Tools** menu on the **Standard** toolbar. Click there and then the **Calendar**. It is feasible to use the Moon's and Sun's right ascensions to estimate their angular separation. This angle indicates how far apart the Sun and Moon appear to be from each other in the sky. On this date the Moon is a little more than 5° south of the Sun, while their right ascensions are nearly the same. However,

© 2011 Cengage Learning. All Rights Reserved. May not be scanned, copied or duplicated, or posted to a publicly accessible website, in whole or in part.

the lunar phase is determined from its ecliptic longitude rather than its right ascension. This can be found using the Find command and locating the Sun and Moon on March 24th at 4:29 AM at the Ball State Observatory.

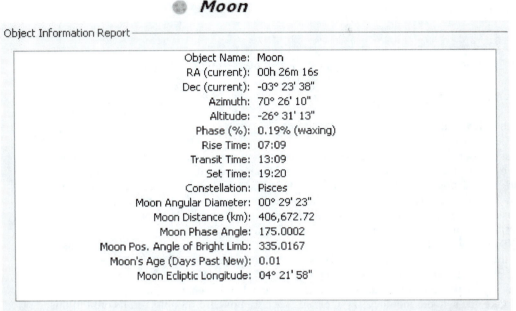

Figure 9-4 Object Information Report Window for the Moon on March 24th

In order to view the Sky window displayed in **Figure 9-2**, follow the instructions provided in **Exercise B**.

Exercise B: Observing the New Phase of the Moon on March 24, 2020

1. Run *TheSkyX*.

2. Open the Sky file named ***NewMoon.skyx***.

3. Click on the Sun or right-click on the mouse button to find the Sun's coordinates and record them here.

    R.A.$_{Sun}$ = _____ H _____ M _____ S   Declination$_{Sun}$ = __ _____ ° _____ ' _____ "

4. Record the Sun's *Geocentric Longitude*. _____ ° _____ '

5. Now, find the Moon and [Center]. Left-click on the mouse button to display its Object Information Report window.

6. Use the same technique as in #3 above, to find the Moon's coordinates and record them here.

    R.A.$_{Moon}$ = _____ H _____ M _____ S   Declination$_{Moon}$ = __ _____ ° _____ ' _____ "

7. Look in the Result Object Information Report window to find the Moon's phase or the the amount of its surface that is illuminated.

    Is the phase = 0.0%? _____.   What is the phase? _____ %

© 2011 Cengage Learning. All Rights Reserved. May not be scanned, copied or duplicated, or posted to a publicly accessible website, in whole or in part.

8. Compare the Moon's *Ecliptic Longitude* _____° _____' to that of the Sun's *Geocentric Longitude* _____° _____'. Are they about the same? _____. In other words, it is the difference between these two angles that roughly determines the phase of the Moon as it journey's around Earth.

9. Compare the Moon's right ascension to that of the Sun's. Sun _____[H] Moon _____[H]

    Is their difference about 0[H]? _____

10. ![Zoom In] to observe the Moon's phase up close as it would appear if in a telescope.

11. What is the age of the Moon? _____ days.

12. Close the file and do not save it.

    The Sun and Moon are both seen in this window, even though the Moon is actually not visible from Earth. The phase that is given is 0.19% (waning). This means that because the Moon is 5° below the Sun, and 0.19% of the Moon might be seen from Earth. However, since the Moon is in the day sky on this date it really is not visible at all from Earth's surface. It is in the direction of the Sun and has its dark (night) side toward Earth. The Sun is much too bright to be able to see the Moon. Do not try looking for the Moon when it is so close to the Sun!

    Figure 9-5 portrays the lunar phases as they appear from Earth's surface. In this figure, "Phase A" represents the New Moon phase and is displayed as a totally black circle at the left side of the diagram. This phase corresponds to the Moon's position at *Point A* in **Figure 9-1**. The lunar cycle begins with the Moon aligned on the Sun at *Point A*, in **Figure 9-1**, or when its phase is a New Moon. It will take one synodic month (29½[D]) for the Moon to move from Point A in its orbit to this point again or from one New Moon to the next New Moon. These lunar images were rendered using *TheSky6 Professional* Edition.

Figure 9-5 Phases of the Moon as Seen from Earth's Surface

First Quarter Moon

    As the Moon moves counterclockwise around Earth in **Figure 9-1**, eastward in the sky as viewed from Earth, more and more of its illuminated surface becomes visible. The Moon eventually reaches Point C in its orbit. It is now 90° east of the Sun. The Moon has moved one-quarter of the distance around its orbit at this point and is called a *First Quarter phase*. The right ascension of the Moon is nearly 6[H] greater (east) than the Sun's right ascension when the Moon is located at this point.

    The time that it has taken for the Moon to move this far in its orbit is a little over a week (7.38[D]). When the phase of the Moon is a First Quarter, it appears to rise and set about 6 hours after the Sun. It is visible in the sky from about midday until just about midnight. From Earth, the Moon appears as a half circle or half-moon in the sky. It is the right half of the Moon that is seen. **Figure 9-5**, "Phase C"

© 2011 Cengage Learning. All Rights Reserved. May not be scanned, copied or duplicated, or posted to a publicly accessible website, in whole or in part.

represents the First Quarter phase, and as one can see it appears as a half circle with the right half illuminated.

Figure 9-6 represents *TheSkyX's* rendition of a First Quarter Moon as seen in Gemini on April 1, 2024, at 5:22 AM DST as viewed from the Ball State Observatory.

Figure 9-7 is an image rendered in *TheSkyX* to show the appearance of a First Quarter phase of the Moon as it looks when viewed through a telescope on this date. It is a zoomed in view of the Moon in **Figure 9-6**. Notice that it resembles "Phase C" in **Figure 9-5**.

Figure 9-6 First Quarter Phase of the Moon in Gemini on April 1, 2024

Figure 9-7 Appearance of the First Quarter Phase of the Moon as Seen in *TheSkyX*

In order to view the sky window displayed in **Figure 9-6**, follow the instructions provided in **Exercise C**.

© 2011 Cengage Learning. All Rights Reserved. May not be scanned, copied or duplicated, or posted to a publicly accessible website, in whole or in part.

Exercise C: Observing the First Quarter Phase of the Moon on April 1, 2020

1. Run *TheSkyX*.

2. Open the Sky file named **FirstQuarter.skyx**.

3. Click on the Sun or right-click on the mouse button to find the Sun's coordinates and record them here.

    R.A.$_{Sun}$ = _____ ^{H} _____ ^{M} _____ ^{S} Declination$_{Sun}$ = __ _____ $^\circ$ _____ ' _____ "

4. Record the Sun's *Geocentric Longitude.* _____ $^\circ$ _____ '

5. Now, find the Moon and [Center]. Left-click on the mouse button to display its Object Information Report window.

6. Use the same technique as in #3 above, to find the Moon's coordinates and record them here.

    R.A.$_{Moon}$ = _____ ^{H} _____ ^{M} _____ ^{S} Declination$_{Moon}$ = __ _____ $^\circ$ _____ ' _____ "

7. Look in the Result Object Information Report window to find the Moon's phase or the the amount of its surface that is illuminated.

    Is the phase = 50.0%? _____.   What is the phase? _____%

8. Compare the Moon's *Ecliptic Longitude* _____ $^\circ$ _____ ' to that of the Sun's *Geocentric Longitude* _____ $^\circ$ _____ '. Are they about the same? _____. In other words, it is the difference between these two angles that roughly determines the phase of the Moon as it journey's around Earth.

9. Compare the Moon's right ascension to that of the Sun's. Sun _____ ^{H} Moon _____ ^{H}

    Is their difference about 6^H? _____

10. [Zoom In] to observe the Moon's phase up close as it would appear if in a telescope.

11. What is the age of the Moon past New? _____ days.

12. Close the file and do not save it.

Full Moon

As the Moon continues its counterclockwise journey around Earth, it travels from Point C to Point E. By the time the Moon reaches Point E, it has traveled halfway (two-quarters) around its orbit. At Point E, the Moon's position in its orbit is exactly 180° from the Sun; that is, its position in the sky is directly opposite the Sun. The right ascension of the Moon at this point is exactly 12^H greater (east) than the Sun's right ascension. The time it has taken the Moon to reach this point in its orbit is a little over two weeks (14.77^D).

From Earth, the Moon now appears as a full circle. Whenever the Moon is located at this point in its orbit, its phase is called a *Full phase*. In **Figure 9-5**, "Phase E" represents the *Full phase* and is displayed as a fully illuminated Moon. When the Moon is a Full Moon it appears to rise when the Sun appears to set and appears to set when the Sun appears to rise. It is visible in Earth's sky throughout the entire night.

© 2011 Cengage Learning. All Rights Reserved. May not be scanned, copied or duplicated, or posted to a publicly accessible website, in whole or in part.

Figure 9-8 displays the Full phase of the Moon in Virgo on April 7, 2020 at 9:37PM DST as viewed by an observer from the Ball State Observatory.

Figure 9-8 Full Moon Phase as Seen in Virgo on April 7, 2020

Figure 9-9 displays *TheSkyX's* rendition of what a Full Moon phase looks like when viewed with a telescope on this date. It is a zoomed in view of the Moon in **Figure 9-8**. If you look at **Figure 9-5**, you will notice that it does resemble "Phase E".

Figure 9-9 Appearance of the Full Phase of the Moon as Seen in *TheSkyX*

In order to view the sky window displayed in **Figure 9-8**, then follow the instructions provided in **Exercise D**.

Exercise D: Observing the Full Phase of the Moon on April 7, 2020

1. Run *TheSkyX*.

© 2011 Cengage Learning. All Rights Reserved. May not be scanned, copied or duplicated, or posted to a publicly accessible website, in whole or in part.

2. Open the Sky file named ***FullMoon.skyx***.

3. Click on the Sun or right-click on the mouse button to find the Sun's coordinates and record them here.

    R.A.$_{Sun}$ = _____ ^{H} _____ ^{M} _____ ^{S} Declination$_{Sun}$ = __ _____ $^{\circ}$ _____ $'$ _____ $''$

4. Record the Sun's *Geocentric Longitude*. _____ $^{\circ}$ _____ $'$

5. Now, find the Moon and | Center | . Left-click on the mouse button to display its Object Information Report window.

6. Use the same technique as in #3 above, to find the Moon's coordinates and record them here.

    R.A.$_{Moon}$ = _____ ^{H} _____ ^{M} _____ ^{S} Declination$_{Moon}$ = __ _____ $^{\circ}$ _____ $'$ _____ $''$

7. Look in the Result Object Information Report window to find the Moon's phase or the the amount of its surface that is illuminated.

    Is the phase = 100.0%? _____.   What is the phase? _____%

8. Compare the Moon's *Ecliptic Longitude* _____ $^{\circ}$ _____ $'$ to that of the Sun's *Geocentric Longitude* _____ $^{\circ}$ _____ $'$. Are they about the same? _____.   In other words, it is the difference between these two angles that roughly determines the phase of the Moon as it journey's around Earth.

9. Compare the Moon's right ascension to that of the Sun's. Sun _____ ^{H} Moon _____ ^{H}

    Is their difference about 12^{H}? _____

10. | Zoom In | to observe the Moon's phase up close as it would appear if in a telescope.

11. What is the age of the Moon past New? _____ days.

12. Close the file and do not save it.

Last Quarter Moon

The Moon continues to move in its orbit from Point E to Point G in **Figure 9-1**. When it reaches Point G, the Moon is once again located 90° from the Sun. It is, however, now 90° west of the Sun or 270° from the point where the cycle began (New Moon). The Moon has moved three-quarters of the distance around its orbit to reach this point. The Moon's phase at Point G is again called a quarter phase.

Its right ascension is now about 18^{H} greater (east) than the Sun's right ascension at this point. Or, it is 6^{H} less than the Sun's right ascension because it is west of the Sun. The time it has taken the Moon to move this far in its orbit is a little over three weeks (22.15^{D}). The Moon's phase at Point G was once called the 3rd Quarter phase but is now known as a *Last Quarter phase* of the Moon. When the Moon is a Last Quarter Moon, it appears to rise and set approximately 6 hours before the Sun (remember it is west). It is visible in Earth's sky from about midnight until approximately midday.

From Earth, the Moon once again appears as a half circle or a half-moon in the sky. But it is the *left half* of the Moon that is seen not the *right half* like the First Quarter phase. The reason for this is that the Moon is now located west of the Sun. In **Figure 9-5**, "Phase G" represents the Last Quarter phase in this figure and is displayed as a half circle with the left half of the Moon illuminated. **Figure 9-10** displays the Last Quarter phase of the Moon in Sagittarius on April 14, 2020, at 5:57 PM DST for an observer located at the Ball State Observatory.

© 2011 Cengage Learning. All Rights Reserved. May not be scanned, copied or duplicated, or posted to a publicly accessible website, in whole or in part.

Figure 9-10 Last Quarter Phase of the Moon in Sagittarius on April 14, 2020

Figure 9-11 is *TheSkyX's* rendition of how a Last Quarter phase of the Moon would look if viewed through a telescope on this date. It is a zoomed in view of the Moon in **Figure 9-10**. Notice that it looks like "Phase G" in **Figure 9-5**.

Figure 9-11 Appearance of the Last Quarter Phase of the Moon as Seen in *TheSkyX*

© 2011 Cengage Learning. All Rights Reserved. May not be scanned, copied or duplicated, or posted to a publicly accessible website, in whole or in part.

In order to view the sky window displayed in **Figure 9-10**, simply follow the instructions provided in **Exercise E**.

Exercise E: Observing the Last Quarter Phase of the Moon on April 14, 2020

1. Run *TheSkyX*.

2. Open the Sky file named ***LastQuarter.skyx***.

3. Click on the Sun or right-click on the mouse button to find the Sun's coordinates and record them here.

    R.A._{Sun} = _____ ^{H} _____ ^{M} _____ ^{S} Declination_{Sun} = __ _____ $^\circ$ _____ ' _____ "

4. Record the Sun's *Geocentric Longitude*. _____ $^\circ$ _____ '

5. Now, find the Moon and [Center]. Left-click on the mouse button to display its Object Information Report window.

6. Use the same technique as in #3 above, to find the Moon's coordinates and record them here.

    R.A._{Moon} = _____ ^{H} _____ ^{M} _____ ^{S} Declination_{Moon} = __ _____ $^\circ$ _____ ' _____ "

7. Look in the Result Object Information Report window to find the Moon's phase or the the amount of its surface that is illuminated.

    Is the phase = 90.0%? _____. What is the phase? _____%

8. Compare the Moon's *Ecliptic Longitude* _____ $^\circ$ _____ ' to that of the Sun's *Geocentric Longitude* _____ $^\circ$ _____ '. Are they about the same? _____. In other words, it is the difference between these two angles that roughly determines the phase of the Moon as it journey's around Earth.

9. Compare the Moon's right ascension to that of the Sun's. Sun _____ ^{H} Moon _____ ^{H}

    Is their difference about 6^H? _____

10. [Zoom In] to observe the Moon's phase up close as it would appear if in a telescope.

11. What is the age of the Moon past New? _____ days.

12. Close the file and do not save it.

   The Moon finally moves from Point G to Point A in **Figure 9-1**, thus completing one lunar cycle about Earth. The total time it has taken the Moon to complete this cycle is 29.531^D with respect to the Sun. This is the Moon's *synodic* or *phase period*! At this point the cycle starts over again.

Other Phases of the Moon in *TheSkyX*

   The phases discussed so far are those that have well-defined measured positions in the Moon's orbit. That is, the angles for these particular phases of the Moon are precisely known with

© 2011 Cengage Learning. All Rights Reserved. May not be scanned, copied or duplicated, or posted to a publicly accessible website, in whole or in part.

respect to Earth and the Sun. The angles, or one might say elongations, are 0°, 90°, 180°, and 270° (New, First Quarter, Full, and Last Quarter). However, during the time that it has taken the Moon to complete one phase cycle around Earth, it has also exhibited a number of other phases in Earth's sky.

For example, as the Moon moves from Point A to Point C the sunlit portion of the Moon's surface gradually becomes more visible to observers on Earth. The Moon appears to go through a series of *crescent phases* in Earth's sky, each larger than the previous night, during its phase cycle. The illuminated areas are visible on the right edge of the Moon as viewed from Earth. In **Figure 9-5**, "Phase B" represents a crescent phase of the Moon as viewed from Earth during this period of time.

Likewise, as the Moon continues moving from Point C to Point E the illuminated portion continues to get larger in Earth's sky as the Moon moves from the First Quarter phase to the Full phase. During this portion of its cycle the Moon's phase is known as a *gibbous phase*. From Earth, the Moon appears larger than a half circle but less than a full circle. The illuminated area is still on the right edge of the Moon as viewed from Earth. In **Figure 9-5**, "Phase D" represents a gibbous phase of the Moon as viewed from Earth while moving from Point C to Point E.

As the Moon continues moving from Point E to Point G the illuminated portion of the Moon becomes less than a full circle but larger than a half circle. The Moon's phase is once again a gibbous phase while moving through this part of its orbit. In Earth's sky, it still appears larger than a half circle again and smaller than a full circle. The illuminated area, however, is now on the left edge now rather than the right edge as was seen earlier. In **Figure 9-5**, "Phase F" represents a gibbous phase of the Moon as it appears from Earth while moving between Point E and Point G.

The Moon finally completes its cycle around Earth while moving from Point G to Point A. The illuminated portion of the Moon begins to appear less than a half circle again. Its appearance in Earth's sky is like that of a crescent phase once again. "Phase H," on the right side in **Figure 9-5**, represents a crescent phase of the Moon as viewed from Earth during this period of time.

One should realize by now that at many places in the Moon's orbit the phases of the Moon are either crescent or gibbous. The fact that the Moon has many of these phases might seem to pose a problem in locating where the Moon is in its orbit. However, there is a way to tell precisely where the Moon is located in its orbit with respect to the Sun.

As the Moon goes around Earth and consequently through its phases, one notices that when the Moon moves between New and Full the illuminated portion becomes larger. By the same token, when the Moon moves between Full and New its illuminated portion becomes smaller.

When the Moon moves between New and Full the phases are referred to as *waxing phases*, because they appear to get larger. When the Moon moves between Full and New the phases are referred to as *waning phases*, because they appear to get smaller.

Waxing Phases of the Moon in *TheSkyX*

At Point B in **Figure 9-1**, the phase of the Moon is referred to as a *waxing crescent*. At Point D in **Figure 9-1**, this phase of the Moon is referred to as a *waxing gibbous* Moon. In **Figure 9-5**, "Phase B" represents the way a waxing crescent Moon appears in Earth's sky, and "Phase D" represents how a waxing gibbous Moon appears. During the waxing stage of the Moon's phase cycle, the right-hand edge of the Moon is visible from Earth. **Figure 9-12** displays the waxing crescent phase of the Moon in Taurus on April 26, 2020, approximately two hours before the apparent sunset as seen from the Ball State Observatory.

© 2011 Cengage Learning. All Rights Reserved. May not be scanned, copied or duplicated, or posted to a publicly accessible website, in whole or in part.

Figure 9-12 Waxing Crescent Phase of the Moon in Taurus on April 26, 2020

Figure 9-13 shows *TheSkyX's* rendition of what a waxing crescent phase of the Moon looks like when viewed through a telescope on this date. It is a zoomed in view of the Moon in **Figure 9-12**. Notice that it resembles "Phase B" in **Figure 9-5**. A waxing crescent Moon appears to rise a few hours after the Sun appears to rise and appears to set a few hours after the Sun. It is visible in Earth's sky in the early evening just after sunset for a few hours.

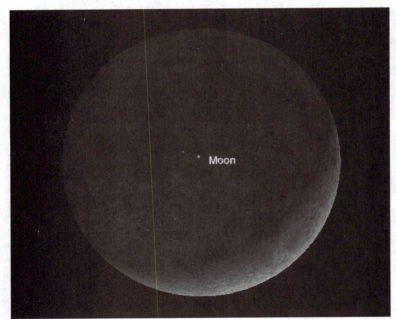

Figure 9-13 Appearance of the Waxing Crescent Phase of the Moon in *TheSkyX*

In order to view the sky window displayed in **Figure 9-12**, then follow the instructions provided in **Exercise F**.

© 2011 Cengage Learning. All Rights Reserved. May not be scanned, copied or duplicated, or posted to a publicly accessible website, in whole or in part.

Exercise F: Observing the Waxing Crescent Phase of the Moon on April 26, 2020

1. Run *TheSkyX*.

2. Open the Sky file named ***WaxingCrescent.skyx***.

3. Click on the Sun or right-click on the mouse button to find the Sun's coordinates and record them here.

    R.A.$_{Sun}$ = _____ ^{H} _____ ^{M} _____ ^{S} Declination$_{Sun}$ = __ _____ $^\circ$ _____ $'$ _____ $''$

4. Record the Sun's *Geocentric Longitude.* _____ $^\circ$ _____ $'$

5. Now, find the Moon and [Center]. Left-click on the mouse button to display its Object Information Report window.

6. Use the same technique as in #3 above, to find the Moon's coordinates and record them here.

    R.A.$_{Moon}$ = _____ ^{H} _____ ^{M} _____ ^{S} Declination$_{Moon}$ = __ _____ $^\circ$ _____ $'$ _____ $''$

7. Look in the Result Object Information Report window to find the Moon's phase or the the amount of its surface that is illuminated.

    Is its Phase greater than 0% and less than 50%? _____.   What is the phase? _____%

8. Compare the Moon's *Ecliptic Longitude* _____ $^\circ$ _____ $'$ to that of the Sun's *Geocentric Longitude* _____ $^\circ$ _____ $'$. Are they about the same? _____.   In other words, it is the difference between these two angles that roughly determines the phase of the Moon as it journey's around Earth.

9. Compare the right ascensions of the Sun and Moon.

    Sun's R.A. = _____ ^{H} _____ ^{M} Moon's R.A. = _____ ^{H} _____ ^{M}

    Is their difference between 0^H and 6^H? _____.

10. [Zoom In] to observe the Moon's phase up close as it would appear if in a telescope.

11. What is the age of the Moon past New? _____ days.

12. Close the file and do not save it.

   Figure 9-14 displays the view of a waxing gibbous Moon in Cancer on April 2, 2020, at 5:08 PM DST. This is approximately two hours before apparent sunset, as seen from the Ball State Observatory.

© 2011 Cengage Learning. All Rights Reserved. May not be scanned, copied or duplicated, or posted to a publicly accessible website, in whole or in part.

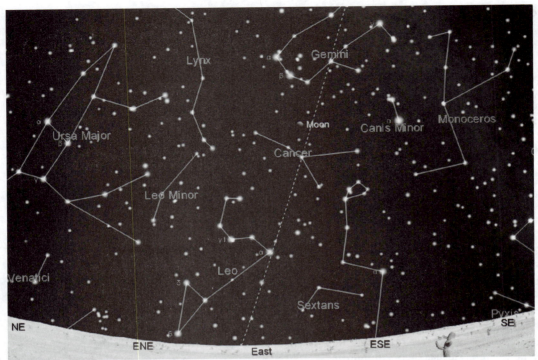

Figure 9-14 Waxing Gibbous Phase of the Moon in Cancer on April 2, 2020

    Figure 9-15 represents *TheSkyX's* rendition of what a waxing gibbous phase of the Moon looks like when viewed through a telescope on this date. It is a zoomed in view of the Moon in **Figure 9-14**. Notice that it is like "Phase D" in **Figure 9-5**. A waxing gibbous Moon appears to rise sometime after midday until just before sunset. It is visible in the eastern sky from the early afternoon until it appears to set sometime after midnight.

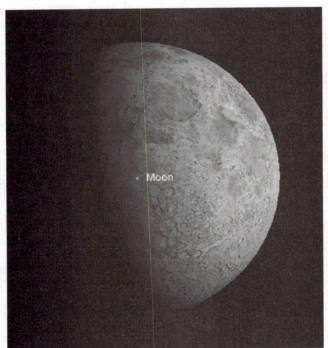

Figure 9-15 Appearance of the Waxing Gibbous Phase of the Moon as Seen in *TheSkyX*

© 2011 Cengage Learning. All Rights Reserved. May not be scanned, copied or duplicated, or posted to a publicly accessible website, in whole or in part.

In order to view the sky window displayed in **Figure 9-14**, then follow the instructions provided in **Exercise G**.

Exercise G: Observing a Waxing Gibbous Phase of the Moon on April 2, 2020

1.   Run *TheSkyX.*

2.   Open the Sky file named ***WaxingGibbous.skyx***.

3.   Click on the Sun or right-click on the mouse button to find the Sun's coordinates and record them here.

   R.A.·_{Sun} = _____ ^H _____ ^M _____ ^S   Declination_{Sun} = __ _____ ° _____ ' _____ "

4.   Record the Sun's *Geocentric Longitude.* _____ ° _____ '

5.   Now, find the Moon and 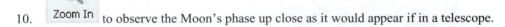 . Left-click on the mouse button to display its Object Information Report window.

6.   Use the same technique as in #3 above, to find the Moon's coordinates and record them here.

   R.A._{Moon} = _____ ^H _____ ^M _____ ^S   Declination_{Moon} = __ _____ ° _____ ' _____ "

7.   Look in the Result Object Information Report window to find the Moon's phase or the the amount of its surface that is illuminated.

   Is its Phase greater than 50% and less than 100%? _____ .   What is the phase? _____%

8.   Compare the Moon's *Ecliptic Longitude* _____ ° _____ ' to that of the Sun's *Geocentric Longitude* _____ ° _____ '. Are they about the same? _____ .   In other words, it is the difference between these two angles that roughly determines the phase of the Moon as it journey's around Earth.

9.   Compare the right ascensions of the Sun and Moon.

   Sun's R.A. = _____ ^H _____ ^M   Moon's R.A. = _____ ^H _____ ^M

   Is their difference between 6^H and 12^H? _____ .

10.   **Zoom In** to observe the Moon's phase up close as it would appear if in a telescope.

11.   What is the age of the Moon past New? _____ days.

12.   Close the file and do not save it.

Waning Phases of the Moon in *TheSkyX*

At Point F in **Figure 9-1**, the phase of the Moon is referred to as a waning gibbous phase. At Point H in **Figure 9-1**, the phase of the Moon is referred to as a waning crescent. In **Figure 9-5**, "Phase F" represents the way a waning gibbous phase of the Moon appears and "Phase H" represents how a waning crescent phase of the Moon appears in Earth's sky. During the waning stage of the Moon's phase cycle, the left-hand edge of the Moon is visible from Earth.

© 2011 Cengage Learning. All Rights Reserved. May not be scanned, copied or duplicated, or posted to a publicly accessible website, in whole or in part.

Figure 9-16 displays a view of a waning gibbous phase of the Moon in Scorpius on April 11, 2020, at 3:15 AM DST as seen from the Ball State Observatory.

Figure 9-16   Waning Gibbous Phase of the Moon in Scorpius on April 11, 2020

Figure 9-17 represents what a waning gibbous phase of the Moon looks like when viewed through a telescope on this date. It is a zoomed in view of the Moon in **Figure 9-16**. Notice that it resembles "Phase F" in **Figure 9-5**. A waning gibbous Moon appears to rise sometime after midnight and is visible in the eastern sky from early morning until it appears to set sometime after midday.

Figure 9-17 Appearance of the Waning Gibbous Phase of the Moon as Seen in *TheSkyX*

© 2011 Cengage Learning. All Rights Reserved. May not be scanned, copied or duplicated, or posted to a publicly accessible website, in whole or in part.

In order to view the sky window displayed in **Figure 9-16**, then follow the instructions provided in **Exercise H**.

Exercise H: Observing the Waning Gibbous Phase of the Moon on April 11, 2020

1. Run *TheSkyX*.

2. Open the Sky file named *WaningGibbous.skyx*.

3. Click on the Sun or right-click on the mouse button to find the Sun's coordinates and record them here.

    R.A.$_{Sun}$ = _____ ^{H} _____ ^{M} _____ ^{S} Declination$_{Sun}$ = __ _____ $^\circ$ _____ $'$ _____ $''$

4. Record the Sun's *Geocentric Longitude*. _____ $^\circ$ _____ $'$

5. Now, find the Moon and [Center]. Left-click on the mouse button to display its Object Information Report window.

6. Use the same technique as in #3 above, to find the Moon's coordinates and record them here.

    R.A.$_{Moon}$ = _____ ^{H} _____ ^{M} _____ ^{S} Declination$_{Moon}$ = __ _____ $^\circ$ _____ $'$ _____ $''$

7. Look in the Result Object Information Report window to find the Moon's phase or the the amount of its surface that is illuminated.

    Is its Phase greater than 50% and less than 100%? _____. What is the phase? _____%

8. Compare the Moon's *Ecliptic Longitude* _____ $^\circ$ _____ $'$ to that of the Sun's *Geocentric Longitude* _____ $^\circ$ _____ $'$. Are they about the same? _____. In other words, it is the difference between these two angles that roughly determines the phase of the Moon as it journey's around Earth.

9. Compare the right ascensions of the Sun and Moon.

    Sun's R.A. = _____ ^{H} _____ ^{M} Moon's R.A. = _____ ^{H} _____ ^{M}

    Is their difference between 12^H and 18^H? _____.

10. [Zoom In] to observe the Moon's phase up close as it would appear if in a telescope.

11. What is the age of the Moon past New? _____ days.

12. Close the file and do not save it.

    Figure 9-18 shows the waning crescent phase of the Moon in Aquarius on April 18, 2020, at 5:28 AM DST, approximately 30 minutes before the apparent sunrise as seen from the Ball State Observatory.

© 2011 Cengage Learning. All Rights Reserved. May not be scanned, copied or duplicated, or posted to a publicly accessible website, in whole or in part.

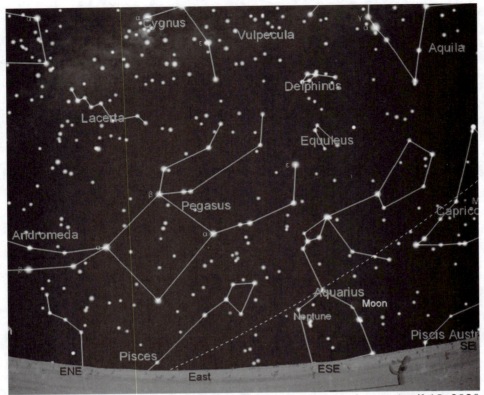

Figure 9-18 Waning Crescent Phase of the Moon in Aquarius on April 18, 2020

Figure 9-19 represents what the waning crescent phase of the Moon looks like when viewed through a telescope on that morning. It is a zoomed in view of the Moon in **Figure 9-17**. Notice that it resembles "Phase H" in **Figure 9-5**. A waning crescent Moon appears to rise a few hours before the Sun and is visible before the Sun appears to rise in the early morning. It appears to set a few hours before the Sun in the late afternoon.

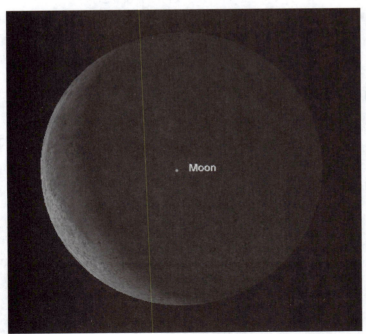

Figure 9-19 Appearance of the Waning Crescent Phase of the Moon as Seen in *TheSkyX*

© 2011 Cengage Learning. All Rights Reserved. May not be scanned, copied or duplicated, or posted to a publicly accessible website, in whole or in part.

In order to view the sky window displayed in **Figure 9-18**, then follow the instructions provided in **Exercise I**.

Exercise I: Observing the Waning Crescent Phase of the Moon on April 18, 2020

1. Run *TheSkyX*.

2. Open the Sky file named ***WaningCrescent.skyx***.

3. Click on the Sun or right-click on the mouse button to find the Sun's coordinates and record them here.

    R.A.$_{Sun}$ = _____ ^{H} _____ ^{M} _____ ^{S} Declination$_{Sun}$ = __ _____ $^\circ$ _____ $'$ _____ $''$

4. Record the Sun's *Geocentric Longitude*. _____ $^\circ$ _____ $'$

5. Now, find the Moon and [Center]. Left-click on the mouse button to display its Object Information Report window.

6. Use the same technique as in #3 above, to find the Moon's coordinates and record them here.

    R.A.$_{Moon}$ = _____ ^{H} _____ ^{M} _____ ^{S} Declination$_{Moon}$ = __ _____ $^\circ$ _____ $'$ _____ $''$

7. Look in the Result Object Information Report window to find the Moon's phase or the the amount of its surface that is illuminated.

    Is its Phase greater than 0% and less than 50%? _____. What is the phase? _____%

8. Compare the Moon's *Ecliptic Longitude* _____ $^\circ$ _____ $'$ to that of the Sun's *Geocentric Longitude* _____ $^\circ$ _____ $'$. Are they about the same? _____. In other words, it is the difference between these two angles that roughly determines the phase of the Moon as it journey's around Earth.

9. Compare the right ascensions of the Sun and Moon.

    Sun's R.A. = _____ ^{H} _____ ^{M} Moon's R.A. = _____ ^{H} _____ ^{M}

    Is their difference between 18^H and 24^H? _____.

10. [Zoom In] to observe the Moon's phase up close as it would appear if in a telescope.

11. What is the age of the Moon past New? _____ days.

12. Close the file and do not save it.

    TheSkyX displays all lunar phases quite accurately and with amazing clarity. It is easy to determine the phase of the Moon for any date, time, or year. The date and time must first be set to the appropriate Date and Time in *TheSkyX*. One then finds the Moon by accessing the Edit menu on the Standard toolbar at the top of the sky window or right-clicking the mouse any where in the sky window. [Zoom In] allows one to observe the phase of the Moon closely and more clearly — a telescopic view so to speak!

© 2011 Cengage Learning. All Rights Reserved. May not be scanned, copied or duplicated, or posted to a publicly accessible website, in whole or in part.

Figure 9-20 illustrates the phases of the Moon, which have been rendered in *TheSky6 Professional Version*. The frames rendered are all zoomed frames so the look is remarkably like that of the real Moon. The phases in the figure are labeled A through H as they were in **Figure 9-5**. The lunar phases seen in **Figure 9-20** portray one complete phase cycle of the Moon from New Moon to New Moon — a 29½-day cycle.

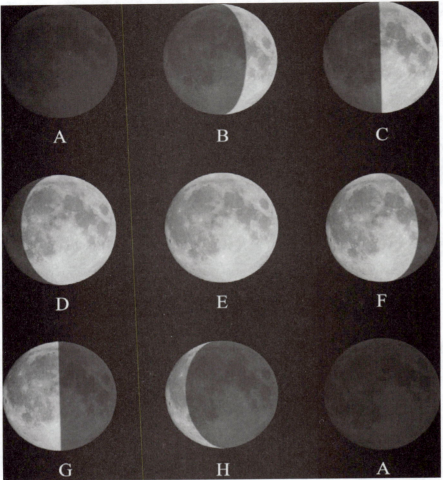

Figure 9-20 Phases of the Moon as Displayed in *TheSky6 Professional Version*

Exercises J, K, L, M, N, and **O** are exercises designed to assist in using *TheSkyX* to determine and identify lunar phases.

Exercise J: Determining the Phase of the Moon

1. Run *TheSkyX* and set as follows:

    Set Date: October 12, 1492
    Set Time: 2:30 AM STD
    Daylight Saving option: "Not Observed"

    Location: San Salvador, Island

    Latitude: 24° 24′ 30″ N
    Longitude: 75° 31′ 53″ W
    Time Zone: -5 hours

© 2011 Cengage Learning. All Rights Reserved. May not be scanned, copied or duplicated, or posted to a publicly accessible website, in whole or in part.

2. Find the Moon and .

3. Sketch its appearance (Zoom in, if necessary). You may wish to use the rotate tool in Orientation menu on the Standard toolbar…click Navigate then Rotation.

4. What is its phase? _____ _____

5. In what constellation is the Moon located? _____

6. What time does it appear to transit the local celestial meridian? _____:_____

7. Close the file and do not save it, or continue with **Exercise K**.

Exercise K: Determining the Phase of the Moon

1. Run *TheSkyX* and set as follows:

    Set Date: July 4, 1776
    Set Time: 9:30 PM STD
    Time Zone: -5 hours
    Daylight Saving option: "Not Observed"

    Location: Philadelphia, Pennsylvania

2. Find the Moon and .

3. Sketch its appearance (Zoom in, if necessary). You may wish to use the rotate tool in Orientation menu on the Standard toolbar…click Navigate then Rotation.

4. What is its phase? _____ _____

5. In what constellation is the Moon located? _____

6. What time does it appear to transit the local celestial meridian? _____:_____

7. Close the file and do not save it, or continue with *Exercise L*.

© 2011 Cengage Learning. All Rights Reserved. May not be scanned, copied or duplicated, or posted to a publicly accessible website, in whole or in part.

Exercise L: Determining the Phase of the Moon

1. Run *TheSkyX* and set as follows:

   Set Date: December 7, 1941
   Set Time: 6:30 AM HST
   Time Zone: -10 hours
   Daylight Saving option: "Not Observed"

   Location: Honolulu, Hawaii

2. Find the Moon and .

3. Sketch its appearance (Zoom in, if necessary). You may wish to use the rotate tool in Orientation menu on the Standard toolbar…click Navigate then Rotation.

4. What is its phase? _____ _____

5. In what constellation is the Moon located? _____

6. What time does it appear to transit the local celestial meridian? _____:_____

7. Close the file and do not save it, or continue with **Exercise M**.

Exercise M: Determining the Phase of the Moon

1. Run *TheSkyX* and set as follows:

   Set Date: July 20, 1969
   Set Time: 8:30 PM DST
   Time Zone: -5.0 hours
   Daylight Saving option: "U.S. and Canada"

   Location: Cape Canaveral, Florida

   Latitude: 28° 29′ 00″ N
   Longitude: 80° 34′ 00″ W

2. Find the Moon and .

3. Sketch its appearance (Zoom in, if necessary). You may wish to use the rotate tool in Orientation menu on the Standard toolbar…click Navigate then Rotation.

© 2011 Cengage Learning. All Rights Reserved. May not be scanned, copied or duplicated, or posted to a publicly accessible website, in whole or in part.

4. What is its phase? _____ _____

5. In what constellation is the Moon located? _____

6. What time does it appear to transit the local celestial meridian? _____:_____

7. Close the file and do not save it, or continue with **Exercise N**.

Exercise N: Determining the Phase of the Moon

1. Run *TheSkyX* and set as follows:

    Set Date: March 21, 5555
    Set Time: 2:30 AM DST
    Time Zone: -5 hours
    Daylight Saving option: "U.S. and Canada"

    Location: New York City, New York

2. Find the Moon and .

3. Sketch its appearance (Zoom in, if necessary). You may wish to use the rotate tool in Orientation menu on the Standard toolbar…click Navigate then Rotation.

4. What is its phase? _____ _____

5. In what constellation is it located? _____

6. What time does it appear to transit the local celestial meridian? _____:_____

7. Close the file and do not save it, or continue with **Exercise O**.

Exercise O: Determining the Phase of the Moon

1. Run *TheSkyX* and set as follows:

    Set Date: December 31, 1947
    Set Time: 2:30 AM STD
    Time Zone: -5.0 hours
    Daylight Saving option: "U.S. and Canada"

    Location: Ball State Observatory

2. Find the Moon and .

© 2011 Cengage Learning. All Rights Reserved. May not be scanned, copied or duplicated, or posted to a publicly accessible website, in whole or in part.

3. Sketch its appearance (Zoom in, if necessary). You may wish to use the rotate tool in Orientation menu on the Standard toolbar…click Navigate then Rotation.

4. What is its phase? _____ _____

5. In what constellation is the Moon located? _____

6. What time does it appear to transit the local celestial meridian? _____:_____

7. Close the file and do not save it.

Understanding why the Moon goes through phases and why it appears the way it does in Earth's sky is all a matter of angles. Many people have the notion that the phases of the Moon are caused by shadows. This is simply not true! Phases of the Moon are a direct result of viewing the Moon at different positions in its orbit from Earth relative to the Sun in the sky.

Is there an easier way to determine the phases of the Moon? Perhaps the simplest way is to consult a typical calendar. This might take away some of the fun of determining the lunar phase. However, *TheSkyX* software is equipped with a lunar phase calendar. It displays the phases of the Moon for the current month and year.

In this mode, the previous or next month's phases can easily be viewed for any day of the month for any year. The day on which the Moon's phase is a New Moon, First Quarter, Full Moon, or Last Quarter is displayed as well as the time it occurs. The time indicated below that particular phase, is the time at which the Moon reaches that point (New, First Quarter, Full, Last Quarter) in its orbit and is expressed in the time at the observer's location.

The Moon Phase Calendar can be printed out, along with the times of the apparent moonrise and moonset. It is a very useful tool provided in the software and helps in planning observing sessions. **Figure 9-21** displays the Moon Phase Calendar in *TheSkyX* for April 2020. If you would like to use this tool, then follow the instructions described in the **Exercises P** and **Q**.

© 2011 Cengage Learning. All Rights Reserved. May not be scanned, copied or duplicated, or posted to a publicly accessible website, in whole or in part.

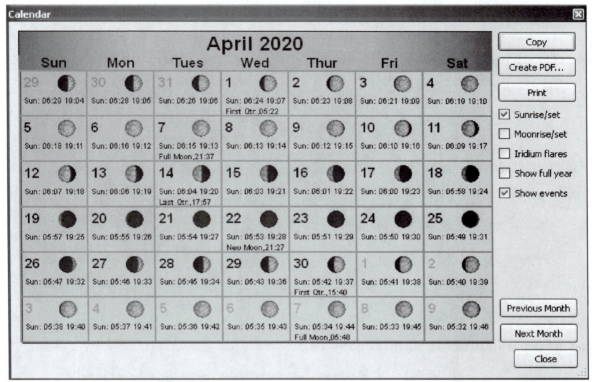

Figure 9-21 Moon Phase Calendar for April 2020

Exercise P: Displaying the Moon Phase Calendar in April 2020

1. Run *TheSkyX* and set as follows:

2. Set Date: April 1, 2020
   Set Time: Any value
   Any Time Zone:

   Set Time: Any value

3. Click Tools on the Standard toolbar at the top of the sky window.

4. Click on Calendar.

5. Make sure the "Sunrise/set" box on the right edge of the calendar is checked.

6. Click the "Show events" box on the right edge of the calendar. The calendar should now appear like the one displayed in **Figure 9-21**.

7. For this month, on what date is the Moon's phase:

   New: _____

   Waxing gibbous: _____

   First Quarter: _____

   Waning crescent: _____

   Full: _____

   Last Quarter: _____

© 2011 Cengage Learning. All Rights Reserved. May not be scanned, copied or duplicated, or posted to a publicly accessible website, in whole or in part.

8. Close the file and do not save it, or continue with **Exercise Q**.

Exercise Q: Displaying the Moon Phase Calendar for December, 9888

1. Run *TheSkyX* and set as follows:

   Set Date: December 1, 9888
   Set Time: Any value
   Any Time Zone
   Daylight Saving Option: "Not Observed"

2. Open the Moon Phase Calendar.

3. On what date is the Moon's phase Full in December? _____

4. Close the file and do not save it.

If you feel that you need more experience, you might go back and redo **Exercises B** through **I** using the Moon Phase Calendar.

Eclipses of the Sun and Moon in *TheSkyX*

Unlike the phase cycle of the Moon, certain astronomical phenomena *are* directly linked to shadows. These, of course, are eclipses. Earth and Moon are large enough, relatively speaking, to cast shadows in space. With Earth in motion around the Sun and the Moon in motion around Earth, there inevitably arises the possibility that these three objects might line up with one another. When this happens, an eclipse is possible. Eclipses of the Moon are known as *lunar eclipses* and those of the Sun are called *solar eclipses*.

Twice a month the Sun, Moon, and Earth align with each other at the time of a New and Full Moon. So why don't we observe a lunar and solar eclipse every two weeks? That is, why doesn't Earth or the Moon cast a shadow on each other every time these alignments occur?

Eclipses involve three aspects of Earth, Moon, and Sun. The first and perhaps the most obvious thing is that eclipses depend on the phase of the Moon. Eclipses can only occur at a time when the Moon's phase is New or Full. However, even when this happens, it does not necessarily guarantee that an eclipse will occur.

The second thing depends on the geometry of Earth, Moon, and Sun with respect to each other. The Earth, Moon, and Sun must be nearly aligned with one another in space before an eclipse can occur. Every two weeks Earth, Moon, and Sun are in alignment with each other. This occurs when the Moon is at the point in its orbit when it is between Earth and the Sun (New) and again when the Moon is 180° from the Sun (Full). Eclipses do not always happen just because Earth, Moon, and Sun are aligned with each other either. The reason is because the Moon's orbit is tilted with respect to Earth's orbit around the Sun.

The last condition has to do with the Moon's orbit itself. Because the Moon's orbit is tilted a little more than 5°, this means that during half of the lunar phase cycle the Moon is above Earth's orbit and during the other half of its phase cycle the Moon is below Earth's orbit. The two places where the Moon crosses Earth's orbit in the plane of the ecliptic are known as *nodes*. One node is located where the Moon moves from below Earth's orbit to above it. This point is defined as the *ascending node*. The other point is located where the Moon moves from above Earth's orbit to below it. This point is defined as the *descending node*. If one draws a line between these two nodes across the Moon's orbit connecting them, this line is defined as the *line of nodes*.

Figure 9-22 illustrates the Moon's orbit around Earth and Earth's orbit around the Sun. The line of nodes is also depicted in this figure as the dotted line. Earth's shadow is shown in this figure as well. In order to have a lunar or solar eclipse occur, the *three conditions* discussed above must be fulfilled simultaneously. That is, all three of these of things must happen at the same time!

© 2011 Cengage Learning. All Rights Reserved. May not be scanned, copied or duplicated, or posted to a publicly accessible website, in whole or in part.

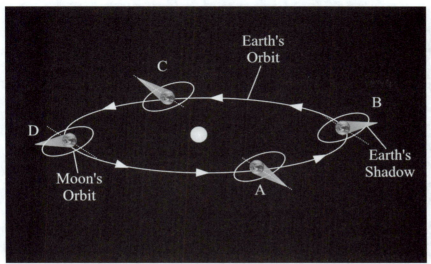

Figure 9-22 Moon's Orbit Around Earth and Earth's Orbit Around the Sun

The first condition is that the Moon's phase *must be* New or Full! The second condition is that the Moon *must be* crossing Earth's orbit. In other words, the Moon must be *at* or *near* a node! Finally, the line of nodes *must be* aligned on the Sun. When these three conditions are satisfied at the same instant in time then it is most probable some type of an eclipse will occur. In **Figure 9-22** the most likely positions for an eclipse to occur are at positions A and C.

Lunar Eclipses

As was mentioned earlier, Earth and Moon both cast shadows in space. These shadows are, of course, three-dimensional shadows (cones). **Figure 9-23** depicts sunlight shining from the left and Earth's shadow being cast into space toward the right. Sunlight passes entirely around Earth and produces two shadows in the direction opposite the Sun. Because light from the left and right edges of the Sun pass around the left and right edges of Earth, respectively, this produces a very dark shadow in space. Light from the left edge of the Sun passes around the right edge of Earth and light from the right edge of the Sun passes around the left edge of Earth, respectively, produces a lighter shadow in space. The geometry of the Sun and Earth in regard to producing these shadows is depicted in **Figure 9-23**. This phenomenon produces one shadow that is quite dark and another one that is much lighter. The darker shadow of Earth, labeled "*A*" in **Figure 9-23**, is defined as Earth's *umbra*. The lighter shadow of Earth, labeled "*B*" in **Figure 9-23**, is defined as Earth's *penumbra*. If the Moon passes through either of these shadows, then people on Earth observe a lunar eclipse. *TheSkyX* displays Earth's umbral and penumbral shadows on the desktop window during the course of a month.

© 2011 Cengage Learning. All Rights Reserved. May not be scanned, copied or duplicated, or posted to a publicly accessible website, in whole or in part.

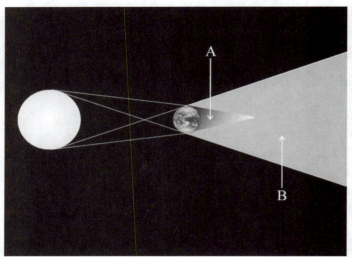

Figure 9-23 Umbra and penumbra of Earth's Shadow in Space

Types of Lunar Eclipses

The type of lunar eclipse observed from Earth's surface depends on the part of Earth's shadow through which the Moon moves. If the Moon moves entirely through Earth's umbra, (A) then a *total eclipse* of the Moon is observed. If the Moon just misses the umbra and moves through the part of the umbra and part of the penumbra, then a *partial eclipse* of the Moon is observed. If the Moon misses the umbral shadow completely and moves only through the penumbra (B) then a *penumbral eclipse* is observed.

Of all these types of lunar eclipses, the most striking is a total eclipse. When the Moon is completely immersed in Earth's umbral shadow a total eclipse occurs. During a total eclipse, the Moon appears to turn a deep reddish color at mid-eclipse. The reason for this is that mostly red light from the Sun makes its way through Earth's atmosphere and falls on the Moon's surface. Unfortunately, *TheSkyX* does not render color images of total eclipses. It does, however, display the location of Earth's umbra and penumbra in the sky. **Figure 9-24** is an image of the totally eclipsed Moon taken July 6, 1982. The photo was taken at mid-eclipse. The authors gratefully acknowledge the source of these eclipse pictures as a courtesy of Dr. Dale Ireland's website: http://www.drdale.com and are used here with his permission.

Figure 9-24 Total Lunar Eclipse of July 6, 1982 (Photo courtesy of Dr. Dale Ireland)

Figure 9-25 displays the March 24, 1997, partial eclipse of the Moon. The photo was taken at mid-eclipse.

© 2011 Cengage Learning. All Rights Reserved. May not be scanned, copied or duplicated, or posted to a publicly accessible website, in whole or in part.

Figure 9-25 Partial Lunar Eclipse of March 24, 1997 (Photo courtesy of Dr. Dale Ireland)

Penumbral eclipses of the Moon are not very exciting to watch. The brightness of the Moon changes so little that the change is virtually undetectable from Earth, penumbral eclipses are rarely of interest to anyone. If one did *not know* the Moon was passing through Earth's penumbra ahead of time, they would be hard-pressed to determine whether or not an eclipse was taking place.

Figure 9-26 displays the penumbral eclipse that took place on November 18, 1994. As you can see in the photograph, it is very difficult to determine whether or not an eclipse is occurring. The photo was taken at mid-eclipse.

Figure 9-26 Penumbral Lunar Eclipse of November 18, 1994 (Photo courtesy of Dr. Dale Ireland)

Solar Eclipses

Figure 9-27 depicts the shadow of the Moon being cast on Earth from space. It shows the path this shadow makes on Earth's surface. Sunlight once again is shining from the left in this figure. As sunlight passes entirely around the Moon, it produces two shadows in the direction opposite the Sun. Like Earth's shadow, the Moon also has a darker shadow and a lighter one. The Moon's *darker shadow*, displayed in the figure, is also called the *umbra* while its *lighter shadow* is called the *penumbra*. If an observer standing on Earth happens to have either Moon's shadow pass over them, they will observe a solar eclipse.

© 2011 Cengage Learning. All Rights Reserved. May not be scanned, copied or duplicated, or posted to a publicly accessible website, in whole or in part.

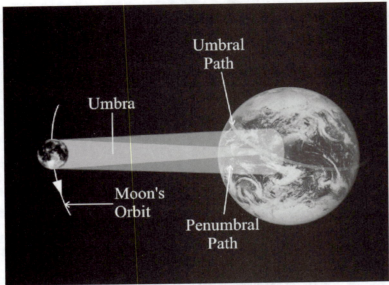

Figure 9-27 The Umbra of the Moon's Shadow.

Types of Solar Eclipses

Like lunar eclipses, the type of solar eclipse observed from Earth depends on which part of the Moon's shadow passes over the observer. If an observer's location lies *completely within* the umbral path as shown in **Figure 9-27**, then anywhere along that path a total solar eclipse is observed. **Figure 9-28** displays the total solar eclipse that took place on February 26, 1998.

Figure 9-28 Total Solar Eclipse of February 26, 1998 (Photo courtesy of Dr. Dale Ireland)

Observing total solar eclipses is quite interesting from a scientific point of view. With the bright visible portion of the Sun (known as the *photosphere*) eclipsed, astronomers can study the outer part of the Sun's atmosphere called the *corona*.

If an observer's location is just *outside* the umbral path, but *anywhere within* the penumbral path, then partial eclipses (of varying degrees) may be seen from many locations on Earth. Scientifically, there is not much interest in partial solar eclipses because the bright photosphere of the Sun is only partially blocked by the Moon. **Figure 9-29** is an image of the partial solar eclipse taken on February 26, 1998. The picture was taken at the time of mid-eclipse.

© 2011 Cengage Learning. All Rights Reserved. May not be scanned, copied or duplicated, or posted to a publicly accessible website, in whole or in part.

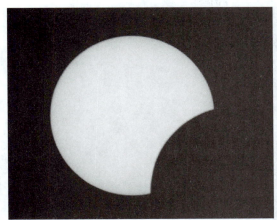

Figure 9-29 Partial Solar Eclipse of February 26, 1998 (Photo courtesy of Dr. Dale Ireland)

In some instances the Moon's umbral shadow does not completely reach Earth's surface. When this situation arises, the last type of solar eclipse occurs. As the Moon's umbra makes its way toward Earth, it sometimes converges several miles above Earth's surface. Eventually it diverges, however, and finally reaches Earth's surface. An observer standing on Earth looking toward the Sun and Moon in this situation observes the Moon's angular size to be smaller than that of the Sun.

As the Moon moves between Earth and the Sun, from Earth's surface an observer sees most of the Moon blocking the sunlight except for a small ring of light that encircles it. This small ring is called an *annulus*. Consequently, this type of solar eclipse is known as an *annular eclipse*. The geometry of an annular eclipse is illustrated **Figure 9-30**.

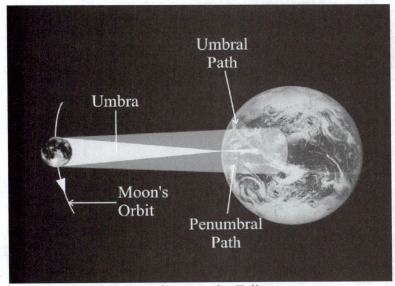

Figure 9-30 Geometry of an Annular Eclipse

Figure 9-31 displays *TheSkyX's* rendition of the annular eclipse that took place on May 10, 1994, as viewed from the Ball State University campus. It is a zoomed in view of the Moon. The time of mid-eclipse was 12:09 PM EST.

© 2011 Cengage Learning. All Rights Reserved. May not be scanned, copied or duplicated, or posted to a publicly accessible website, in whole or in part.

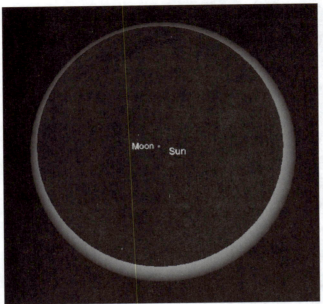

Figure 9-31 *TheSkyX's* Rendition of the Annular Solar Eclipse on May 10, 1994

Figure 9-32 is an actual photograph of the eclipse taken on May 10, 1994 from the Ball State University campus at the time of mid-eclipse by Dr. Jordan.

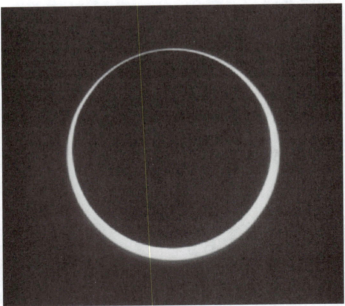

Figure 9-32 Photograph of the Annular Solar Eclipse on May 10, 1994 (Photo Courtesy of Dr. Jordan)

In order to view the sky window displayed in **Figure 9-31**, then follow the instructions provided in **Exercise R**.

Exercise R: Displaying the Annular Eclipse on May 10, 1994

1. Run *TheSkyX* and set as follows:

    Set Date: May 10, 1994
    Set Time: 12:09 PM EST

© 2011 Cengage Learning. All Rights Reserved. May not be scanned, copied or duplicated, or posted to a publicly accessible website, in whole or in part.

Location: Ball State Observatory
Daylight Saving option: "Not Observed"

2. Find the Sun or the Moon then 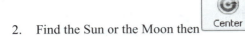 .

3. Now, Zoom In on the Sun or Moon.

4. What are the equatorial coordinates of the Sun and Moon?

Sun: R.A.: _____^H _____^M _____^S   Declination: __ _____° _____' _____"

Moon: R.A.: _____^H _____^M _____^S   Declination: __ _____° _____' _____"

5. Close the file and do not save it.

If an observer's location is outside the penumbral path, then *no* eclipses are observed. *TheSkyX* does an excellent job of finding and rendering eclipses for any epoch. In fact, the software contains another tool package that finds and illustrates eclipses for any given year.

Exercises S, **T**, **U**, **V** and **W** are all designed to assist in using this tool and determining when and where solar or lunar eclipses have or will occur. Note, however, that finding an eclipse with the Eclipse Finder does not guarantee that the eclipse will be visible from a given observer's location.

In the case of solar eclipses, a feature in the Eclipse Finder allows one to observe where the path of totality is located on Earth. Clicking on this feature renders a graphical image of Earth showing where the Moon's path is traced over the area to be eclipsed. In fact, it may be necessary to change one's location on Earth in order to view the eclipse. **Exercises S**, **T**, and **U** are designed to assist in determining solar eclipses that have or will occur. **Exercises V** and **W** are designed to assist in determining when lunar eclipses will occur.

Exercise S: Displaying the Partial Solar Eclipse on December 25, 2000

1. Run *TheSkyX* and set as follows:

    Set Date: January 1, 2000
    Location: Ball State Observatory
    Daylight Saving option: "Not Observed"

2. Click Tools on the Standard toolbar at the top of the sky window.

3. Now, click on the "Solar and Lunar Eclipse Viewer."

4. Click on the "Eclipses" tab.

5. Now, click on the "Partial Solar" eclipse on "12/25/2000" as is shown in **Figure 9-33**. Details concerning the eclipse are given in the "Local Circumstances" box.

© 2011 Cengage Learning. All Rights Reserved. May not be scanned, copied or duplicated, or posted to a publicly accessible website, in whole or in part.

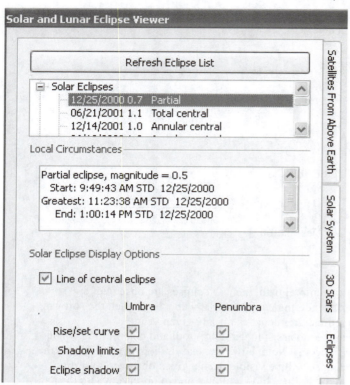

Figure 9-33 Selecting the Partial Solar Eclipse on December 25, 2000

6. Set the Date: December 25, 2000
    Set the Time: 09:49:43 STD
    Daylight saving option: "Not Observed"

7. Find the Sun or Moon and [Center].

8. When viewing this eclipse in TheSkyX, the Moon appears just right (west) and above (north) of the Sun.

9. The Moon eclipses the northern half of the Sun as shown in **Figure 9-34**. Notice too that the Sun and Moon are both in the constellation of Sagittarius.

© 2011 Cengage Learning. All Rights Reserved. May not be scanned, copied or duplicated, or posted to a publicly accessible website, in whole or in part.

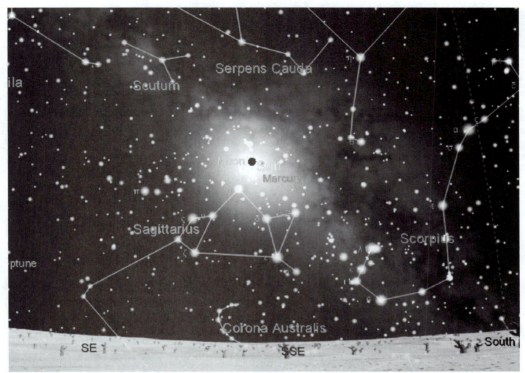

Figure 9-34 Mid-eclipse of the Partial Solar Eclipse on December 25, 2000

10.   Close the file and do not save it, or continue with **Exercise T**.

Exercise T: Displaying the Total Solar Eclipse on August 21, 2017

1.   Run *TheSkyX*.

2.   Set Date: January 1, 2017.
      Location: Ball State Observatory
      Daylight Saving option: "U.S. and Canada"

3.   Click Tools on the Standard toolbar at the top of the sky window.

4.   Now, click on the "Solar and Lunar Eclipse Viewer" once again.

5.   Click on the "Eclipses" tab.

6.   Now, click on the "Partial Solar" eclipse on "08/21/2017" as is shown in **Figure 9-35**.
      Details concerning the eclipse are displayed in the "Local Circumstances" box once again.
      If the list of Solar Eclipses is not opened, then click on ⊞ Solar Eclipses button to expand
      the list.

© 2011 Cengage Learning. All Rights Reserved. May not be scanned, copied or duplicated, or posted to a publicly accessible website, in whole or in part.

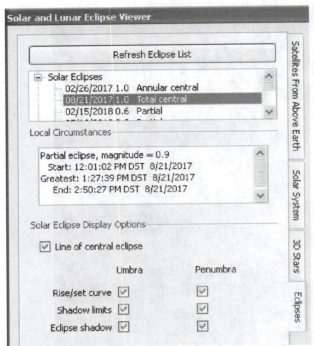

Figure 9-35 Solar and Lunar Eclipse Viewer Window for January 1, 2017

7. After selecting the Total eclipse on August 21, 2017, in addition to information displayed in the "Local Circumstances" box, an Earth globe appears like the one shown in **Figure 9-36**.

8. Notice too that when one clicks the "Line of Central Eclipse" a yellow line appears on the figure and runs East-West through the middle of the United States. This is the path of totality through the United States in **Figure 9-36**. Next, if one clicks the "Shadow Limits," a red line appears to show where one has to be on Earth to observe any eclipse at all. In this instance the eclipse is visible from all sites in the U.S. However, when you click on the "Eclipse Shadow" for the umbra and penumbra, a white oval-shaped line appears. One should notice that the Ball State Observatory only observes a partial eclipse not a total one on this date.

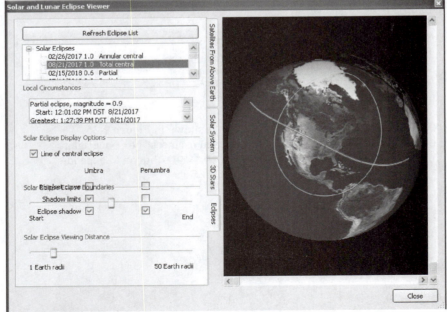

Figure 9-36 Solar and Lunar Eclipse Viewer Window for August 21, 2017

© 2011 Cengage Learning. All Rights Reserved. May not be scanned, copied or duplicated, or posted to a publicly accessible website, in whole or in part.

9. Close the Solar Lunar Eclipse Viewer window and right click on your mouse. Make sure the time is set to 11:15 AM DST and the date is set to August 21, 2017. Now, find the Sun and

.

10. The screen should appear like the one shown in **Figure 9-37**.

Figure 9-37 *TheSkyX's* Rendition of the Total Solar Eclipse on August 21, 2017 from the Ball State Observatory

11. Zoom To an approximate 25° field of view using the Zoom In button. Set the Time Flow rate to a 1^M increment.

12. Click Go Forward to put the Moon in motion in order to observe the eclipse. Click Stop. Try to stop the time at approximately 2:28 PM DST…this is when mid-eclipse occurs at the Ball State Observatory. Reset the time to 11:15 AM and Zoom To an 8° field of view. Now, click Step Forward to watch the eclipse at 1^M increments. This slows the motion of the Moon as illustrated in **Figure 9-37**.

13. After zooming to an 8° field of view, one should notice that at mid-eclipse only a very small portion of the top of the Sun is visible. As stated earlier, the eclipse is actually *not* a total eclipse at the Ball State Observatory but a partial. This is shown in **Figure 9-38**.

© 2011 Cengage Learning. All Rights Reserved. May not be scanned, copied or duplicated, or posted to a publicly accessible website, in whole or in part.

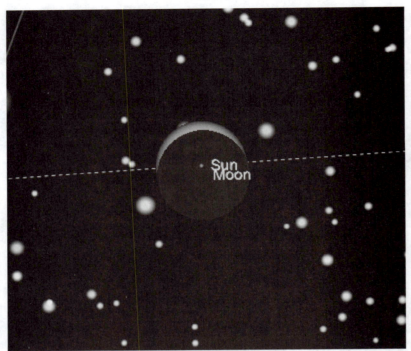

Figure 9-38 Mid-eclipse on August 21, 2017 Viewed from the Ball State Observatory

14. In order to view complete totality one must change their latitude.

15. Click "Input" on the Standard toolbar and then Location.

16. Change the latitude to 36° 00' N then press [X].

17. Reset the time to 12:00 Noon DST and the screen will appear as the one displayed in **Figure 9-39**.

Figure 9-39 Total Eclipse on August 21, 2017 from Latitude of 36° North

18. Click [Go Forward] to replay the eclipse sequence. In other words, one would have to be near a line from Kansas City, Kansas through Chattanooga, Tennessee to observe this total eclipse. But all of the United States will observe some type of eclipse.

© 2011 Cengage Learning. All Rights Reserved. May not be scanned, copied or duplicated, or posted to a publicly accessible website, in whole or in part.

19. **Figure 9-40** displays how the eclipse appears from latitude 36° N at the time of mid-eclipse.

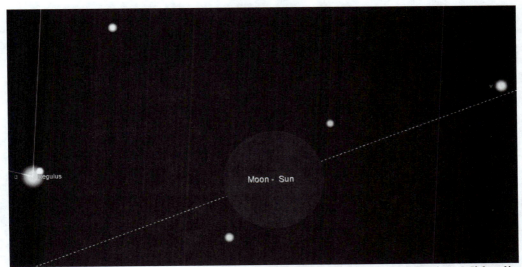

Figure 9-40 Total Eclipse on August 21, 2017 from Latitude of 36° North at Mid-eclipse

20. Close the file and do not save it, or continue with **Exercise U**.

Exercise U: Displaying the Annular Solar Eclipse on May 9, 2013

1. Run *TheSkyX* and set as follows:

   Set Date: January 1, 2013
   Set Time: 10:00 PM STD

   Location: Ball State Observatory
   Daylight Saving option: "U.S. and Canada"

2. Click Tools on the Standard toolbar at the top of the sky window.

3. Click Solar and Lunar Eclipse Viewer once again. A window like the one displayed in **Figure 9-41** appears.

© 2011 Cengage Learning. All Rights Reserved. May not be scanned, copied or duplicated, or posted to a publicly accessible website, in whole or in part.

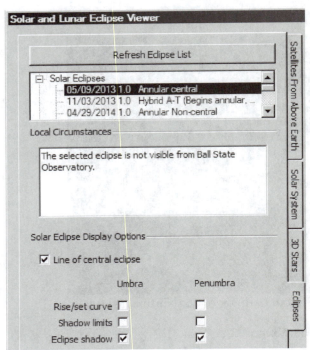

Figure 9-41 Solar and Lunar Eclipse Viewer Window for January 1, 2013

4. If not opened, expand the menu of "Solar eclipses" and click the "Annular central" eclipse on "05/09/2013."

5. Just below the window containing the index of eclipses, in the "Local Circumstances" window, it states: "The selected eclipse is not visible from the Ball State Observatory."

6. In the right-half of the Solar and Lunar Eclipse View window the "Path of Totality" (yellow line) and the penumbral shadow (white line) are shown in **Figure 9-42**.

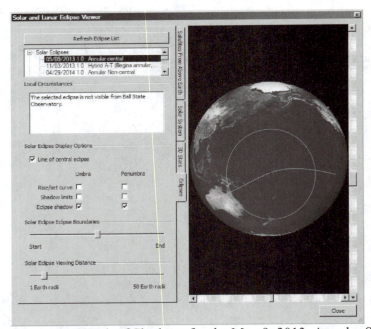

Figure 9-42 Path of Shadows for the May 9, 2013, Annular Solar Eclipse

7. The path of totality is across the continent of Australia.

© 2011 Cengage Learning. All Rights Reserved. May not be scanned, copied or duplicated, or posted to a publicly accessible website, in whole or in part.

8. If one had set *TheSkyX* to the date and time of this eclipse to observe it, you would have noticed that the Moon does not move in front of the Sun. The reason, of course, for this is that your location is probably not set to Australia.

9. If you wish to view this eclipse, then continue. Otherwise, close the file and do not save it.

10. Change your latitude, longitude, date, and time to the following:

    Date: May 10, 2013
    Time: 6:00 AM
    Time Zone: 10 hours

    Latitude: 11° S
    Longitude: 151° E
    Daylight Saving option: "Not Observed"

11. Find the Sun and [Center]. Zoom To an 8° field of view. You may have to Navigate in order to position the Moon left of the Sun.

12. Set the Time Flow rate to a 1^M increment.

13. Now, click [Go Forward] to watch the eclipse. It is interesting to note here that the Moon Moon appears to move to the right. Why? _____.

14. Close the file and do not save it, or continue with **Exercise V**.

Exercise V: Observing a Total Lunar Eclipse in 2014

1. Run *TheSkyX* and set as follows:

    Set Date: January 1, 2014
    Location: Oakland, CA.
    Set Time: 12:00 Noon STD
    Time Zone: -8 hours
    Daylight Saving option: "U.S. and Canada"

2. Click Tools menu on the Standard toolbar at the top of the sky window.

3. Click on the Solar and Lunar Eclipse Viewer again. This opens a window that appears like the one shown in **Figure 9-43**.

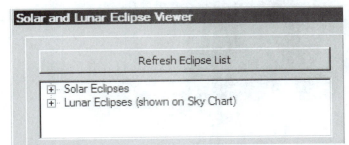

Figure 9-43 Solar Lunar Eclipse Viewer Window for 2014 Lunar Eclipses

© 2011 Cengage Learning. All Rights Reserved. May not be scanned, copied or duplicated, or posted to a publicly accessible website, in whole or in part.

4. Click the ⊞ Lunar Eclipses button to expand the list. Now, find the next Total Lunar Eclipse and click the "Total Lunar" eclipse on "04/15/2014" as displayed in **Figure 9-44**.

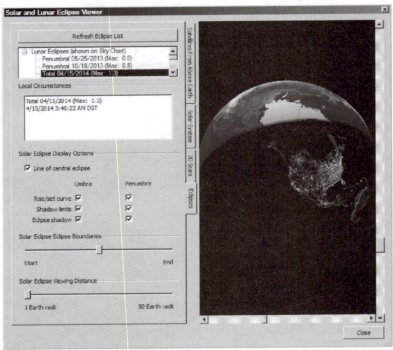

Figure 9-44 Total Lunar Eclipse on April 15, 2014

5. The Region of the Earth that can observe this eclipse is shown in **Figure 9-45**.

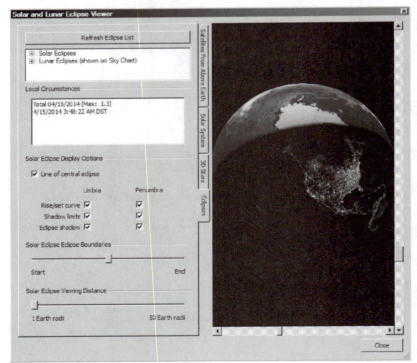

Figure 9-45 Lunar Eclipse Viewer Window Showing Regions of Totality

© 2011 Cengage Learning. All Rights Reserved. May not be scanned, copied or duplicated, or posted to a publicly accessible website, in whole or in part.

6. You may change the view orientation using the slide bars above and below the Earth. Just explore the features in this window.

7. Now, close this window and reset *TheSkyX* as follows if not set already.

   Set Date: April 15, 2014
   Set Time: 12:48 AM DST
   Daylight Saving option: "U.S. and Canada"
   Zoom To: 10°

8. The time set is to mid-eclipse in Oakland, CA. Notice too that the Moon is completely immersed in Earth's umbral shadow. The eclipse has been rendered in *TheSkyX* and is displayed in **Figure 9-46**.

   However, when an observer goes to different locations across the United States it necessary to note here that *TheSkyX* automatically adjusts the observer's time to the time of mid-eclipse at that particular location.

   If on the other hand, clocks were adjusted to the time at a particular location in regard to Oakland's mid-eclipse time, then one would see differences in the amount of how much of the Moon is eclipsed.

   In other words, the time of totality in Oakland is not the same time as it would be for an observer in New York City. Remember there is a three-hour time difference between the two locations. If we adjust our clocks to 12:48 AM at each location, then observers would see how longitude and latitude affects how much of the Moon is seen during the eclipse.

   This is the purpose of this exercise!

Figure 9-46 Mid-eclipse of the Total Lunar Eclipse on April 15, 2014 from Oakland, CA

© 2011 Cengage Learning. All Rights Reserved. May not be scanned, copied or duplicated, or posted to a publicly accessible website, in whole or in part.

9. Changing an observer's location on Earth allows one to view latitudinal and longitudinal effects of the eclipse. For example, change the location to Phoenix, Arizona.

    Set Date: April 14, 2014
    Set time: 11:48 PM STD
    Set Location: Phoenix, Arizona
    Set Daylight Savings option to "Not Observed"

    Phoenix is on Standard Time all year round, a correction in time of 1^H is necessary.

10. Find the Moon and [Center] . Does anything appear to change from this location? _____.

11. Is there any difference in the amount of the Moon eclipsed? _____.

12. [Step Forward] to change the time. Does this change anything for the eclipse? _____.
    After making the correction for time, does the eclipse appear to be the same? _____.

13. Now, change the location to Chicago, Illinois and you will have to make some necessary corrections.

    Set Daylight Savings option to "U.S. and Canada"
    Set time: 12:48 AM DST
    Chicago is on Daylight Savings Time this time of year.

14. Does anything appear to change from this location? _____

15. Was there any difference in the amount of the Moon eclipsed? _____.

16. [Step Forward] to change the time. Does this change anything for the eclipse? _____.
    After making the correction for time, does the eclipse appear to be the same?
    _____.

17. What time does mid-eclipse occur in Chicago, Illinois? _____:_____.

18. Finally, change the location to New York City and you will also have to make necessary corrections here too.

    Set Daylight Savings option to "U.S. and Canada"
    Set time: 12:48 AM DST

19. Does anything appear to change from this location? _____.

20. Was there any difference in the amount of the Moon eclipsed? _____.

21. [Step Forward] to change the time. Does this change anything for the eclipse? _____.
    After making the correction for time, does the eclipse appear to be the same? _____.

22. What time does mid-eclipse occur in New York City? _____:_____

© 2011 Cengage Learning. All Rights Reserved. May not be scanned, copied or duplicated, or posted to a publicly accessible website, in whole or in part.

23. What conclusions might you make about the time at which mid-eclipse occurs for observers at the four different locations? _____

    _____.

24. Close the file and do not save it, or continue with **Exercise W**.

Exercise W: Displaying a Partial Lunar Eclipse in 2017

1. Run *TheSkyX* and set as follows:
   Set Date: January 1, 2017
   Location: Ball State Observatory
   Time: Anytime
   Daylight Saving option: "U.S. and Canada"

2. Click Tools on the Standard toolbar at the top of the sky window.

3. Click on the Solar and Lunar Eclipse Viewer.

4. Click the "Partial Lunar" eclipse that occurs on "08/07/2017." The Region of the Earth that is able to observe this eclipse is shown in **Figure 9-47**.

Figure 9-47 Partial Lunar Eclipse Viewer Window Showing Regions of the Partial Eclipse

5. At first, one should notice that the partial eclipse occurs at 2:53 **PM** DST for an observer at the Ball State Observatory. This means that it is *not* visible from this location.

6. However, it is still possible to observe this eclipse in *TheSkyX*. Simply click on the eclipse date in the Solar and Lunar Eclipse Viewer. If necessary, Find the Moon and [Center].

© 2011 Cengage Learning. All Rights Reserved. May not be scanned, copied or duplicated, or posted to a publicly accessible website, in whole or in part.

7. Close the eclipse viewer window. The desktop then will display the eclipse in question and is shown in **Figure 9-48**. Notice that the Moon moves through both the umbra and penumbra of the Earth's shadow as illustrated in the figure.

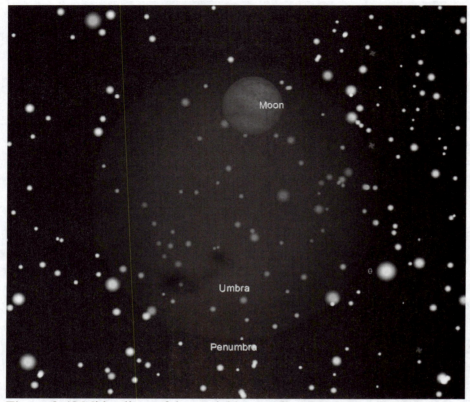

Figure 9-48 Mid-eclipse of the partial lunar eclipse on August 7, 2017

8. To find out when this eclipse occurs at a particular location on Earth, one needs to know when a Full Moon occurs in August 2017. This information can be found on the Moon Phase Calendar.

9. Reset *TheSkyX* as follows:

    Set Date: August 1, 2017
    Set Time: 8:00 AM DST
    Daylight Saving option: "U.S. and Canada"

10. Click Tools on the toolbar at the top of the sky window.

11. Click the "Calendar" to display it. It is displayed in **Figure 9-49** for August 2017.

© 2011 Cengage Learning. All Rights Reserved. May not be scanned, copied or duplicated, or posted to a publicly accessible website, in whole or in part.

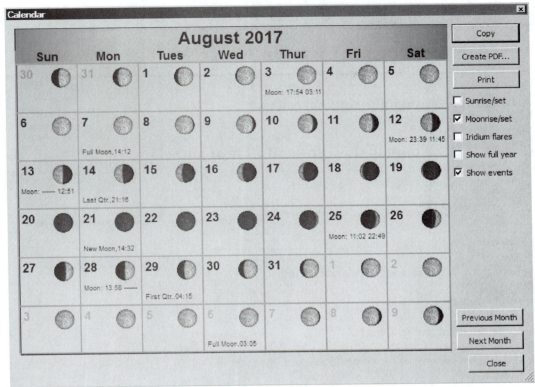

Figure 9-48 Moon Phase Calendar for August, 2017

12. Notice that the Full Moon occurs on the 7th at 2:12 pm DST for an observer at the Ball State Observatory. This time, of course, is adjusted accordingly for the time zone in which one is located as well as any corrections for Daylight Savings Time.

13. There is one thing to note here. Even though a partial eclipse of the Moon takes place on this date it is not visible from the Ball State Observatory since it is afternoon and the Sun is above the horizon.

14. When does this eclipse occur in Los Angeles, California? _____:_____. Is it visible from this location? _____.

15. When does this eclipse occur in Memphis, Tennessee? _____:_____. Is it visible from this location? _____.

16. When does this eclipse occur in Bangor, Maine? _____:_____. Is it visible from this from this location? _____.

17. Where do you have to be on Earth to observe this partial lunar eclipse? _____

18. Close the file and do not save it.

© 2011 Cengage Learning. All Rights Reserved. May not be scanned, copied or duplicated, or posted to a publicly accessible website, in whole or in part.

Chapter 9

TheSkyX Review Exercises

Use *TheSkyX* to complete the following exercises:

TheSkyX Exercise 1: Determining the Type and Date of an Eclipse

1. What type of eclipse first occurs in the year 2015 C.E.? _____

2. Is it a total, partial, penumbral, or annular eclipse? _____

3. On what date does the eclipse occur? _____

4. Approximately what time does this eclipse occur? _____:_____

5. Is it visible from New York City? _____

6. Is it visible from Edmonton, Canada? _____

7. Is it visible from any part of the United States or Canada on this date? _____
   If so, where? _____

8. If not, where do you have to go on Earth to observe it?

   Latitude = _____°   Longitude = _____°

TheSkyX Exercise 2: Determining the Date and Time of a Lunar Eclipse

1. On what date does the last lunar eclipse occur in 2015 C.E.? _____

2. Is it a total, partial, or penumbral eclipse? _____

3. Approximately what time does this eclipse occur? _____:_____

4. Is it visible from Memphis, Tennessee? _____

5. Is it visible from any part of the United States or Canada? _____

6. If so, where? _____

7. If not, where do you have to go on Earth to observe it?

   Latitude = _____°   Longitude = _____°

© 2011 Cengage Learning. All Rights Reserved. May not be scanned, copied or duplicated, or posted to a publicly accessible website, in whole or in part.

TheSkyX Exercise 3: Determining the Date and Time of a Solar Eclipse

1. On what date did the first solar eclipse take place in the year 1776 C.E.? _____

2. Approximately what time did this eclipse occur? _____:_____

3. Was it a total, partial, or annular eclipse? _____

4. Was it visible from Philadelphia, Pennsylvania? _____

5. Was it visible from any part of the United States or Canada? _____

6. If so, where? _____

7. If not, where did you have to go on Earth to observe it?

   Latitude = _____°   Longitude = _____°

TheSkyX Exercise 4: Determining the Date and Time of a Lunar Eclipse

1. On what date did the last lunar eclipse occur in the year 1776 C.E.? _____

2. Was it a total, partial, or penumbral eclipse? _____

3. Was it visible from Philadelphia, Pennsylvania? _____

4. Approximately what time did this eclipse occur? _____:_____

5. Was it visible from any part of the United States or Canada? _____

6. If so, where? _____

7. If not, where did you have to go on Earth to observe it?

   Latitude = _____°   Longitude = _____°

TheSkyX Exercise 5: Determining the Type and Date of an Eclipse

1. On what date does the first eclipse occur in the year 5555 C.E.? _____

2. Is it a lunar or solar eclipse? _____

3. Is it a total, partial, annular, or penumbral eclipse? _____

4. Approximately what time does this eclipse occur? _____:_____

5. Is it visible from the United States or Canada? _____

6. If so, where? _____

7. If not, where do you have to go on Earth to observe it?

   Latitude = _____°   Longitude = _____°

© 2011 Cengage Learning. All Rights Reserved. May not be scanned, copied or duplicated, or posted to a publicly accessible website, in whole or in part.

Chapter 9

TheSkyX Review Questions

a

Review Questions:

1. The phase of the Moon when it is aligned with the Sun in the sky is called the
   _____ Moon.

     (a) Full   (b) Last Quarter   (c) First Quarter   (d) New

2. The phase of the Moon when it is located 180° from the Sun is called the
   _____ Moon.

     (a) Full   (b) Last Quarter   (c) First Quarter   (d) New

3. The phase of the Moon when it is located 90° east of the Sun is called the
   _____ Moon.

     (a) Full   (b) Last Quarter   (c) First Quarter   (d) New

4. As the Moon moves between a New and a Full phase, the illuminated portion of it
   becomes larger and larger. The phases during this stage of the Moon's phase cycle
   are known as the _____ phases.

     (a) waving   (b) waning   (c) waxing   (d) wearing   (e) none of these

5. As the Moon moves between a Full and a New phase, the illuminated portion of it
   becomes smaller and smaller. The phases during this stage of the Moon's phase cycle
   are known as the _____ phases.

     (a) waving   (b) waning   (c) waxing   (d) wearing   (e) none of these

6. The two shadows cast by both Earth and the Moon into space are called the
   _____ and the _____.

7. What three conditions are necessary for a lunar or solar eclipse to occur?

     (a) _____

     (b) _____

     (c) _____

8. When the Moon moves entirely through the umbra of Earth's shadow, an observer on
   Earth observes a _____ lunar eclipse.

     (a) partial   (b) annular   (c) penumbral   (d) total   (e) sesquicentennial

9. When the Moon moves partly through the umbra and penumbra of Earth's shadow, an
   observer on Earth observes a _____ lunar eclipse.

     (a) partial   (b) annular   (c) penumbral   (d) total   (e) sesquicentennial

© 2011 Cengage Learning. All Rights Reserved. May not be scanned, copied or duplicated, or posted to a publicly accessible website, in whole or in part.

10. If an observer is standing on Earth and the Moon's umbral shadow passes over them, then that observer observes a _____ solar eclipse.

    (a) partial   (b) annular   (c) penumbral   (d) total

11. If an observer is standing on Earth and the moon's penumbral shadow passes over them, then that observer observes a _____ solar eclipse.

    (a) partial   (b) annular   (c) penumbral   (d) total

12. The type of eclipse that occurs when the moon's umbral shadow doesn't quite reach the ground is called a (n) _____ eclipse of the Sun.

    (a) partial   (b) annular   (c) penumbral   (d) total

© 2011 Cengage Learning. All Rights Reserved. May not be scanned, copied or duplicated, or posted to a publicly accessible website, in whole or in part.

Appendix A

The Brightest Stars

Star	Name	Position 2000.0 R.A.		Dec.		Apparent Magnitude	Spectral Type[*]	Absolute Magnitude	Approximate Distance (LY)
Sol	Sun	----------		----------		-26.7	G2V	+4.85	----------
α CMa A	Sirius	06	45.1	-16	43	-1.44	A0V	+1.44	8.6
α Car	Canopus	06	24.0	-52	42	-0.62	F0Ib	-2.50	95.9
α Boo	Arcturus	14	15.7	+19	11	-0.05	K2IIIp	-0.20	36.7
α Cen A	Rigil Kentaurus	14	39.6	-60	50	-0.01	G2V	+4.37	4.39
α Lyr	Vega	18	36.9	+38	47	+0.03	A0V	+0.60	25.3
α Aur	Capella	05	16.7	+46	00	+0.08	G6III	+0.40	42.2
β Ori A	Rigel	05	14.5	-08	12	+0.18	B8Ia	-8.10	772.9
α CMi A	Procyon	07	39.3	+05	14	+0.40	F5IV-V	+2.70	11.4
α Ori	Betelgeuse	05	55.2	+07	24	+0.45	M2Iab	-7.20	427.5
α Eri	Achernar	01	37.7	-57	14	+0.45	B3Vp	-1.30	143.8
β Cen AB	Hadar	14	03.8	-60	22	+0.61	B1III	-4.40	525.2
α Aql	Altair	19	50.8	+08	52	+0.76	A7IV-V	+2.30	16.8
α Tau A	Aldeberan	04	35.9	+16	31	+0.87	K5III	-0.30	65.1
α Vir	Spica	13	25.2	-11	10	+0.98	B1V	-3.20	262.2
α Sco A	Antares	16	29.4	-26	26	+1.06	M1Ib	-5.20	604.0
α PsA	Fomalhaut	22	57.6	-29	37	+1.17	A3V	+2.00	25.1
β Gem	Pollux	07	45.3	+28	02	+1.15	K0IIIvar	+0.70	33.7
α Cyg	Deneb	20	41.4	+45	17	+1.25	A2Ia	-7.20	3229.3
β Cru	Beta Crucis	12	47.7	-59	41	+1.25	B0.5III	-4.70	352.6
α Leo A	Regulus	10	08.4	+11	58	+1.36	B7V	-0.30	77.5
α Cru A	Acrux	12	26.6	-63	06	+0.77	B0.5IV	-4.20	320.7
ε CMa A	Adhara	06	58.6	-28	58	+1.50	B2II	-4.80	430.9
λ Sco	Shaula	17	33.6	-37	06	+1.62	B1.5IV	-3.50	702.9
γ Ori	Bellatrix	05	25.1	+06	21	+1.64	B2III	-3.90	243.0
β Tau	Elnath	05	26.3	+28	36	+1.65	B7III	-1.50	131.0

[*]Spectral Types: I = Supergiants, II = Bright Giants, III = Giants, IV = Subgiants, V = Dwarfs

© 2011 Cengage Learning. All Rights Reserved. May not be scanned, copied or duplicated, or posted to a publicly accessible website, in whole or in part.

Appendix B

Constellation Names and Their Abbreviations

Latin Name	Possessive Form	Abbreviation	Translation
Andromeda	Andromedae	And	Andromeda*
Antlia	Antliae	Ant	Pump
Apus	Apodis	Aps	Bird of Paradise
Aquarius	Aquarii	Aqr	Water Bearer
Aquila	Aquilae	Aql	Eagle
Ara	Arae	Ara	Altar
Aries	Arietis	Ari	Ram
Auriga	Aurigae	Aur	Charioteer
Boötes	Boötis	Boo	Herdsman
Caelum	Caeli	Cae	Chisel
Camelopardalis	Camelopardalis	Cam	Giraffe
Cancer	Cancri	Cnc	Crab
Canes Venatici	Canum Venaticorum	Cvn	Hunting Dogs
Canis Major	Canis Majoris	CMa	Big Dog
Canis Minor	Canis Minoris	CMi	Little Dog
Capricornus	Capricorni	Cap	Goat
Carina	Carinae	Car	Ship's Keel**
Cassiopeia	Cassiopeiae	Cas	Cassiopeia*
Centaurus	Centauri	Cen	Centaur*
Cepheus	Cephei	Cep	Cepheus*
Cetus	Ceti	Cet	Whale
Chamaeleon	Chamaeleonis	Cha	Chameleon
Circinus	Circini	Cir	Compass
Columba	Columbae	Col	Dove
Coma Berenices	Comae Berenices	Com	Berenice's Hair*
Corona Australis	Coronae Australis	CrA	Southern Crown
Corona Borealis	Coronae Borealis	CrB	Northern Crown
Corvus	Corvi	Crv	Crow
Crater	Crateris	Crt	Cup
Crux	Crucis	Cru	Southern Cross
Cygnus	Cygni	Cvg	Swan
Delphinus	Delphini	Del	Dolphin
Dorado	Doradus	Dor	Swordfish
Draco	Draconis	Dra	Dragon
Equuleus	Equulei	Equ	Little Horse
Eridani	Eri	River Eridanus*	
Fornax	Fornacis	For	Furnace
Gemini	Geminorum	Gem	Twins
Grus	Gruis	Gru	Crane
Hercules	Herculis	Her	Hercules*

© 2011 Cengage Learning. All Rights Reserved. May not be scanned, copied or duplicated, or posted to a publicly accessible website, in whole or in part.

Latin Name	Possessive Form	Abbreviation	Translation
Horologium	Horologii	Hor	Clock
Hydra	Hydrae	Hya	Hydra*(Water Monster)
Hydrus	Hydri	Hyi	Sea Serpent
Indus	Indi	Ind	Indian
Lacerta	Lacertae	Lac	Lizard
Leo	Leonis	Leo	Lion
Leo Minor	Leonis Minoris	LMi	Little Lion
Lepus	Leporis	Lep	Hare
Libra	Librae	Lib	Scales
Lupus	Lupi	Lup	Wolf
Lynx	Lyncis	Lyn	Lynx
Lyra	Lyrae	Lyr	Lyre (Harp)
Mensa	Mensae	Men	Table (Mountain)
Microscopium	Microscopii	Mic	Microscope
Monoceros	Monocerotis	Mon	Unicorn
Musca	Muscae	Mus	Fly
Norma	Normae	Nor	Carpenter's Square
Octans	Octantis	Oct	Octant
Ophiuchus	Ophiuchi	Oph	Ophiuchus*(Serpent Bearer)
Orion	Orionis	Ori	Orion* (The Hunter)
Pavo	Pavonis	Pav	Peacock
Pegasus	Pegasi	Peg	Pegasus*(Winged Horse)
Perseus	Persei	Per	Perseus
Phoenix	Phoenicis	Phe	Phoenix
Pictor	Pictoris	Pic	Easel
Pisces	Piscium	Psc	Fishes
Piscis Austrinus	Piscis Austrini	PsA	Southern Fishes
Puppis	Puppis	Pup	Ship's Stern**
Pyxis	Pyxidis	Pyx	Ship's Compass**
Reticulum	Reticuli	Ret	Net
Sagitta	Sagittae	Sge	Arrow
Sagittarius	Sagittarii	Sgr	Archer
Scorpius	Scorpii	Sco	Scorpion
Sculptor	Sculptoris	Scl	Sculptor
Scutum	Scuti	Sct	Shield
Serpens	Serpentis	Ser	Serpent
Sextans	Sextantis	Sex	Sextant
Taurus	Tauri	Tau	Bull
Telescopium	Telescopii	Tel	Telescope
Triangulum	Trianguli	Tri	Triangle

© 2011 Cengage Learning. All Rights Reserved. May not be scanned, copied or duplicated, or posted to a publicly accessible website, in whole or in part.

Latin Name	Possessive Form	Abbreviation	Translation
Triangulum Australe	Trianguli Australis	TrA	Southern Triangle
Tucana	Tucanae	Tuc	Toucan
Ursa Major	Ursae Majoris	UMa	Big Bear
Ursa Minor	Ursae Minoris	UMi	Little Bear
Vela	Velorum	Vel	Ship's Sails**
Virgo	Virginis	Vir	Virgin
Volans	Volantis	Vol	Flying Fish
Vulpecula	Vulpeculae	Vul	Little Fox

*Proper Names.

**Use to form the constellation Argo Navis, the Argonauts' Ship

© 2011 Cengage Learning. All Rights Reserved. May not be scanned, copied or duplicated, or posted to a publicly accessible website, in whole or in part.

The Greek Alphabet

Letter	Upper case	Lower Case
Alpha	A	α
Beta	B	β
Gamma	Γ	γ
Delta	Δ	δ
Epsilon	E	ε
Zeta	Z	ζ
Eta	H	η
Theta	Θ	θ
Iota	I	ι
Kappa	K	κ
Lambda	Λ	λ
Mu	M	μ

Letter	Upper Case	Lower Case
Nu	N	ν
Xi	Ξ	ξ
Omicron	O	ο
Pi	Π	π
Rho	P	ρ
Sigma	Σ	σ
Tau	T	τ
Upsilon	Y	υ
Phi	Φ	φ
Chi	X	χ
Psi	Ψ	ψ
Omega	Ω	ω

© 2011 Cengage Learning. All Rights Reserved. May not be scanned, copied or duplicated, or posted to a publicly accessible website, in whole or in part.

Appendix D

The Messier Catalogue

M#	NGC#	R. A. (2000)	Dec. (2000)	m_v	Description
1	1952	05 34.5	+22 01	11.3	Crab Nebula in Taurus
2	7089	21 33.5	-00 49	6.5	Globular Cluster in Aquarius
3	5272	13 42.2	+28 23	6.4	Globular Cluster in Canes Venatici
4	6121	16 23.6	-26 32	5.9	Globular Cluster in Scorpius
5	5904	15 18.6	+02 05	5.8	Globular Cluster in Serpens
6	6405	17 40.1	-32 13	5.2	Open Cluster in Scorpius
7	6475	17 53.9	-34 49	3.3	Open Cluster in Scorpius
8	6523	18 03.8	-24 23	5.8	Lagoon Nebula in Sagittarius
9	6333	17 19.2	-18 31	7.9	Globular Cluster in Ophiuchus
10	6254	16 57.1	-04 06	6.6	Globular Cluster in Ophiuchus
11	6705	18 51.1	-06 16	5.8	Open Cluster in Scutum
12	6218	16 47.2	-01 57	6.6	Globular Cluster in Ophiuchus
13	6205	16 41.7	+36 28	5.9	Globular Cluster in Hercules
14	6402	17 37.6	-03 15	7.6	Globular Cluster in Ophiuchus
15	7078	21 30.0	+12 10	6.4	Globular Cluster in Pegasus
16	6611	18 18.8	-13 47	6.0	Open Cluster and Nebula in Serpens
17	6618	18 20.8	-16 11	7.0	Omega Nebula in Sagittarius
18	6613	18 19.9	-17 08	6.9	Open Cluster in Sagittarius
19	6273	17 02.6	-26 16	7.2	Globular Cluster in Ophiuchus
20	6514	18 02.6	-23 02	8.5	Trifid Nebula in Sagittarius
21	6531	18 04.6	-22 30	5.9	Open Cluster in Sagittarius
22	6656	18 36.4	-23 54	5.1	Globular Cluster in Sagittarius
23	6494	17 56.8	-19 01	5.5	Open Cluster in Sagittarius
24	6603	18 16.9	-18 29	4.5	Open Cluster in Sagittarius
25	IC4725	18 31.6	-19 15	4.6	Open Cluster in Sagittarius
26	6694	18 45.2	-09 24	8.0	Open Cluster in Scutum
27	6853	19 59.6	+22 43	8.1	Planetary (Dumbbell Nebula)
28	6626	18 24.5	-24 52	6.9	Globular Cluster in Sagittarius
29	6913	20 23.9	+38 32	6.6	Open Cluster in Cygnus
30	7099	21 40.4	-23 11	7.5	Globular Cluster in Capricornus
31	224	00 42.7	+41 16	3.4	Galaxy in Andromeda (Sb)
32	221	00 42.7	+40 52	8.2	Elliptical Galaxy (Companion M31)
33	598	01 33.9	+30 39	5.7	Spiral Galaxy in Triangulum (Sc)
34	1039	02 42.0	+42 47	5.2	Open Cluster in Perseus
35	2168	06 08.9	+24 20	5.1	Open Cluster in Gemini
36	1960	05 36.1	+34 08	6.0	Open Cluster in Auriga
37	2099	05 52.4	+32 33	5.6	Open Cluster in Auriga
38	1912	05 28.7	+35 50	6.4	Open Cluster in Auriga
39	7092	21 32.2	+48 26	4.6	Open Cluster in Cygnus
40	———	12 22.4	+58 05	8.0	Double Star Winnecke 4 Separation of 50"(UMa)

© 2011 Cengage Learning. All Rights Reserved. May not be scanned, copied or duplicated, or posted to a publicly accessible website, in whole or in part.

M#	NGC#	R. A. (2000)	Dec. (2000)	m$_v$	Description
41	2287	06 47.0	-20 44	4.5	Open Cluster in Canis Major
42	1976	05 35.4	-05 27	4.0	Nebula in Orion
43	1982	05 35.6	-05 16	9.0	Smaller Part of the Nebula in Orion
44	2632	08 40.1	+19 59	3.1	Open Cluster in Cancer (Beehive)
45	——	03 47.0	+24 07	1.2	Open Cluster in Taurus (Pleiades)
46	2437	07 41.8	-14 49	6.1	Open Cluster in Puppis
47	2422	07 36.6	-14 30	4.4	Open Cluster in Puppis, West of M46
48	2548	08 13.8	-05 48	5.8	Open Cluster in Hydra
49	4472	12 29.8	+08 00	8.4	Bright Elliptical Galaxy in Virgo
50	2323	07 03.2	-08 20	5.9	Open Cluster in Monoceros
51	5194/95	13 29.9	+47 12	8.1	Whirlpool Galaxy in CnV
52	7654	23 24.2	+61 35	6.9	Open Cluster in Cassiopeia
53	5024	13 12.9	+18 10	7.7	Globular Cluster in Coma Berenices
54	6715	18 55.1	-30 29	7.7	Globular Cluster in Sagittarius
55	6809	19 40.0	-30 58	7.0	Globular Cluster in Sagittarius
56	6779	19 16.6	+30 11	8.2	Globular Cluster in Lyra
57	6720	18 53.6	+33 02	9.0	Planetary Nebula; Ring Nebula in Lyra
58	4579	12 37.7	+11 49	9.8	Bright Barred-Spiral Galaxy in Virgo
59	4621	12 42.0	+11 39	9.8	Bright Elliptical Galaxy in Virgo
60	4649	12 43.7	+11 33	8.8	Bright Elliptical Galaxy in Virgo
61	4303	12 21.9	+04 28	9.7	Spiral Galaxy in Virgo
62	6266	17 01.2	-30 07	6.6	Globular Cluster in Ophiuchus
63	5055	13 15.8	+42 02	8.6	Spiral Galaxy in Canes Venatici
64	4826	12 56.7	+21 41	8.5	Spiral Galaxy in Coma Berenices
65	3623	11 18.9	+13 05	9.3	Spiral Galaxy in Leo
66	3627	11 20.2	+12 59	9.0	Spiral Galaxy in Leo
67	2682	08 50.4	+11 49	6.9	Open Cluster in Canes Venatici
68	4590	12 39.5	-26 45	8.2	Globular Cluster in Hydra
69	6637	18 31.4	-32 21	7.7	Globular Cluster in Sagittarius
70	6681	18 43.2	-32 18	8.1	Globular Cluster in Sagittarius
71	6838	19 53.8	+18 47	8.3	Globular Cluster in Sagitta
72	6981	20 53.5	-12 32	9.4	Globular Cluster in Aquarius
73	6994	20 58.9	-12 38	—	Open Cluster in Aqr (Group of 4 Stars)
74	628	01 36.7	+15 47	9.2	Spiral Galaxy in Pisces
75	6864	20 06.1	-21 55	8.6	Globular Cluster in Sgr (59,000 LY)
76	650/51	01 42.4	+51 34	11.5	Planetary Nebula in Perseus
77	1068	02 42.7	-00 01	8.8	Spiral Galaxy in Cetus
78	2068	05 46.7	+00 03	8.0	Small Reflection Nebula in Orion
79	1904	05 24.5	-24 33	8.0	Globular Cluster in Lepus
80	6093	16 17.0	-22 59	7.2	Globular Cluster in Scorpius

© 2011 Cengage Learning. All Rights Reserved. May not be scanned, copied or duplicated, or posted to a publicly accessible website, in whole or in part.

M#	NGC#	R. A. (2000)	Dec. (2000)	m_v	Description
81	3031	09 55.6	+69 04	6.8	Bright Spiral Galaxy in UMa
82	3034	09 55.8	+69 41	8.4	Irregular Galaxy in Uma (Exploding)
83	5236	13 37.0	-29 52	10.1	Spiral Galaxy in Hydra
84	4374	12 25.1	+12 53	9.3	Elliptical Galaxy in Virgo
85	4382	12 25.4	+18 11	9.3	Elliptical Galaxy in Coma Berenices
86	4406	12 26.2	+12 57	9.2	Elliptical Galaxy in Virgo
87	4486	12 30.8	+12 24	8.6	Elliptical (Peculiar) Galaxy in Vir
88	4501	12 32.0	+14 25	9.5	Spiral Galaxy in Coma Berenices
89	4552	12 35.7	+12 33	9.8	Elliptical Galaxy in Virgo
90	4569	12 36.8	+13 10	9.5	Spiral Galaxy in Virgo
91	4548	12 35.4	+14 30	10.2	Barred Spiral Galaxy in Com (M58?)
92	6341	17 17.1	+43 08	6.5	Another Globular Cluster in Hercules
93	2447	07 44.6	-23 52	6.2	Open Cluster in Puppis
94	4736	12 50.9	+41 07	8.1	Spiral (Peculiar) Galaxy in CVn
95	3351	10 44.0	+11 42	9.7	Barred Spiral Galaxy in Leo
96	3368	10 46.8	+11 49	9.2	Spiral Galaxy in Leo close to M95
97	3587	11 14.8	+55 01	11.2	Planetary Nebula in UMa (Owl Nebula)
98	4192	12 13.8	+14 54	10.1	Spiral Galaxy in Coma Berenices
99	4254	12 18.8	+14 25	9.8	Spiral Galaxy in Coma Berenices
100	4321	12 22.9	+15 49	9.4	Spiral Galaxy in Coma Berenices
101	5457	14 03.2	+54 21	7.7	Spiral Galaxy (Face-on) in Ursa Major
102	5866	15 06.5	+55 46	10.0	M102 = M101? Possible Duplication
103	581	01 33.2	+60 42	7.4	Open Cluster in Cassiopeia
104	4594	12 40.0	-11 37	8.3	Spiral Galaxy in Vir (Sombrero Galaxy)
105	3379	10 47.8	+12 35	9.3	Elliptical Galaxy in Leo (Near M95 & M96)
106	4258	12 19.0	+47 18	8.3	Spiral Galaxy in Canes Venatici
107	6171	16 32.5	-13 03	8.1	Globular Cluster in Ophiuchus
108	3556	11 11.5	+55 40	10.0	Spiral Galaxy in Ursa Major
109	3992	11 57.6	+53 23	9.8	Spiral Galaxy in Ursa Major
110	205	00 40.4	+41 41	8.0	Elliptical Galaxy in Andromeda (Companion M31)

© 2011 Cengage Learning. All Rights Reserved. May not be scanned, copied or duplicated, or posted to a publicly accessible website, in whole or in part.

Appendix E

Using Menus, Shortcuts, Tools, and Preferences in *TheSkyX*

This appendix is devoted to help set up *TheSkyX* desktop window. It explains how to use the Menus, a variety of shortcuts, Tools and Preferences available in *TheSkyX*. The user will be given a sampling of these features in the menus listed at the top of the Standard toolbar in *TheSkyX* desktop window. The Standard toolbar is shown in **Figure E-1**.

| File | Edit | Display | Orientation | Input | Tools | Telescope | Help |

Figure E-1 Standard Toolbar in *TheSkyX* Desktop Window

TheSkyX's databases are complete catalogues. From these catalogues different types of objects may be selected and displayed on the desktop. Other windows determine the celestial objects and the object types that one may want to display in *TheSkyX* window. Objects that users are not interested in displaying may be hidden. This, in effect, reduces the clutter in what is displayed on the desktop window.

Using Preference windows allow users to create new or personal sky windows or charts. Essentially *TheSkyX* window or a sky chart may be customized to one's own personal taste or needs. The font styles, colors, lines, fill colors, patterns may be changed as well using Preference menus. All of the preference files in *TheSkyX* folder have an extension "svp." Once the sky window has been customized, then the "Save As" option is used to save the file. Preference files are located in the *TheSkyX's* folder "…Software Bisque\TheSkyX Student Edition\user\SVP." In addition to these items, we also describe the use and function of the Tool menu located on the Standard toolbar in *TheSkyX*.

The Display Menu in *TheSkyX*

Clicking Display on the Standard toolbar at the top of the sky window opens a menu like the one shown in **Figure E-2**. This window contains many sub-menus that allow the user to configure *TheSkyX* window to one's own personal settings. That is, it allows one to display as many or as few objects on the computer screen. Object preferences may be changed at any time. The background to *TheSkyX* desktop window may be changed from "Photo Like" to Chart Like" depending on the users needs. These options are the first two options at the top of the "Display" menu window shown in **Figure E-2**. At the bottom of Display menu window in **Figure E-2** are command shortcut windows (Find, Date And Time, etc.). They too will be used extensively in *TheSkyX*.

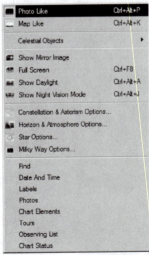

Figure E-2 The Display Menu

© 2011 Cengage Learning. All Rights Reserved. May not be scanned, copied or duplicated, or posted to a publicly accessible website, in whole or in part.

Objects that are listed in the "Celestial Objects" window are listed by type – as stellar or non-stellar objects – not alphabetically as is shown in **Figure E-3**. Any "Celestial Object" may be displayed or hidden using this shortcut.

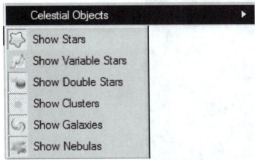

Figure E-3 The Celestial Objects Menu

Everything can be displayed on the desktop if so desired simply click all of the boxes next to the objects in the menu. This is illustrated in **Figure E-3**. This perhaps the best place to start. As one looks at the list of celestial objects, it must be decided at first what to display or not to display. Clicking the box next to the object will display the object not clicking the box next to the celestial object temporarily removes this object from the computer screen.

Begin by selecting a few celestial objects at first. This is perhaps the wisest thing to do when using *TheSkyX*. This makes the desktop less busy and easier to locate stars, planets, and constellations. Typically, the stars, Sun, Moon, planets, and constellation figures should be displayed at first. Later on, as the use of *TheSkyX* becomes more familiar then add or remove other items as needed.

Setting the Options for the Stars with the Display Menu in *TheSkyX*

For example, let's start with something simple. Click off all of the boxes except the box next to "Show Stars." The desktop window should now look like the one displayed in **Figure E-4**. If the box is clicked again, then the stars disappear as shown in **Figure E-5**. Click "Show Stars" once again and they reappear on the desktop.

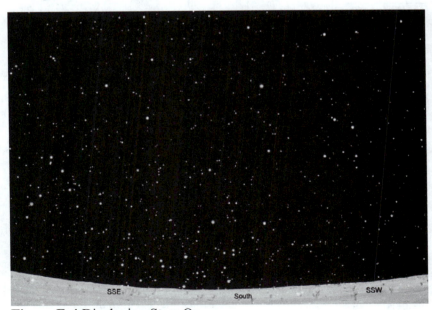

Figure E-4 Displaying Stars On

© 2011 Cengage Learning. All Rights Reserved. May not be scanned, copied or duplicated, or posted to a publicly accessible website, in whole or in part.

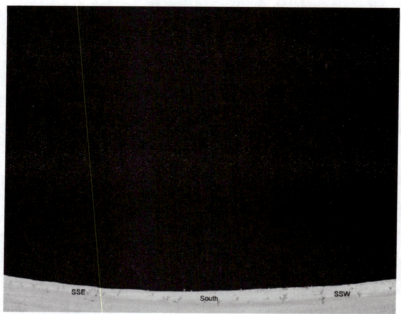

Figure E-5 Displaying the Stars Off

In the "Star Options" menu the magnitude limits for the stars may be set by clicking on the "Appearance" tab. The default setting for stars in *TheSkyX* is shown in **Figure E-6**. These settings, of course, may not seem realistic for your location. For instance Sirius, the brightest star in the sky, has a magnitude of -1.5. The Sun, Moon, and planets are displayed regardless of the magnitude settings.

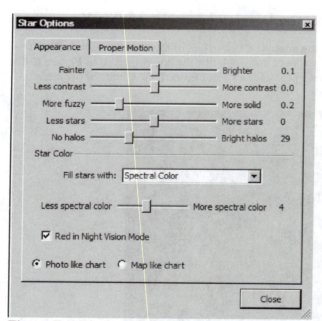

Figure E-6 Displaying the Stars Options Default Menu

The faintest object that can be seen easily with the naked-eye from most medium-sized cities such as Muncie, Indiana is about +3.5 - 6th magnitude is the faintest seen from a mountain top. Most non-stellar objects are fainter than 6th magnitude but most of them can be displayed in *TheSkyX*. To change the magnitude settings, slide the "Fainter to Brighter" bar in the direction of the magnitudes that one wishes to change. **Figure E-7** displays *TheSkyX* window using the "Star Options" Default menu settings.

© 2011 Cengage Learning. All Rights Reserved. May not be scanned, copied or duplicated, or posted to a publicly accessible website, in whole or in part.

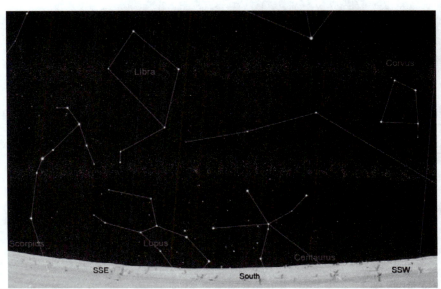

Figure E-7 Displaying the Stars in the Star Options Default Setting

In **Figure E-8** the default settings have been changed slightly to reflect a more realistic view of the sky from a rural location. **Figure E-9** displays *TheSkyX* window after making changes to the default settings.

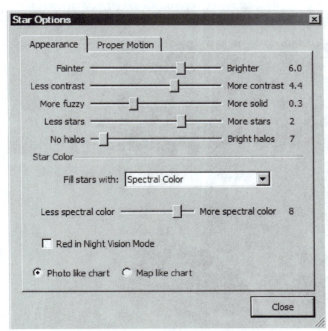

Figure E-8 Slightly Changed Stars Options Menu

© 2011 Cengage Learning. All Rights Reserved. May not be scanned, copied or duplicated, or posted to a publicly accessible website, in whole or in part.

Figure E-9 The Sky After Changing the Star Options Setting

Differences between **Figures E-8** and **E-9** are very apparent. There are far fewer stars in **Figure E-8**, as is typically seen in the night sky from a medium-size city. Changing the magnitude limits or the number of stars may be necessary in order to more accurately represent what is seen at one's specific location. Just go outside and observe and adjust the sky window accordingly using the instructions in the previous example.

Figure E-10 shows how Dr. Jordan's sky window appears on his desktop. If needed, objects can be easily added or removed. In fact, the preferences can be set to suit any taste. It is better and less confusing if you configure the sky in a simple fashion at first.

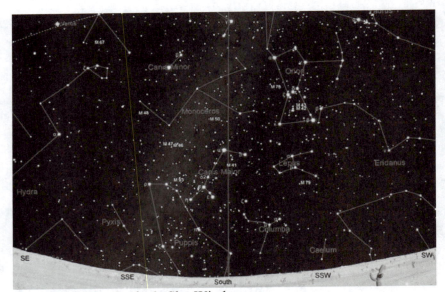

Figure E-10 Dr. Jordan's Sky Window

We suggest that the beginner display the images in *TheSkyX* window too. This enhances one's experience while using *TheSkyX*. Displaying images in *TheSkyX* is accomplished by clicking on the celestial object. Whenever an object in TheSkyX window is clicked on, a Find window opens. The window displays information about the celestial object in question. In that window is the "Show Photo Option" (   ). This option allows the user to display images of non-stellar objects automatically from *TheSkyX's* database. As an example, **Figure E-11** displays the Owl Nebula, a planetary nebula. In the case of stars, however, an HR Diagram appears indicating the location of that star on the diagram. This is explained in more detail later on in ***Displaying***

© 2011 Cengage Learning. All Rights Reserved. May not be scanned, copied or duplicated, or posted to a publicly accessible website, in whole or in part.

Spectral Colors in TheSkyX section. This option is displayed for the star α Leonis, classified as a B7V star, in **Figure E-12**.

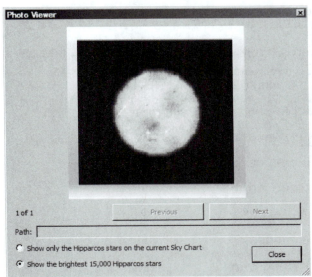

Figure E-11 Photo View of the Owl Nebula

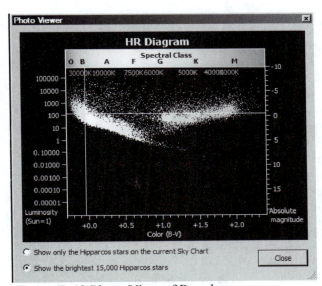

Figure E-12 Photo View of Regulus

It is also possible to display images of Solar System objects. Simply click on the object (Sun, Moon, or planet) in question and then the ⬛ button in the Find window. Accessing stunning images of many of the astronomical objects makes for a pleasant experience in using *TheSkyX*. It may be impossible to see some of the non-stellar objects with the naked-eye, but the images are no less beautiful to look at and appreciate.

Displaying Reference Lines with the Display Menu in *TheSkyX*

There are other options in the Display menu on the Standard toolbar at the top of the sky window. One such option is to display, hide, or change the attributes of Reference Lines in *TheSkyX*. In order to make these changes, one opens the Chart Elements window. It is there that settings can be made which are necessary to help in visualizing the orientation of and viewing

© 2011 Cengage Learning. All Rights Reserved. May not be scanned, copied or duplicated, or posted to a publicly accessible website, in whole or in part.

portions of the sky that one is interested in viewing. The style and colors of these reference lines can be changed by clicking on the "Edit Attributes" bar at the bottom of the Chart Elements window.

The following procedure outlines how to change the display of the reference lines in the sky window.

1. Click Display on the Standard toolbar at the top of the sky window, then on "Chart Elements." The Chart Elements menu is displayed like the one shown in **Figure E-13**.

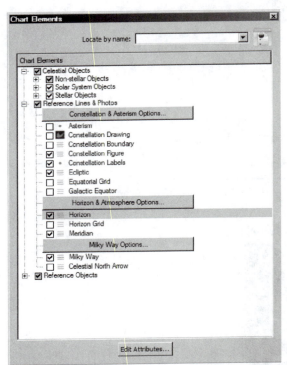

Figure E-13 Setting Reference Lines in the Chart Elements Window

2. When setting up the sky window, it is helpful to have several Reference Lines displayed. These might include the Constellation Figures and Constellations Labels, the Ecliptic, and the Milky Way. Other reference lines that are very useful to have are Horizon-based lines. This obviously includes the Horizon but also should include the Local Celestial Meridian. Make sure that these items are checked in the Chart Elements menu.

3. It is also convenient to have one's horizon customized as well as being displayed in the sky window. Use the "Horizon & Atmosphere Options" bar (Horizon & Atmosphere Options...) to set up the horizon in one of two fashions. After clicking on this bar, the menu displayed in **Figure E-14** opens.

© 2011 Cengage Learning. All Rights Reserved. May not be scanned, copied or duplicated, or posted to a publicly accessible website, in whole or in part.

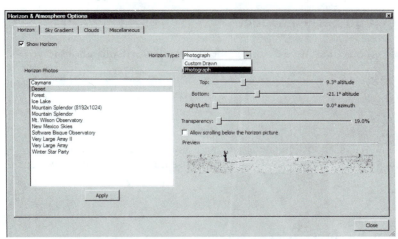

Figure E-14 The Horizon & Atmosphere Menu Window

It is in this menu that one can change or customize their horizon. There are a variety of "Horizon Photos" available to choose from in this menu window. One may also download other "Horizon Photos" from Software Bisque's website.

There are a couple of options that one can choose in order to represent the horizon in *TheSkyX*. The first option is a completely transparent one that allows seeing objects that are below one's local horizon. The second option is an opaque one which resembles what one actually sees when outside observing. Depending upon the situation, it is sometimes useful and convenient to switch from one option to the other.

The transparency may be changed from 0%, totally opaque, to 100%, totally transparent. In order to change the transparency of the horizon slide, the bar in the "Horizon and Atmosphere Options" menu window as shown in **Figure E-15**. Typically, 35% – 40% transparency is sufficient to allow one to "see" below the horizon from any location on Earth. Be sure to click the box to "Allow scrolling below the horizon picture."

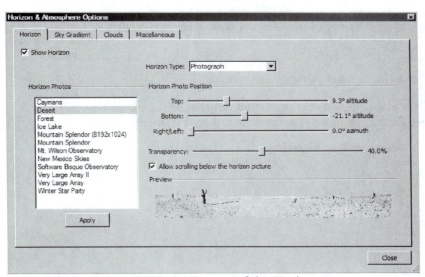

Figure E-15 Setting the Transparency of the Horizon

Displaying Mirror Image –

The Mirror Image option in the Display menu flips the sky window in a left-right fashion. This is the way it would appear in an astronomical telescope (East-West flipped).

© 2011 Cengage Learning. All Rights Reserved. May not be scanned, copied or duplicated, or posted to a publicly accessible website, in whole or in part.

Displaying *TheSkyX* in the Full Screen Mode –

The Full Screen mode is sometimes used to look at the sky window without all of the menus displayed.

Displaying the Daytime Sky –

The Show Daylight option is used to simulate the daytime sky. Usually, this option is chosen to observe the apparent short-term motion of the Sun. To use this mode, a time increment must be set beforehand.

Displaying the Night Vision Mode –

The Night Vision Mode is essential if you decide to use *TheSkyX* outside or in an observatory. It redraws the toolbars in red to minimize the loss of dark adaptation.

Displaying the Photo Like Mode –

The Photo Like Mode draws the sky window with the sky background as black and the stars as colored-dots. It makes the window look more realistic and what would be seen in the night sky.

Displaying the Map Like Mode –

The Map Like Mode redraws the sky window with the sky background white and the stars as colored-dots. It makes the window look more like what would be seen on a star chart or in a book of star atlases. It displays the way the sky window looks when it is printed.

Displaying the Constellations and Asterisms –

This menu window allows the user to display the constellation names and figures in a variety of ways. **Figure E-16** displays the Constellation & Asterism Options window.

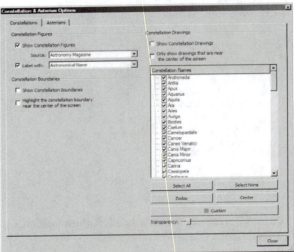

Figure E-16 The Constellation & Asterism Options Window

© 2011 Cengage Learning. All Rights Reserved. May not be scanned, copied or duplicated, or posted to a publicly accessible website, in whole or in part.

Figure E-17 shows the Constellation and Asterism Options window displaying the "Label with:" option displayed. The default setting is the Astronomical Name with the "Source" being *Astronomy Magazine*. My preference is *Sky and Telescope Magazine*.

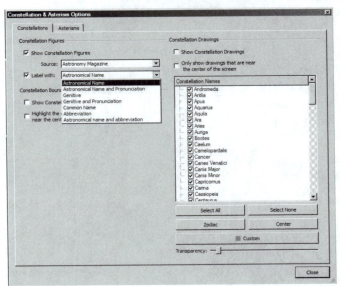

Figure E-17 The Constellation & Asterism Options Window Displaying Name Options

In addition to displaying constellations and asterisms this window allow one to display "Constellation Drawings." This makes *TheSkyX* window look more like what is seen in a star atlas of old. One can customize these drawings as well. Just explore this menu for yourself.

Displaying the Horizon & Atmosphere Options –

This option is used to change the settings for the observer's horizon as well as a few other things. It is in this menu window changing the appearance of the horizon from a "Photograph" to a "Custom Draw" is a simple task. The authors have chosen the "Desert" photo throughout this entire workbook. The horizon is an individual choice and may be changed at any time.
In addition to setting the horizon attributes one may change the "Sky Gradient," whether or not to display "Clouds" is another personal option to consider. The "Horizon &Atmosphere Options" window is displayed in **Figure E-18**.

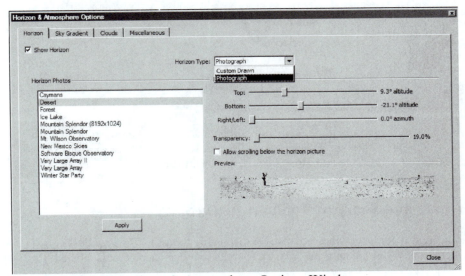

Figure E-18 The Horizon & Atmosphere Options Window

© 2011 Cengage Learning. All Rights Reserved. May not be scanned, copied or duplicated, or posted to a publicly accessible website, in whole or in part.

In the "Miscellaneous" tab the Sun and Moon's Halo may also be changed to suit the user as well as being able to simulate meteor showers. This is shown in **Figure E-19**.

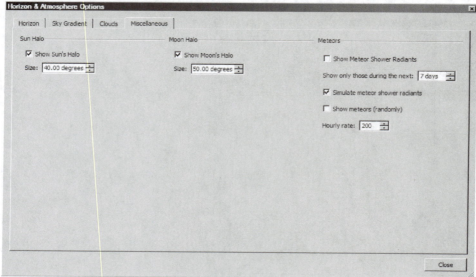

Figure E-19 The Horizon & Atmosphere Options Window

Displaying the Milky Way Options –

The Milky Way may be displayed in only two different modes. One is in Black and White and the other is in color. The color mode makes the appearance of the Milky Way much more appealing to the eye in the desktop window of *TheSkyX*.

Displaying "Command" Shortcuts with the Display Menu in *TheSkyX*

There are eight different "Command" menu windows that may be displayed in *TheSkyX* software. These windows are the ones that are most often used in *TheSkyX*. They are…

Find
Date And Time
Labels
Photos
Chart Elements
Tours
Observing List
Chart Status

We believe it best not to go through these menus at this time. There is ample time to learn how to use these as you progress through this workbook. The most often used "Command" windows in *TheSkyX* are the Find, Date and Time, Labels, and Chart Elements.

Displaying Preferences and Shortcut Toolbars in *TheSkyX*

There is no general Preference window per se for displaying things in *TheSkyX* desktop window as there were in previous versions of *TheSky*. There is only one menu that is labeled

© 2011 Cengage Learning. All Rights Reserved. May not be scanned, copied or duplicated, or posted to a publicly accessible website, in whole or in part.

"Preferences" and this is located in the "Tools" menu on the Standard toolbar at the top of the desktop window.

When TheSkyX is initially installed there are many icons displayed across the top of your desktop window…just under the Standard toolbar at the top of the desktop window. As was discussed in Chapter 2, these shortcut toolbars may be customized by the user at any time.

In order to customize these shortcut toolbars go to "Tools" on the Standard toolbar at the top of the sky window. Scroll down to "Preferences." It is in the "Preference" menu that one can address how to display the shortcut "Toolbars," the "Status Windows," the "Report Setup," and the "Advanced" setting in *TheSkyX*.

One or all of the Preferences on these shortcut toolbars may be customized. When customizing, one has the choice of adding or removing a variety of shortcuts to any of these toolbars. The following example demonstrates how one can add a shortcut to the "File Toolbar" menu.

1. Click on "Tools" on the Standard toolbar at the top of the desktop window.

2. **Figure E-20** displays the Preference menu window with the "Toolbars" icon highlighted.

3. In the left-hand pane one may choose from a list of toolbars to display in the *TheSkyX* window. In the right-hand pane are all of the toolbars that may be displayed on the desktop window.

4. Below the list of toolbars are three choices: Show All, Hide All, or Customize…" Currently, all of these toolbars with their shortcuts are displayed on the desktop window.

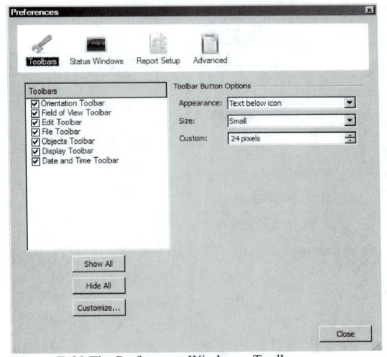

Figure E-20 The Preferences Window – Toolbars

5. Let's customize the "File Toolbar." Click the "Customize" button below the list. A window such as the one in **Figure E-21** appears. Notice that the "File Toolbar" appears in the upper right-hand panel of this menu window but is not highlighted.

© 2011 Cengage Learning. All Rights Reserved. May not be scanned, copied or duplicated, or posted to a publicly accessible website, in whole or in part.

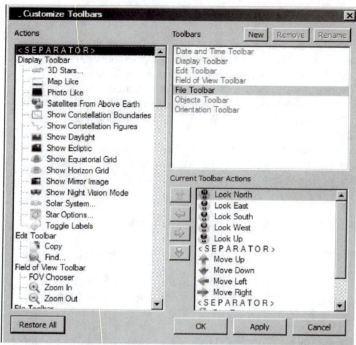

Figure E-21 The Preferences Window – Customize File Toolbar

6. Left-click with the mouse to highlight this toolbar as shown in **Figure E-22**.

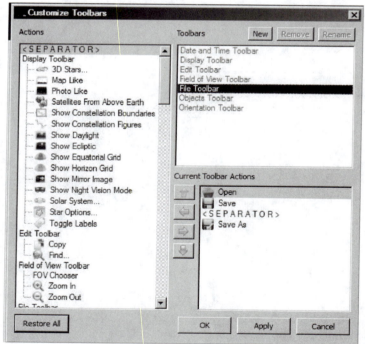

Figure E-22 The Preferences Window – File Toolbar Highlighted

7. Notice that there are already three items listed in the File Toolbar.

8. If you look back into Chapter 2 you will see that there were other shortcuts, 10 in all, displayed in the File Toolbar. They were added using this procedure.

9. Let's add the shortcut "Exit."

© 2011 Cengage Learning. All Rights Reserved. May not be scanned, copied or duplicated, or posted to a publicly accessible website, in whole or in part.

10. Use the scroll bar on the right-hand side of the left panel and scroll to "All Other Commands."

Now, find the 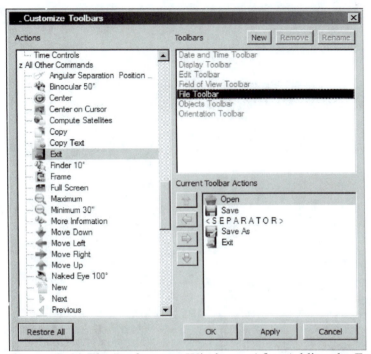 **Exit** icon and highlight it. Use the ⇨ in order to transfer this shortcut to the File Toolbar.

11. The customized File Toolbar now appears like the one shown **Figure E-23**.

Figure E-23 The Preferences Window – After Adding the Exit Icon

12. Click "Apply" and then "OK." That's all there is to it. The toolbars in *TheSkyX* window may be changed at any time depending upon what is needed or the user's preferences. In Chapter 7 there is an example of how to customize the Display in the desktop window. It provides information on how to customize and add important information to the Chart Status window and display it on the desktop.

In *TheSkyX* "preferences" are imbedded in different menu windows as attributes. These attributes may also be edited at any time. They give the user the flexibility to change the appearance (colors, fonts, symbols, etc.) of objects or reference lines in *TheSkyX*.

For example, being an observational astronomer and knowing that the Sun is a yellow star, Dr. Jordan prefers the ecliptic to be represented as a yellow dashed-line, rather than a blue colored one (default setting in *TheSkyX*). It is simple to change this feature in *TheSkyX* by following through the exercise that follows.

Setting the Ecliptic to a Yellow-Dashed Line with the Display Menu

1. Click Display on the Standard toolbar at the top of the sky window and then click "Chart Elements." A window is displayed like the one in **Figure E-24**.

2. Expand the "Reference Lines and Photos" menu by clicking on the ⊞ next to it. The menu now appears like the one shown in **Figure E-25**.

© 2011 Cengage Learning. All Rights Reserved. May not be scanned, copied or duplicated, or posted to a publicly accessible website, in whole or in part.

3. In the Constellation & Asterism Options menu click on the Ecliptic as is shown in **Figure E-25**.
4. Now click on "Edit Attributes" button at the bottom of the window. A window like the one shown in **Figure E-26** appears.

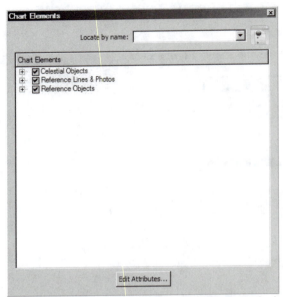

Figure E-24 The Chart Elements Menu Window

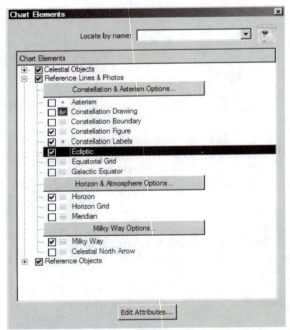

Figure E-25 The Expanded View of Reference Lines and Photos

© 2011 Cengage Learning. All Rights Reserved. May not be scanned, copied or duplicated, or posted to a publicly accessible website, in whole or in part.

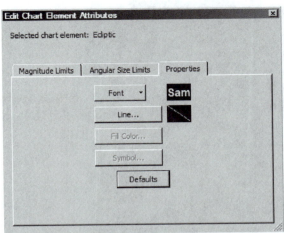

Figure E-26 The Edit Chart Element Attributes Window

5. Click the "Line…" button then the "Style." A menu like the one displayed in **Figure E-27** appears.

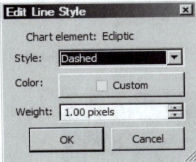

Figure E-27 Using Edit Line Style to Change Style/Color of Ecliptic

6. This menu allows one to change the Style, Color, and Weight (thickness) of a line.

7. To change the line style, click "Style" and change it to a dashed-line, and click OK. This returns you back to the Attributes menu. Now, click the "Color" and change it to bright yellow or any other color you desire. Click the desired color (yellow) this again returns you back to the Attributes menu. Click "OK." This will close the window.

8. The ecliptic in your desktop sky window should now appear as a yellow-dashed line as shown in **Figure E-28**.

© 2011 Cengage Learning. All Rights Reserved. May not be scanned, copied or duplicated, or posted to a publicly accessible website, in whole or in part.

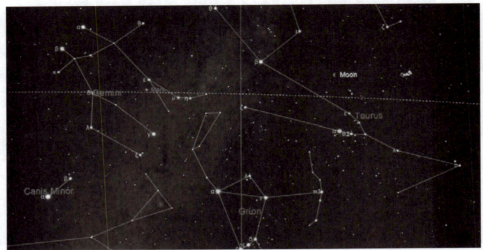

Figure E-28 Appearance of *TheSkyX* Desktop Window after Changing the Line Style

9. When the Attributes menu is displayed once more, click "Font" then "Properties." This menu allows one to change the Style and Size of the font. After changing the style of the fonts just click "Ok." Now, click "Font" again and then "Color." A palette of colors opens from which the user may choose from a variety of colors for their desktop window.

10. **Figure E-29** also illustrates the changes that were made to the ecliptic font in *TheSkyX* window. The color is now yellow and the font size is larger.

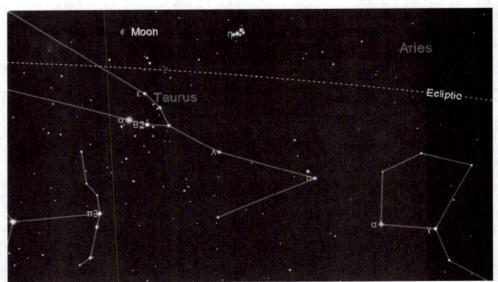

Figure E-29 Appearance of *TheSkyX* Desktop Window after Changing the Font Style

One can change the appearance of the sky window with this menu at any time. In fact, we recommend it. However, change the preferences to suit your tastes! If you do not like the change, then start again.

Setting the Common Names in *TheSkyX* with the Display Menu in *TheSkyX*

Probably the feature that is most helpful in the Display menu…and it must be active…is the Common Names menu. For example, if *TheSkyX* is running and the computer screen is filled with stars and constellations and no names or designations are displayed, this might seem like a severe

© 2011 Cengage Learning. All Rights Reserved. May not be scanned, copied or duplicated, or posted to a publicly accessible website, in whole or in part.

problem. There is a simple solution. **Figure E-30** illustrates the sky with no names or designations displayed.

Figure E-30 Desktop Window **Without** the Common Names Toggled-On

In order to display the common names, one must make sure that the common names are toggled on so that they will be displayed in the sky window. The following procedure explains how to toggle on the common names in the desktop sky window.

1. Click Display on the Standard toolbar at the top of the sky window.

2. Next click "Labels." This opens the Labels menu window as shown in **Figure E-31**.

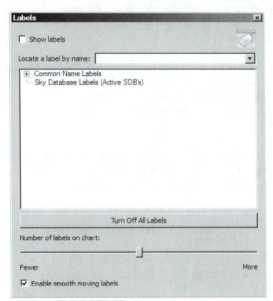

Figure E-31 *TheSkyX*

3. Next click the expand button,⊞ next to the "Common Name Labels." This displays the entire menu as shown in **Figure E-32**.

© 2011 Cengage Learning. All Rights Reserved. May not be scanned, copied or duplicated, or posted to a publicly accessible website, in whole or in part.

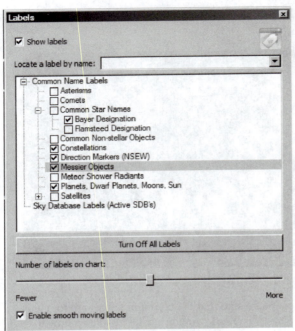

Figure E-32 The Expanded Label Menu.

4. Check the list to see what labels have been checked. At first, it might be wise to display just a few "Common Name Labels."

5. It is suggested, for now, that in the "Labels" menu only the "Bayer Designations" and "Constellations" boxes be checked. This makes the desktop sky window easier to view and without a lot of clutter.

    One should also click the "Direction Markers" (NESW) box so that they are displayed on the horizon and perhaps the "Messier Objects" box might be checked too. Finally, click the "Planets, Dwarf Planets, Moons, Sun" box. Of course, one may choose as many or as few labels as to display on the desktop window by using the slide bar below the menu. This displays more or less star labels on the desktop.

    Click the "Enable smooth moving labels" box near the bottom-left of the menu and make sure that the box at the upper-left is clicked too…"Show Labels." Afterwards, the constellation names, star designations, and Messier Objects designations appear in the sky window as shown in **Figure E-33**.

6. Another, perhaps faster way, to show the Labels of astronomical objects on the desktop window is to simply click ![Toggle Labels] on the shortcut menu.

© 2011 Cengage Learning. All Rights Reserved. May not be scanned, copied or duplicated, or posted to a publicly accessible website, in whole or in part.

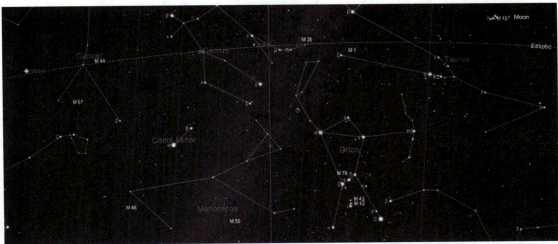

Figure E-33 Desktop Window **With** the Common Names Toggled-On

One can also select the names of other types of astronomical objects to be displayed in the sky window. To accomplish this you must access the Label menu again. Simply click Display on the Standard toolbar at the top of your sky window, then on "Labels." The menu like the one displayed in **Figure E-34** appears once again.

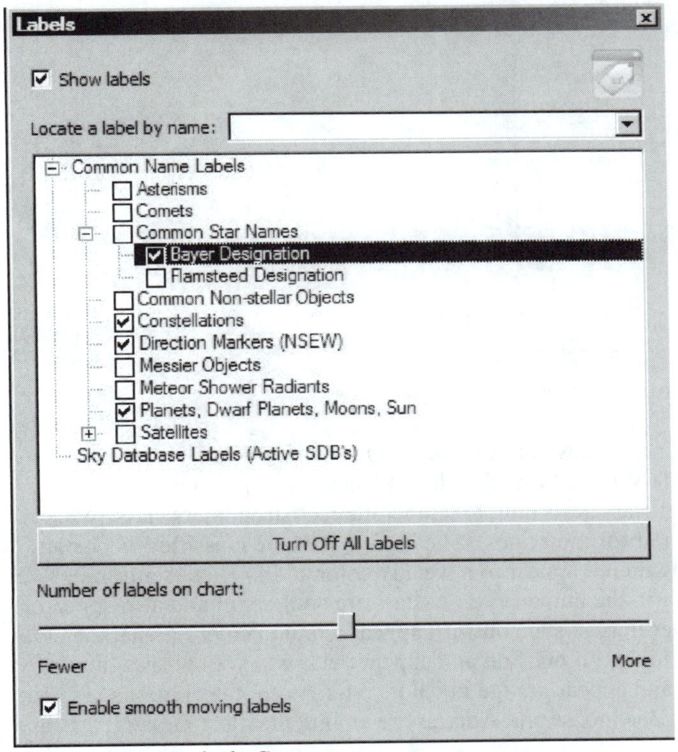

Figure E-34 Labels Setup

This menu is very versatile. In addition to the Constellations, Messier Objects, and Bayer Letter designations, one can also display the Flamsteed designations, asterisms, comets, and common non-stellar objects...keeping in mind, the more names displayed, the more busy and confusing the sky window becomes.

© 2011 Cengage Learning. All Rights Reserved. May not be scanned, copied or duplicated, or posted to a publicly accessible website, in whole or in part.

Displaying Spectral Colors With the Display Menu in *TheSkyX*

Whenever the mouse is clicked anywhere in *TheSkyX* desktop window, a Find Result window always appears. Usually, the Result tab indicates the location of the "Mouse click position" and displays information in the "Object Information Report" section. However, when one clicks on a particular star, the star's name and information about the star appears in the Object Information Report. One such piece of information is its "Spectral" characteristic or color.

When *TheSkyX* was first installed this option for the stars was already preset. In other words, the stars on the desktop window are displayed with different hues of color. This is done in order to simulate their approximate temperatures. This option, however, may be changed to reflect more or less color of the stars to be displayed. Enhancing the colors of stars to reflect their temperatures is accomplished by opening the "Star Options" menu window. This is displayed in **Figure E-35**. The slide bar allows the user to increase or decrease the "Spectral Color" of the stars on their desktop window. Simply explore this option until you've decided what works best for you.

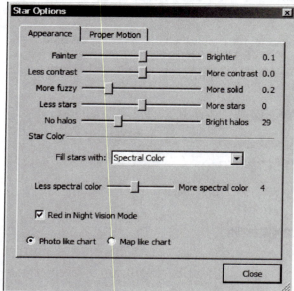

Figure E-35 The Star Options Menu Window

The colors, however, are based on the laws of radiation…specifically Wien's Law. It is left as an exercise for the user to investigate Wien's Law as well as Planck's Law.

There are seven major classes of stars according to temperature. Astronomers classify the stars by their spectral characteristics or absorption lines. The hottest stars are classified as O-stars and appear white in color. In fact, they do not appear to have any color at all. B-stars are somewhat cooler and appear bluish-white in color to the human eye. A-stars are cooler still and also appear bluish-white in color. F-stars are cooler than A-stars but still appear white in color. G-stars, on the other hand, have temperatures similar to that of our Sun and appear yellow to yellowish-white in color. K-stars are cooler than G-stars and appear orange in color. M-stars are the coolest stars and appear reddish orange to red in color. Betelgeuse and Antares are among the most famous stars in this group of stars.

Figure E-36 lists the seven classes of stars with their approximate temperatures and colors. As one can see, different classes have temperature ranges. Therefore, subclasses were later added to account for the differences in temperatures. In **Figure E-36**, the temperatures expressed are degrees on the Kelvin scale. There are ten subclasses for each spectral type except for the O and M stars. These subclasses range from 0 to 9 with 0 the hottest and 9 the coolest within each type. The Sun, for example, is classified as a G2 star because of its spectrum (temperature). *TheSkyX* will indicate the temperature of the star by giving its "Spectral" type in the Find Result window.

© 2011 Cengage Learning. All Rights Reserved. May not be scanned, copied or duplicated, or posted to a publicly accessible website, in whole or in part.

Spectral Type:	Temperature:	Color:
O	> 25,000 K	Blue - White
B	11,000 – 25,000 K	Blue – White
A	7,500 – 10,000 K	Blue - White
F	6,000 – 7,500 K	White
G	5,000 – 6,000 K	Yellow - White
K	3,500 – 5,000 K	Orange - Red
M	< 3,500 K	Red

Figure E-36 The Spectral Classes of Stars.

In addition to the spectral type, *TheSkyX* displays the luminosity type for stars. The luminosity of a star is the total amount of electromagnetic radiation (γ–rays through radio) that a star radiates away from its surface per unit area per second. This in essence is the total power output of the star. The Sun, for example, radiates 3.83×10^{26} Watts of energy from its surface every second!

Stars are then classified by their luminosity. The major types of stars classified according to their luminosity are the following:

I – Supergiants

II – Bright Giants

III – Giants

IV – Sub-giants

V – Main Sequence (dwarfs)

It will be left as an exercise for the user to investigate these specific properties for the stars.

Changing the Attributes in *TheSkyX* with the Display Menu in *TheSkyX*

As one progresses through the chapters and exercises in this workbook it becomes very apparent that the sky windows presented are not like the default (what was installed) windows. Most of the fonts have different styles, sizes, and colors. Even the lines used in many of the windows are different as well. This section guides the user through the process of how to change these attributes in *TheSkyX*.

It is a simple procedure to change most or all of the attributes in *TheSkyX*. It can be done in a couple of ways. Probably the simplest way is the using the Chart Elements menu in the Display window. **Figure E-37** displays all of the Elements available in *TheSkyX* program.

© 2011 Cengage Learning. All Rights Reserved. May not be scanned, copied or duplicated, or posted to a publicly accessible website, in whole or in part.

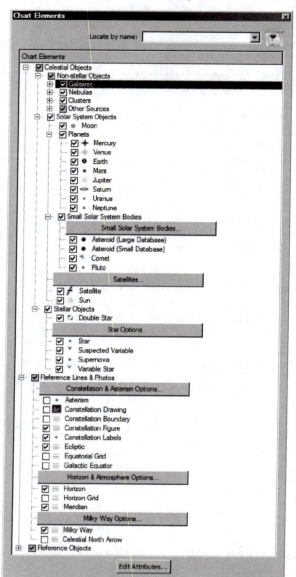

Figure E-37 Chart Elements Menu with All Elements Displayed

Suppose we would like to change the attributes of galaxies. Notice that "Galaxies" is highlighted in the Chart Elements in **Figure E-37**. If you click on "Edit Attributes…" at the bottom of the window a menu opens like the one displayed in **Figure E-38**.

© 2011 Cengage Learning. All Rights Reserved. May not be scanned, copied or duplicated, or posted to a publicly accessible website, in whole or in part.

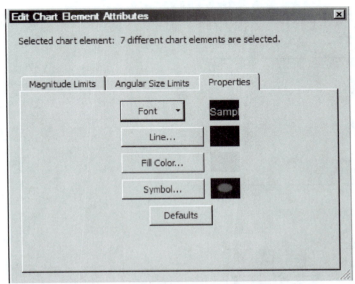

Figure E-38 The Chart Element Attributes Window

If we wanted to change the font properties (style, size, etc.) of "Galaxies," first, then we would click "Font" then "Properties." **Figure E-39** displays this window after clicking on "Font." If we wanted to change the font color of the text describing "Galaxies," then we would click "Color."

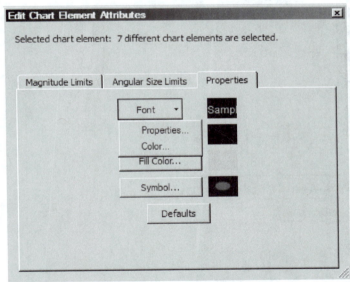

Figure E-39 Selecting "Font" in the Edit Chart Element Attributes Window

After clicking "Properties" in the Edit Chart Element Attributes window, a menu like the one shown in **Figure E-40** appears. In addition to changing the font style, size, and color it is possible the change the chart symbol for objects displayed in *TheSkyX* desktop window. Usually, the symbol is not changed but the symbol's size may be changed for easier recognition. The author has chosen to change the symbol sizes for the planets for illustrative purposes only. It is evident in many of the exercise windows found in this workbook. **Figure E-41** displays the Edit Chart Symbol window in *TheSkyX*.

© 2011 Cengage Learning. All Rights Reserved. May not be scanned, copied or duplicated, or posted to a publicly accessible website, in whole or in part.

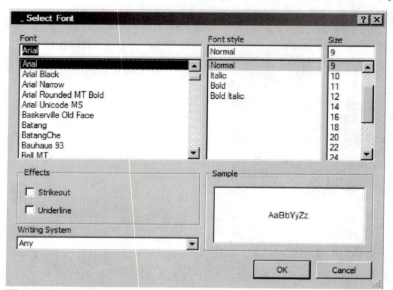

Figure E-40 The Select Font Window

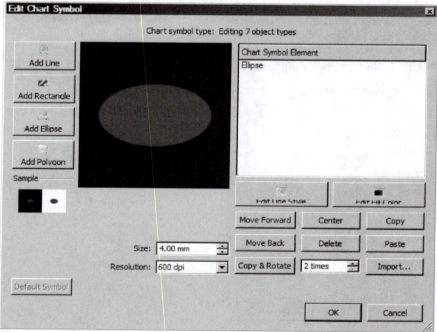

Figure E-41 The Edit Chart Symbol Window

As stated at the beginning of this section, there are a couple of ways for one to change the Chart attributes in *TheSkyX*. The simplest way is to change an object's attribute in the desktop window is to right-mouse click on the object. For example, if one wishes to change the color of the lines in the constellations, put the cursor on a constellation line and right-mouse click on it. This opens a menu window such as the one displayed in **Figure E-42**.

© 2011 Cengage Learning. All Rights Reserved. May not be scanned, copied or duplicated, or posted to a publicly accessible website, in whole or in part.

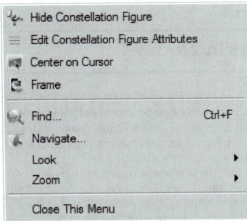

Figure E-42 The Edit Constellation Figure Attributes

Now, click "Edit Constellation Figure Attributes." This opens a menu window like the one that was displayed in **Figure E-26** and is shown in **Figure E-43**. This menu, as before, allows the user to change the font's color, style, and size and as well as the line's color, style, and thickness.

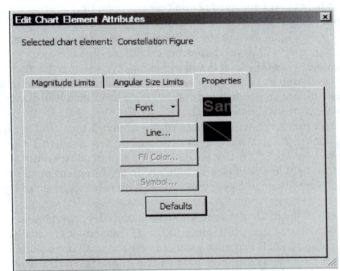

Figure E-43 The Edit Constellation Figure Attributes

With these preferences and attribute changes one can make your sky window as personal as you wish. The main thing is to enjoy the sky above and use *TheSkyX*.

May all of you have Clear Skies!

© 2011 Cengage Learning. All Rights Reserved. May not be scanned, copied or duplicated, or posted to a publicly accessible website, in whole or in part.

Appendix F

Stellar Magnitudes in *TheSkyX*

After a casual survey of the nighttime sky, it becomes quite apparent that not all stars appear to have the same brightness. On any clear night, for example, one might be able to see about 3000 to 5000 stars. This, of course, includes the entire sky, both Northern and Southern Hemispheres, and would require observing every night for an entire year! The number of stars seen depends on several things, such as sky conditions, the brightness of the stars, and the observer's vision.

The Greek astronomer Hipparachus developed the first brightness system for stars in the second century before the Common Era (C.E.). He divided stars into six separate groups and assigned numbers to them to represent their apparent brightnesses. To the brightest stars he assigned the number 1, and to the faintest stars, the number 6. Other stars between these two extremes he assigned numbers 2, 3, 4, and 5, respectively. It is a *reversed numbering* system. That is, the *brighter* the star the *smaller* the number, and the *fainter* the star the *larger* the number.

Today we refer to this system as the *magnitude system*. The one in use today differs slightly from the system developed by Hipparchus 22 centuries ago. A magnitude is simply a number assigned to a celestial object that represents the object's apparent brightness.

The brightness is that amount of light energy received per unit area per unit time from a particular object. It is the visible radiant flux coming to us from the object. Some objects only reflect light, such as planets, whereas others generate their own light, such as stars. Regardless of what object(s) one may see, as long as light is received from them, they will have an apparent brightness or magnitude.

The brightness of stars is determined by two factors: the star's luminosity, which is the total amount of electromagnetic energy radiated from its surface into space per unit area per unit time, and its distance from Earth. If one can measure the apparent magnitude of a star and determine its distance from Earth, then its total power output or luminosity can be calculated.

There is a relationship in nature regarding the human senses known as the "Weber-Fechner Law of Psychophysics" (c. 1860). It states that: *"Equal increments of sensations are associated with equal increments of the logarithm of the stimulus."* This law is said to apply within certain limits to all human sense organs. The eye responds to light in a nonlinear fashion due to the variations in the intensity or brightness of light sources. Simply said, the eye is able to see a tremendous range of light variation.

This system is essentially the basis of the modern magnitude scale adopted by astronomers today. The astronomer N. E. Pogson (c. 1860) suggested that a difference of five (5) magnitudes (actually the difference between a first-magnitude star and a sixth-magnitude star) correspond to a ratio of luminous flux or brightness of 100 to 1. A ratio of brightness corresponding to a step of one magnitude should be the fifth root of 100, which is approximately 2.512.

Because differences of magnitude correspond to ratios of brightness, *differences in magnitude* are *added together* whereas the corresponding *ratios of brightness* are *multiplied together*. If one is acquainted with logarithms, the reason for this rule is clear. This can be easily verified using the values in the Magnitude Scale and multiplying the ratio of brightness for the difference of 3.0 magnitudes by the ratio of brightness for a difference of 5.0 magnitudes.

Magnitude difference of 3.0	$\rightarrow$	15.85:1
Magnitude difference of 5.0	$\rightarrow$	100.0:1
Their product	$\rightarrow$	1585.0

The product of these two ratios of brightness is 1585.0. This is the same value as that of a magnitude difference of 8.0 magnitudes. This is also true for any combination of eight magnitudes such as 3.0 + 5.0, or 4.0 + 4.0, or 6.0 + 2.0, and 7.0 + 1.0 while their corresponding ratios of brightness are multiplied together...try it!

© 2011 Cengage Learning. All Rights Reserved. May not be scanned, copied or duplicated, or posted to a publicly accessible website, in whole or in part.

Magnitude Scale

Difference of Magnitude	Ratio of Brightness	Difference in Magnitudes	Ratio of Brightness
0.1	1.096	1.0	2.512
0.2	1.202	2.0	6.31
0.3	1.318	3.0	15.85
0.4	1.445	4.0	39.81
0.5	1.585	5.0	100.00
0.6	1.738	6.0	251.20
0.7	1.905	7.0	631.00
0.8	2.089	8.0	1585.00
0.9	2.291	9.0	3981.00
		10.0	10,000.00
		15.0	1,000,000.00

The application of the Magnitude Scale may not be apparent (no pun intended) at first glance. The following examples using the Magnitude Scale will help make working with stellar magnitudes more clear. Astronomers use these numbers to describe many astronomical objects quantitatively.

Suppose one observes two stars, let's say Star A and Star B for example. Looking at a star chart or in *TheSkyX* you determine that the magnitude of Star A is +4.5 and the magnitude of Star B is +2.6. The question arises, which star is brighter?

If the answer to the question above was Star B, this is correct! Remember, the smaller the magnitude number, the brighter the star appears in the night sky. That was easy!

Now here is a more practical question. *How much* brighter is Star B than Star A? The first thing that must be determined is the difference in their magnitudes. We apply the rule of subtraction (smaller from the larger) and find that +4.5 minus +2.6 is 1.9. That is, the difference in magnitudes of Star B and Star A is 1.9. Once the difference in the two stars' magnitudes is known, the corresponding ratios of brightness can be determined.

In the Magnitude Scale, the difference of magnitude of 1.9 is not listed. We must look for a combination of two differences in magnitude that add up to 1.9. Adding 1.0 and 0.9 together gives us the difference in magnitudes needed. From these two differences in magnitudes, the corresponding ratios of brightness are 2.512 and 2.291. When we multiply these two ratios together we obtain the answer which is 5.75. One could also use the difference in magnitudes of 0.3 + 0.3 + 0.3 + 1.0 since these too add up to 1.9. Then one would have to multiple 1.318 times 1.318 times 1.318 times 2.512 which equals 5.75 too. Any combination of differences in magnitudes work! So Star B is 5.75 times brighter than Star A. Or, one could also say that Star A is 5.76 times fainter than Star B. Easy? Of course it is!

Here is another and more difficult problem to consider. Suppose Star A is 525 times brighter than Star B, what are the two stars' magnitudes? If the difference in magnitudes of Star A and Star B is known and the magnitude of the one of the stars is known, then the other star's magnitude can be determined. However, neither star's magnitude is known in this case. What is known in this example is simply the ratio of brightness between the two stars. The only thing that we know for sure in this example is that the number associated with the magnitude of Star A is smaller than that of Star B.

We still can make some progress in determining something about the magnitudes of the two stars in this example, however. The first thing we must do is to determine the difference in magnitudes between the two stars. Since the ratio of brightness of the two stars is known (525), we can easily find their difference in magnitudes corresponding to this ratio of brightness. Looking at the Magnitude Scale we notice that the brightness ratio of 525 is not listed. We must find the next smallest ratio of brightness to it (namely 251.2). By dividing the ratio of brightness of 525 by 251.2, yields the ratio of brightness of 2.089. Verify this by multiplying 2.089 by 251.2…when rounded off equals 525! The product of these two ratios of brightness yields the original ratio of brightness of 525. What we have found in this instance are two numbers in the ratio of brightness column whose product is the given ratio of brightness of the two stars.

© 2011 Cengage Learning. All Rights Reserved. May not be scanned, copied or duplicated, or posted to a publicly accessible website, in whole or in part.

In the difference in magnitude columns, opposite these two multiplicands, are two differences in magnitude when added together correspond to the magnitude difference between Star A and Star B. That is, for a ratio of brightness of 525 the corresponding differences in magnitudes are 6.0 and 0.8. Adding these two numbers together gives the difference in magnitudes between Star A and Star B as 6.8. Now that the magnitude difference is known, it is possible to find the magnitude of the two stars provided that the magnitude of one of the stars is known.

If the magnitude of *Star B* is known, then the *difference in magnitudes* of 6.8 is *subtracted* from the magnitude of Star B to obtain the magnitude of Star A. The reason, Star A is brighter than Star B. The magnitude of Star A is represented by a smaller number and is obtained from the subtraction of the difference in magnitudes from the magnitude of Star B. If, on the other hand, the magnitude of *Star A* is known, then the *difference in magnitudes* 6.8 is *added* to the magnitude of Star A to obtain the magnitude of Star B. The reason, Star B is fainter than Star A. The magnitude of Star B is represented by a larger number and is obtained from the addition of the difference in magnitudes to the magnitude of Star A.

So far, only the discussion of apparent magnitudes has been given. The *apparent magnitude* of a star is defined as the *brightness of a star as seen from Earth*. The apparent magnitude of a star does not tell us much about any of the physical characteristics of stars themselves.

Remember the two things that influence a star's apparent brightness—the star's luminosity and distance. If we know a star's distance, then the star's total energy output (luminosity) can be determined. A star's luminosity is determined by two unique properties of the star itself, its temperature and size. Stellar sizes and temperatures are determined from radiation laws and spectroscopic analysis. Astronomers determine the luminosities of stars and compare them. To accomplish this, astronomers visualize all the stars at the same distance from Earth and then re-estimate their apparent brightnesses.

The inverse square law of light is used for these determinations. The farther away a light source is from the observer, the fainter it appears by a factor of one over the distance squared. The brightness determined for stars at this so-called standard distance is defined as the star's absolute magnitude. The *absolute magnitude* of a star is defined as the *star's apparent brightness as seen from a distance of 10 parsecs*. A parsec is a distance equal to 3.26 light years or 206,265 AUs. It also equals about $3.1 \cdot 10^{13}$ kilometers or about $1.9 \cdot 10^{13}$ miles. Once the absolute magnitude of a star is known, calculations like those that were done with apparent magnitudes may be done. However, in this case, the ratios of luminosities are compared instead of ratios of brightness!

In *TheSkyX* apparent magnitudes for most astronomical objects are provided in the Result box in the Find window. Unless the magnitude of the object is brighter than sixth magnitude (naked-eye limit), it is next to impossible to see without a pair of binoculars or a small telescope. Some of the Messier objects are naked-eye objects, many are not. Unless one has an excellent dark site to make one's observations, many Messier objects cannot be seen without the aid of some type of astronomical instrument.

TheSkyX allows observers to set magnitude limits on their desktop in the *Chart Elements* window. Simply click on *Display*, at the top of the Standard toolbar in the sky window then, "Chart Elements." A window such as the one displayed in **Figure F-1** appears.

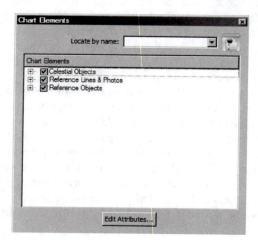

© 2011 Cengage Learning. All Rights Reserved. May not be scanned, copied or duplicated, or posted to a publicly accessible website, in whole or in part.

Figure F-1 The Chart Elements Window in *TheSkyX*
The Celestial Objects menu contains both *Non-stellar* and *Stellar* objects. The magnitude limits for all objects observed in *TheSkyX* can be set with this menu.

To set the magnitude limits of celestial objects in *TheSkyX*, highlight "Celestial Objects" as shown in **Figure F-2**. Now, double-click on the "Edit Attributes" button at the bottom of the window. A window like the one in **Figure F-3** displaying the "Magnitude Limits" tab appears. Notice that in the Magnitude Limits tab that one has the option to choose the magnitude limit in the "Hide fainter than:" description window. One may either "Hide all" or "Show all" the stars in the desktop window. The current setting is to "Show all" magnitudes of celestial objects fainter than 30.0. This number ranges from +30.0 to -6.0. Typically, the magnitude setting is done by sliding the bar to about half-way.

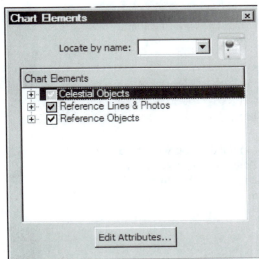

Figure F-2 Changing Magnitude Limits for Celestial Objects

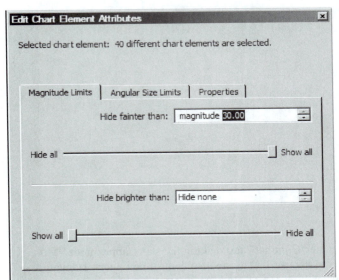

Figure F-3 Edit Chart Element Attributes in the Chart Elements Window

This will display most celestial objects that are brighter than 8th magnitude in the desktop window as shown in **Figure F-4**. This is typically the magnitude range for binoculars. Remember that the faintest star that can be seen from mountain tops is a 6th magnitude star. Make sure that the slide bar at the bottom of the window is set to the "Show all" setting.

© 2011 Cengage Learning. All Rights Reserved. May not be scanned, copied or duplicated, or posted to a publicly accessible website, in whole or in part.

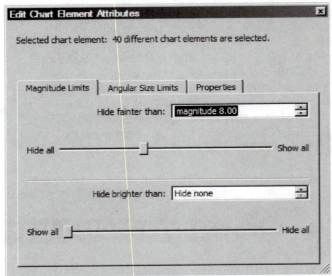

Figure F-4 Setting the Magnitude Limits the Chart Elements Menu

In order to make changes in how or how many stars appear in the desktop window one has to expand the "Stellar Objects" list by clicking on the ⊞ next to it. After clicking to expand the list of Stellar Objects a window like the one displayed in **Figure F-5** opens. Click the "Star Options" bar in **Figure F-5** to open the "Star Options" menu shown in **Figure F-6**.

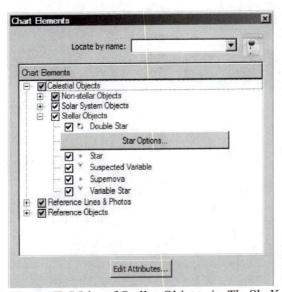

Figure F-5 List of Stellar Objects in *TheSkyX*

The features in this window include setting the "Contrast" and making the star's appearance "More fuzzy" or "More solid" depending on the preference of the user. One can also control how many stars are displayed on the desktop by using the slide bar between "Less stars" and "More stars." One also has the option to change the "Star Color" in this window by using the slide bar between "Less spectral color" and "More spectral color." Using this option adds color to the stars on your desktop. The spectral colors of stars are discussed in detail in *Appendix E*.

© 2011 Cengage Learning. All Rights Reserved. May not be scanned, copied or duplicated, or posted to a publicly accessible website, in whole or in part.

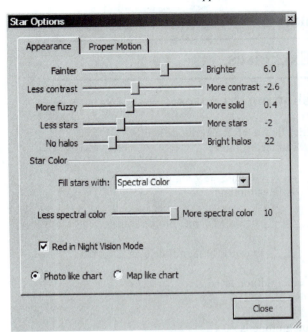

Figure F-4 The Star Options Window in *TheSkyX*

This menu is very versatile in allowing the user to make changes to the preferences in TheSkyX window. One may display the sky window in any fashion they choose. Explore the possibilities!!

© 2011 Cengage Learning. All Rights Reserved. May not be scanned, copied or duplicated, or posted to a publicly accessible website, in whole or in part.

Appendix F

Review Exercises

Use the Magnitude Scale to answer the following exercises

1. A +4.2 magnitude star is _____ times brighter than a +5.7 magnitude star.
   (Answer: 3.98 times brighter)

2. A −1.7 magnitude star is _____ times _____ than a −2.7 magnitude star.
   (Answer: 2.51 times fainter)

3. A +0.8 magnitude star is _____ times _____ than a +1.3 magnitude star.

4. A 3.7 magnitude star is _____ times _____ than −1.2 magnitude star.

5. Star A is 400 times brighter than star B whose magnitude is +6.3.
   The magnitude of star A is _____ (Answer: −0.2)

6. Star A is 63 times fainter than star B whose magnitude is −0.1.
   The magnitude of star A is _____

7. Venus at its greatest brilliancy has a magnitude of −4.6. Polaris, however, has an apparent magnitude of only +2.0.

   Is Venus brighter or fainter than Polaris? _____
   How many times brighter or fainter? _____

8. What significance does this have as we view Venus and Polaris? _____
   _____

9. How much brighter is the Sun than a full Moon? _____

© 2011 Cengage Learning. All Rights Reserved. May not be scanned, copied or duplicated, or posted to a publicly accessible website, in whole or in part.

Appendix F

Review Questions

Answer the following Review Questions about Magnitudes:

1. What does the magnitude of a star measure? _____.

2. What is the apparent magnitude of a star? _____
   _____.

3. What is the absolute magnitude of a star? _____
   _____.

4. Is a star of magnitude of −2 brighter or fainter than a star of magnitude +2?
   _____.

5. How much brighter is a first-magnitude star than a sixth-magnitude star? _____ times.

6. What are the apparent magnitudes of the brightest and faintest stars which can be seen with the naked eye, respectively in the nighttime sky? _____ and _____ magnitude.

© 2011 Cengage Learning. All Rights Reserved. May not be scanned, copied or duplicated, or posted to a publicly accessible website, in whole or in part.